# Bacteriology Research Developments

nova
science publishers

www.novapublishers.com

# Bacteriology Research Developments

**Biofilms: Advances in Research and Applications**
Shane Rowland (Editor)
2021. ISBN: 978-1-68507-062-5 (Hardcover)
2021. ISBN: 978-1-68507-089-2 (eBook)

**Bacteriology: Structure, Reproduction, Plant Diseases and Management**
P.C. Trivedi, Ph.D. (Author), Tripti Agarwal, PhD (Author)
2021. ISBN: 978-1-53619-613-9 (Hardcover)
2021. ISBN: 978-1-53619-678-8 (eBook)

**The Encyclopedia of Bacteriology Research Developments (11 Volume Set)**
Tod Barton (Editor)
Doyle Ortiz (Editor)
2021. ISBN: 978-1-53619-240-7 (Hardcover)
2021. ISBN: 978-1-53619-316-9 (eBook)

**Biodegradation of Cellulose Fibers**
Barbara Simonèiè (Author)
Brigita Tomšiè (Author)
Boris Orel (Author)
Ivan Jerman (Author)
2021. ISBN: 978-1-61668-154-8 (Softcover)
2021. ISBN: 978-1-61324-211-7 (eBook)

**Recent Developments in Enterobacter Research**
Grégoire Roux (Editor)
2020. ISBN: 978-1-53618-615-4 (Hardcover)
2020. ISBN: 978-1-53618-688-8 (eBook)

More information about this series can be found at
https://novapublishers.com/product-category/series/bacteriology-research-developments/

**Rajeshwar P. Sinha**
Editor

# Cyanobacteria

## Life History, Ecology and Impact on Humans

nova
science publishers

www.novapublishers.com

DOI: https://doi.org/10.52305/STKS9486



## NOTICE TO THE READER

The Publisher has taken reasonable care in the preparation of this book, but makes no expressed or implied warranty of any kind and assumes no responsibility for any errors or omissions. No liability is assumed for incidental or consequential damages in connection with or arising out of information contained in this book. The Publisher shall not be liable for any special, consequential, or exemplary damages resulting, in whole or in part, from the readers' use of, or reliance upon, this material. Any parts of this book based on government reports are so indicated and copyright is claimed for those parts to the extent applicable to compilations of such works.

Independent verification should be sought for any data, advice or recommendations contained in this book. In addition, no responsibility is assumed by the Publisher for any injury and/or damage to persons or property arising from any methods, products, instructions, ideas or otherwise contained in this publication.

This publication is designed to provide accurate and authoritative information with regard to the subject matter covered herein. It is sold with the clear understanding that the Publisher is not engaged in rendering legal or any other professional services. If legal or any other expert assistance is required, the services of a competent person should be sought. FROM A DECLARATION OF PARTICIPANTS JOINTLY ADOPTED BY A COMMITTEE OF THE AMERICAN BAR ASSOCIATION AND A COMMITTEE OF PUBLISHERS.

Additional color graphics may be available in the e-book version of this book.

**Library of Congress Cataloging-in-Publication Data**

ISBN: 979-8-88697-169-9

*Published by Nova Science Publishers, Inc. † New York*

*This book is dedicated to Mrs. Madhumita Sinha,*
*Mrs. Ajita Sinha, Mr. Abhinav Sinha,*
*Ms. Trichi Shreya Sinha and Mr. Swapnil Sinha*
*for their continuous love and support.*

# Contents

# Preface

I have great pleasure in bringing forward the book titled Cyanobacteria: Life History, Ecology and Impact on Humans. Cyanobacteria are the prokaryotic photosynthetic microorganisms having ubiquitous distribution on Earth ranging from the Arctic to Antarctic regions. They also form a prominent component of microbial populations in almost all ecosystems, especially in rice fields where they contribute significantly to fertility as natural biofertilizers due to their inherent capacity to fix atmospheric nitrogen.

Cyanobacteria are a rich source of several novel metabolites having the potential for drug development. They are promising bioresource for application in sustainable agriculture, medicine, and green nanotechnology. Certain natural compounds of cyanobacteria such as mycosporine-like amino acids (MAAs) and scytonemin, having UV-absorbing/screening and antioxidative properties, could be biotechnologically exploited by the pharmaceutical industries for the development of natural sunscreens as well as antiaging medicines.

This book encompasses the most recent knowledge and modern trends in the field of cyanobacterial research and their impact on humans. It provides a guide for beginners as well as advanced researchers in the field of cyanobacterial study. I believe that undergraduate, post-graduate and research scholars of several disciplines such as Chemistry, Botany, Microbiology, Biochemistry, Biomedical, Biotechnology, Bioinformatics, Life sciences and Bionanotechnology would also be immensely benefited from this book.

I express my sincere gratitude to all the contributors. Thanks are also due to Ms. Varsha K. Singh for her excellent technical assistance during the editing of this book. I am also thankful to NOVA Science Publishers Inc. for taking such great efforts to complete the publication of this book so efficiently.

Rajeshwar P. Sinha
June, 2022

# Chapter 1

# Evolution and Ecological Significance of Cyanobacteria

**Varsha K. Singh and Rajeshwar P. Sinha***
Laboratory of Photobiology and Molecular Microbiology,
Centre of Advanced Study in Botany, Institute of Science,
Banaras Hindu University, Varanasi, India

## Abstract

Cyanobacteria constitute one of the first forms of life on Earth. During their long evolutionary history, they have participated in several significant events. Particularly, these phototrophic microorganisms are original builders of oxygenic atmosphere and left an indelible footprint in the evolution of photosynthetic organisms. Cyanobacteria are characterized by the ability to produce an array of bioactive compounds having antibacterial, antifungal, antiviral, anti-cancerous, anti-inflammatory and antialgal properties that are of pharmaceutical and cosmeceutical significance. Several strains of cyanobacteria are seen playing role in reclamation of usar (unproductive) soils. They are considered as "soil conditioners" and are significant in improving and restoring the soil fertility. The biotechnological advancements involving cyanobacteria have prompted researchers to come up with well-defined applications of cyanobacteria in a range of industrial fields. This volume is an attempt to enlist useful facts about cyanobacteria and their possible role in resolving the emerging challenges in our daily life.

**Keywords:** bioactive compounds, biotechnological advancements, evolution, soil conditioners

---

* Corresponding Author's Email: rpsinhabhu@gmail.com.

In: Cyanobacteria
Editor: Rajeshwar P. Sinha
ISBN: 979-8-88697-169-9

## Introduction

Cyanobacteria are the essential illustrative of every geographical section on the planet Earth due to their phenomenal adaptive capacities. They owe a high level of physiological flexibility which enables them to scatter on almost all habitable ecosystems. Cyanobacteria are seen inhabiting a range of habitats like marine, freshwater, biological soil crusts, snow, cryoconites, etc. Also, they are found in symbiotic association with different hosts and occur in extreme stressed conditions like volcanic ash, salted soils, and anthropogenically disturbed areas.

Cyanobacteria played a major role in early biogeochemical fluxes and in life and Earth evolution (Pathak et al., 2021). They are the only prokaryotic organisms that perform oxygenic photosynthesis, and are thus generally held responsible for the rise of oxygen in the atmosphere and oceans around 2.4 Ga, during the so-called Great Oxidation Event (GOE) (Bekker et al., 2004; Hannah et al., 2004; Demoulina et al., 2019), facilitated by geological processes (Kasting 2013). Despite the importance of cyanobacteria in the early evolution of Earth and life, fundamental questions about their origin, the timing and pattern of their diversification and the origin of oxygenic photosynthesis remains unexplored (Fischer et. al 2016).

In most cyanobacterial phylogenetic analyses that are using a non-cyanobacterial out-group to root the tree, the reference strain *Gloeobacter violaceus* PCC7421 has a basal position (Hoffmann et al., 2005; Gupta and Mathews 2010). Consequently, the phylogenetic node bearing *Gloeobacter* and the rest of modern cyanobacterial lineages serves as calibration for the origin of cyanobacteria in several studies (Blank and Sanchez-Baracaldo 2010; Demoulina et al., 2019). Microfossils interpreted as cyanobacteria have been reported in rocks as old as the early Archean, but their biogenicity and interpretation is highly debated (Demoulina et al., 2019). The necessity of reliable criteria for the study of microfossils is well illustrated with the famous controversy surrounding the "microfossils" from the 3.45 Ga Apex Chert, Australia (Schopf and Packer 1987). These traces interpreted as fossil cyanobacteria based on their morphology and the geological context (Schopf and Packer 1987) were subsequently reassessed either as pseudofossils (Brasier et al., 2005; Pinti et al., 2009; Bower et al., 2016), contaminants (Altermann and Kazmierczak 2003), mineral artefacts (Javaux and Knoll 2017), or microfossils (Schopf 1993; De Gregorio et al., 2009; Schopf et al., 2017) and the geological context was revised (Brasier et al., 2005). The evolutionary trends and life history patterns of cyanobacteria have been discussed (cf. Chapter 2 for details, this volume).

## The Modern Taxonomy- Beginning of a "Polyphasic Approach"

During 1980's, a new school of thought started to emerge, where a combined total evidence approach began to spread its roots in cyanobacterial taxonomy. Initially it was only the morphological characters along with some ecological and biochemical traits, which were used in this system of classification. Subsequently, with more refinements and clarity, regarding the usage of the *16S rRNA* gene as a molecular marker, cyanobacterial taxonomy improved further. The universal nature and high information content help to unravel and understand major evolutionary changes in cyanobacteria (Woese and Fox 1977). At that time, major reforms were in queue as the botanical as well as bacteriological codes were taken into account for the classification of cyanobacteria. Major revisions were also carried out at that point of time (Komárek and Anagnostidis 1989). Thus, reconciliation efforts were designed to unite both the codes of nomenclature with an ultimate aim to bring up a taxonomic system which could help to understand and study cyanobacteria in a new perspective.

Notably, the use of 16S-23S ITS region in cyanobacterial taxonomy has been an another step that further helped to differentiate closely related genera using the polyphasic approach (Iteman et al., 2000; Boyer et al., 2021). Eventually, it was Komárek et al., (2014) who proposed a new system of classification based solely on the polyphasic approach which as of now constitutes the spirit of the present-day taxonomy of cyanobacteria.

In spite of the universal consensus and agreement in usage of the polyphasic approach, there are still several problematic cases which need to be addressed. The lack of correct naming and improper typification of several taxa has brought about a lot of confusing scenarios in cyanobacterial taxonomy. Another important aspect is the polyphyletic nature of several taxa such as *Nostoc, Leptolyngbya, Calothrix, Trichormus, Anabaena* and many others which needs to be solved, so as to attain monophyly in these large groups (Komárek 2018). A brief introduction to cyanobacterial ecology and systematics has been described (cf. Chapter 3 for details, this volume).

## Potentials of Cyanobacteria in Agricultural and Environmental Sustainability

Cyanobacteria are beneficial microorganisms, having diversified applications in the agricultural and environmental sector (Whitton 2002; Deviram et al., 2013). These are a source of several amino acids, auxins, gibberellins, cytokinins and plant growth promoting factors. They also produce vitamins (e.g., vitamin $B_{12}$ and biotin), polysaccharides and polypeptides (Sood et al., 2011). All these compounds enhance the availability of nutrients and help the plants in taking up nutrients and increase the growth of plant and improve agricultural productivities.

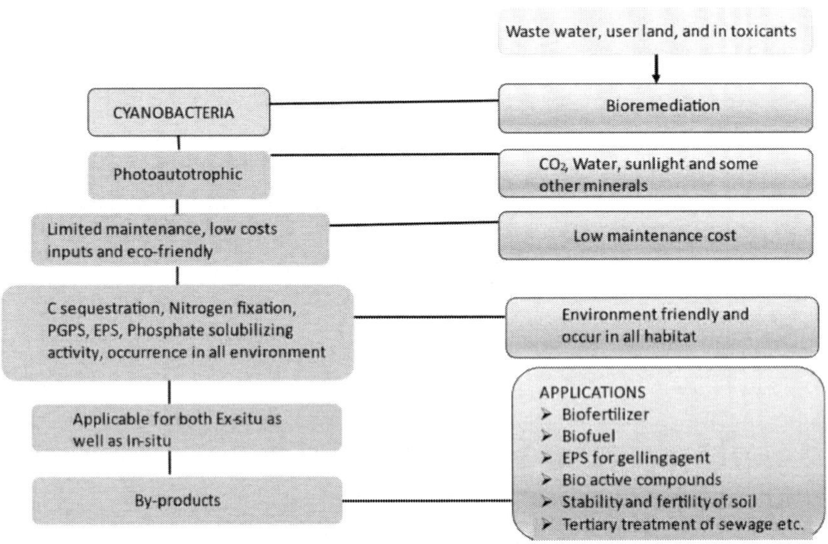


**Figure 1.** Overview of cyanobacteria showing ecological significance.

Cyanobacterial species have a significant role in humus formation as they are involved in two major biological processes i.e., fixation of carbon through photosynthesis and fixation of atmospheric nitrogen. These are the diazotrophs that are useful for the generation of eco-friendly biofertilizers to be made available at a less expensive cost. They mainly function to regulate the nitrogen deficiency in plants, improve the aeration of the soil, increase water holding capacity of soil and add B vitamins to the plants (Song et al., 2005).

Cyanobacteria are emerging as useful organisms within the field of soil restoration and are considered as soil modifiers (Figure 1). Cyanobacteria when applied to the soil will amend the soil stabilization, mineralization, fertility, nourishment and moisture-holding capacity (Rossi et al., 2017). Cyanobacteria produce a range of bioactive metabolites having antibacterial, antiviral, antifungal and anticancerous properties that are eco-friendly and should be suitably applied to the control of phytopathogens (Dahms et al., 2006). Several species of cyanobacteria are known to secrete antibiotics resulting in the death of pathogens like *Nostoc* is thought to secrete an antibiotic that is referred to as bacteriocin that can kill related strains of alga. Also, *Scytonema* sp. is thought to produce cyanobacterin, chlorine-containing gamma lactone (Pignantello et al., 1983).

Cyanobacteria are helpful in land reclamation as these are the primary colonizers of marshlands. These are significant in ecological succession. Soils with increased salinity and alkaline conditions are generally unsuitable for raising crops but cyanobacteria are helpful in the reclamation of such soils. These organisms are now emerging as an efficient bio-agent since various biochemical processes can be applied to them thereby enhancing their aptitude to produce a variety of bioenergy fuels such as bio-diesel, cellulosic ethanol, bio- or syngas, bio-hydrogen, and directly converted fuels etc., (Kumar and Singh 2016). Utilities of cyanobacteria as an efficient bio-agent have been identified and discussed (cf. Chapter 4 for details, this volume).

## Cyanobacteria: A Potential Candidate for Green Nanotechnology

Nanobiotechnology is emerging as one of the important tools in the field of agriculture, industry as well as medicine (Dubchak et al., 2010). At present, several nanoproducts in various forms are commercially available in world markets as paints varnishes, cosmetics, medicines, textile etc., (Roco 2010). The term "nano" refers to particles having dimension in range of 1-100 nm. Nanoparticles (NPs) differ considerably from bulk materials with regard to their physical, chemical and biological properties. These variations are primarily accounted to their high surface area to volume ratio, which leads to substantial differences in their mechanical properties, optical absorption, conductivity, etc., (Nanotechnology 2008; Thakkar et al., 2010; Schmid 2011). The interest has now shifted toward the biological or green synthesis of NPs, as there are various disadvantages associated with the physical and chemical synthesis of NPs (Khanna et al., 2019).

Cyanobacteria are the most preferred candidates for NPs synthesis as they possess various advantages like ability to accumulate heavy metals, contain a number of bioactive compounds like pigments and enzymes that help in stabilizing and act as reducing agents. The cyanobacteria-based NPs also display a large number of chemical, physical and biological properties that further increases the range of its applicability. These NPs have been found to exhibit various properties like antibacterial, antifungal, antialgal, anticancer and photocatalytic activities (MubarakAli et al., 2012). Thus, the use of cyanobacteria to synthesize NPs holds a remarkable value as it is a safe and eco-friendly synthesis, involves low cost, time and energy saving approach and produces novel NPs with various shapes and sizes. Hence by developing and enhancing the biosynthetic processes using cyanobacteria may results in discovery of novel biosynthetic NPs with exclusive properties that would be widely applicable and would be useful in large number of sectors. The applications of cyanobacteria in the field nanobiotechnology to boost agriculture, industrial and pharmaceutical production have been discussed (cf. Chapter 5 for details, this volume).

## Mycosporine-Like Amino Acids in Marine Ecosystems

Generally, the marine organisms, cyanobacteria in particular, have been shown to be the great suppliers of mycosporine-like amino acids produced *via* the Shikimate pathway. The role of MAAs as a UV protective filter has been investigated in cyanobacteria (Sinha et al., 2001; Vega et al., 2020), coral-associated dinoflagellates (Banaszak et al., 2006), marine macro algae (Sun et al., 2020), and a variety of crustaceans.

Marine microbial community is known to harbour a multitude of metabolites and toxins produced as a response to the environmental condition they are exposed to (Nunnery et al., 2010). Even though the presence of MAAs is more expected in lower organisms, other marine microbes are relatively in the rear-end when compared to the production of MAAs in blue-green algal community. Major MAA compounds reported from seaweeds, especially red algae, includes mycosporine-glycine, porphyra-334, asterina-330, shinorine, palythinol, palythine, and palythene (Carreto and Carignan 2011). When environmental circumstances start to deteriorate, the coral zooxanthellae synthesize MAAs along with other metabolites, as a survival tactic. The presence, variation and stress relieving functions including the photoprotective nature of MAAs in marine ecosystems have been discussed (cf. Chapter 6 for details, this volume).

## Potential Uses of MAAs and Scytonemin for Skin Care

In addition to being UV-screening compounds, MAAs have been shown to possess several biological properties, such as antioxidative and DNA-protective properties. MAAs also act as promising inhibitors against protein glycation. MAAs are known to exhibit inhibitory effects on collagenase, which is associated with the skin aging process. Several MAAs have been shown to inhibit collagenase activity *in vitro* (Tarasuntisuk et al., 2018; Orfanoudaki et al., 2019). The DNA-protecting activities of MAAs are also thought to be important for skin care. As DNA can be damaged by ROS generated through UV exposure, the antioxidative activities of MAAs could protect DNA. In fact, mycosporine-2-glycine was found to prevent DNA damage induced by oxidative stress in the A375 human melanoma cell line (Cheewinthamrongrod et al., 2016).

Scytonemin has good prospects in cosmetic and biomedical fields. Assessment of the photoprotective efficacy of scytonemin revealed that scytonemin can prevent up to 90% of solar UV radiation from entering cyanobacterial cells (Derikvand et al., 2017). Along with UV-screening property, scytonemin has antioxidative property as well. It was found that the topical application of scytonemin inhibited 12-*O*-tetradecanoylphorbol-13-acetate (TPA)-induced skin inflammation in mice. It was also demonstrated that scytonemin suppressed LPS-induced expression of tumor necrosis factor-α (TNF-α) and inducible nitric oxide synthase (iNOS) in a mouse macrophage cell line. The further applicable bioactivities of MAAs and scytonemin are being described (cf. Chapter 7 for details, this volume).

## Cyanobacterial Photopigments

Phycobilisomes (PBSs) are the large light-harvesting accessory complexes made up of vividly colored biliproteins (BPs). The major pigments for light absorption in cyanobacteria are phycobiliproteins, whereas carotenoids play an important role in non-photochemical quenching (NPQ) as a barrier against saturating light and a quencher of reactive oxygen species (ROS) (Masojídek et al., 2013). When it comes to carotenoids, these terpenoids are required for the survival of all photosynthetic organisms and are found throughout them. Carotenoids are a broad group of pigments (about 600) that may be split into two categories: carotenes (such as α- and β-carotene) and xanthophylls (oxygenated derivatives of carotenes) such as zeaxanthin and echinenone.

Carotenoids can be found in cyanobacteria's reaction centre, with chlorophyll molecules, or in a protein complex containing a single carotenoid molecule, which is a crucial component for photoprotection—the OCP (Zhang et al., 2014). The OCP is a 35-kDa water-soluble protein that contains a single molecule of hydroxyechinenone in an inactive (orange) form that is activated by blue-green light and transformed to an active (red) form.

Lutein (β, ε-carotene-3,3'-diol) is a xanthophyll that has only 10 conjugated double bonds, giving it a brighter yellowish-green colour than other carotenoids (Torregrosa-Crespo et al., 2018). Lutein, along with astaxanthin and canthaxanthin, is one of the industrially important xanthophylls (Breithaupt 2007).

Lycopene (ψ,ψ-carotene) is a carotenoid with a highly unsaturated hydrocarbon chain made up of 11 conjugated and 2 unconjugated double bonds and no β-ionone ring at the end. It can occur in both cis- and trans-isomeric forms due to the presence of double bonds. However, it is mostly found in the all-trans isomeric form in nature. Under thermal processing and storage settings, lycopene is extremely stable (Torregrosa-Crespo et al., 2018). Various types of cyanobacterial photopigments as well as the most important properties for biotechnological applications, with an emphasis on their bioactivities have been discussed (cf. Chapter 8 for details, this volume).

## Computational Analysis with Cyanobacterial Secondary Metabolites

Cyanobacterial secondary metabolites (CSM) range from low molecular weight compounds such as mycosporine-like amino acids to more complex compounds such as cyanotoxin (Rastogi and Sinha 2009). Secondary metabolites from cyanobacteria show exciting biological activities such as antibiotics, anticancer, antiviral, anti-inflammatory, anti-HIV (human immunodeficiency virus), antibacterial, antifungal, antimalarial and antiprotozoal activities. Metabolites secreted by cyanobacteria have excellent durability and it is therefore, high time to explore CSM and their potential.

A molecular docking study was carried out between the cyanobacterial drug, cryptophycin-F and different types of cancer target proteins. This study compared the docking score between available commercial drugs cabazi-taxel and cryptophycin-F where both the cryptophycin-F and cabazitaxel showed more or less similar docking scores with the target cancer protein. This study confirmed cryptophycin-F can be a good alternative to cabazitaxel without any side effects. However, the study needs further validation through in vitro/in vivo studies (Sangeetha et al., 2014). Recently numerous computational studies have been performed to know the potential of CSM against the viral disease SARS-CoV-2.

Computational method is not only used to discover new drug lead/bioactive molecules against a particular target but it can be also applicable to find metabolic pathways of those bioactive compounds. *In silico* metabolic analysis provides experimental information which can be used to modify a given bioactive compound to increase its biological activity. This method guides to improvement of experimental procedures to get more comprehensive biological information (Yamazaki et al., 2004). Various computational studies which can be used for *in silico* drug reformulation and the role of cyanobacteria in drug development against various diseases through computational studies have been discussed in detail (cf. Chapter 9 for details, this volume).

## Conclusion

It is imperative for the healthy agro-ecosystem to gain sustainability in the true sense in order that it conserves the nature and natural resources, and also maintains the complexity and diversity of ecosystems. Cyanobacteria in this context can be very effective for enriching soil organic carbon and nitrogen and enhancing phosphorous bioavailability to the plants.

Cyanobacteria are excellent accumulators or degraders of various environmental contaminants such as heavy metals, pesticides, and oil containing compounds. Such ubiquitous bio-agents can also be used for capturing and storing of $CO_2$ that may also lead to climate change mitigations through photosynthesis and biological calcification. They are also the ideal source of variety of bioactive compounds with marked antagonistic properties.

## Acknowledgment

Varsha K. Singh (09/0013(12862)/2021-EMR-I) is thankful to the Council of Scientific and Industrial Research, New Delhi, India, for the financial support in the form of fellowship.

## References

Altermann W, Kazmierczak J (2003). Archean microfossils: a reappraisal of early life on Earth. *Res. Microbiol.* 154: 611-617.

Banaszak A T, Santos M G B, LaJeunesse T C, Lesser M P (2006). The distribution of mycosporine-like amino acids (MAAs) and the phylogenetic identity of symbiotic dinoflagellates in cnidarian hosts from the Mexican Caribbean. *J. Exp. Mar. Biol. Ecol.* 337(2): 131-146.

Bekker A, Holland H D, Wang P, IiiD R, Stein H J, Hannah J L, Coetzee L L, Beukes N J (2004). Dating the rise of atmospheric oxygen. *Nature* 427: 117-120.

Bower D M, Steele A, Fries M D, Green O R, Lindsay J F (2016). Raman imaging spectroscopy of a putative microfossil from the ~3.46 Ga Apex chert: insights from quartz grain orientation. *Astrobiology* 16: 169-180.

Boyer S L, Flechtner V R Johansen, J R (2001). Is the 16S–23S rRNA internal transcribed spacer region a good tool for use in molecular systematics and population genetics? A case study in cyanobacteria. *Mol. Biol. Evol.* 18: 1057-1069.

Brasier M D, Green O R, Lindsay J F, McLoughlin N, Steele A, Stoakes C (2005). Critical testing of Earth's oldest putative fossil assemblage from the ~3.5 Ga Apex chert, Chinaman Creek, Western Australia. *Precambrian Res.* 140: 55-102.

Carreto J I, Carignan M O (2011). Mycosporine-like amino acids: relevant secondary metabolites. Chemical and ecological aspects. *Mar. Drugs* 9(3): 387-446.

Cheewinthamrongrod V, Kageyama H, Palaga T, Takabe T, Waditee-Sirisattha R (2016). DNA damage protecting and free radical scavenging properties of mycosporine-2-glycine from the Dead Sea cyanobacterium in A375 human melanoma cell lines. *J. Photochem. Photobiol.* 164: 289-295.

Dahms H U, Xu Y, Pfeiffer C (2006). Antifouling potential of cyanobacteria: a mini-review. *Biofouling* 22: 317-327.

De Gregorio B T, Sharp T G, Flynn G J, Wirick S, Hervig R L (2009). Biogenic origin for Earth's oldest putative microfossils. *Geology* 37: 631-634.

Demoulina C F, Laraa Y J, Corneta L, Françoisa C, Baurainb D, Wilmottec A, Javauxa E J (2019). Cyanobacteria evolution: Insight from the fossil record. *Free Radic. Biol. Med.* 140: 206-223.

Derikvand P, Llewellyn C A, Purton S (2017). Cyanobacterial metabolites as a source of sunscreens and moisturizers: a comparison with current synthetic compounds. *Eur. J. Phycol.* 52: 43-56.

Deviram G, Pant G, Prasuna R (2013). Biochemical changes induced by fungicides in nitrogen fixing *Nostoc* sp. *J. Environ. Sci. Eng.* 55: 81.

Dubchak S, Ogar A, Mietelski J, Turnau K (2010). Influence of silver and titanium nanoparticles on arbuscular mycorrhiza colonization and accumulation of radiocaesium in *Helianthus annuus*. *Spanish J. Agri. Res.* 8(1):103-108.

Fischer W W, Hemp J, Johnson J E (2016). Evolution of oxygenic photosynthesis. *Annu. Rev. Earth Planet Sci.* 44: 647-683.

Gupta R S, Mathews D W (2010). Signature proteins for the major clades of cyanobacteria. *BMC Evol. Biol.* 10: 1-20.

Hannah J L, Bekker A, Stein H J, Markey R J, Holland H D (2004). Primitive Os and 2316 Ma age for marine shale: implications for Paleoproterozoic glacial events and the rise of atmospheric oxygen. *Earth Planet Sci. Lett.* 225: 43-52.

Hoffmann L, Komárek J, Kaštovský J (2005). System of cyanoprokaryotes (cyanobacteria)-state in 2004. *Arch. Hydrobiol. Suppl. Algol. Stud.* 117: 95-115.

Iteman I, Rippka R, de Marsac N T, Herdman M (2000). Comparison of conserved structural and regulatory domains within divergent 16S rRNA–23S rRNA spacer sequences of cyanobacteriaThe GenBank accession numbers for the sequences reported in this paper are AF180968 and AF180969 for ITS-L and ITS-S, respectively. *Microbiol.* 146(6): 1275-1286.

Javaux E J, Lepot K, Van Zuilen M A, Melezhik V A, Medvedev P V (2013). Palaeoproterozoic microfossils. In: Melezhik VA, editor. *Reading the Archive of Earth's Oxygenation:* Springer, New York, 1352-1370.

Kasting J F (2013). What caused the rise of atmospheric $O_2$? *Chem. Geol.* 362: 13-25.

Khanna P, Kaur A, Goyal D (2019). Algae-based metallic nanoparticles: synthesis, characterization and applications. *J. Microbiol. Methods* 163: 105656.

Komárek J, Anagnostidis K (1998). *Freshwater Flora of Central Europe*, Vol 19(1): Cyanoprokaryota: Chroococcales (Süßwasserflora von Mitteleuropa, Bd. 19(1): Cyanoprokaryota: Chroococcales) (H. Ettl, J. Gerloff, H. Heynig & D. Mollenhauer, Eds.). Süßwasserflora von Mitteleuropa.

Komárek J (2018). Several problems of the polyphasic approach in the modern cyanobacterial system. *Hydrobiologia* 811(1): 7-17.

Kumar A, Singh J S (2016). Microalgae and cyanobacteria biofuels: a sustainable alternate to crop-based fuels. In: Singh J S, Singh D P, editors. *Microbes and Environmental Management*, New Delhi: Studium Press Pvt. Ltd. 1-20.

MubarakAli D, Gopinath V, Rameshbabu N, Thajuddin N (2012). Synthesis and characterization of CdS nanoparticles using C-phycoerythrin from the marine cyanobacteria. *Mater. Lett.* 74: 8-11.

Nanotechnology AOo. *National Nanotechnology Strategy Annual Report* 2008: 2007-2008.

Nunnery J K, Mevers E, Gerwick W H (2010). Biologically active secondary metabolites from marine cyanobacteria. *Curr. Opin. Biotechnol.* 21(6): 787-793.

Orfanoudaki M, Hartmann A, Alilou M, Gelbrich T, Planchenault P, Derbré S, Schinkovitz A, Richomme P, Hensel A, Ganzera M (2019). Absolute configuration of mycosporine-like amino acids, their wound healing properties and in vitro anti-aging effects. *Mar. Drugs* 18: 35.

Pathak J, Singh P R, Sinha R P, Rastogi R P (2021). Evolution and Distribution of Cyanobacteria. In: Rastogi RP, editor. *Ecophysiology and Biochemistry of Cyanobacteria*: Springer, Singapore. pp. 1.

Pignantello J J, Porwoll J, Carson R E, Xavier A, Gleason F K, Wood J M (1983). Structure of the antibiotic cyanobacterin, a chlorine containing γ-lactone from the freshwater cyanobacterium *Scytonema hofmanni*. *J. Org. Chem.* 48: 4035-4038.

Pinti D L, Mineau R, Clement V (2009). Hydrothermal alteration and microfossil artefacts of the 3,465-million-year-old Apex chert. *Nat. Geosci.* 2: 640-643.

Rastogi R P, Sinha R P (2009). Biotechnological and industrial significance of cyanobacterial secondary metabolites. *Biotechnol. Adv.* 27(4): 521-539.

Roco M C. *The Long View of Nanotechnology Development: The National Nanotechnology Initiative at 10 Years*. Springer; 2011.

Rossi F, Li H, Liu Y, De Philippis R (2017). Cyanobacterial inoculation (cyanobacterisation): perspectives for the development of a standardized multifunctional technology for soil fertilization and desertification reversal. *Earth-Sci. Rev.* 171: 28-43.

Sangeetha M, Menakha M, Vijayakumar S (2014). *In silico* prediction of anticancer cyanobacterial drug from *Nostoc*. *Biomed. Prev. Nutr.* 4(1): 71-73.

Schmid G (2011). *Nanoparticles: From Theory to Application*: John Wiley & Sons.

Schopf J W (1993) Microfossils of the early archean Apex chert: new evidence of the antiquity of life. *Science* 260: 640-646.

Schopf J W, Kitajima K, Spicuzza M J, Kudryavtsev A B, Valley J W (2017). SIMS analyses of the oldest known assemblage of microfossils document their taxon correlated carbon isotope compositions. *Proc. Natl. Acad. Sci. Unit. States Am.* 115: 53-58.

Schopf J W, Packer B M (1987). Early archean (3.3-3.5-billion-year-old) microfossils from warrawoona group, Australia. *Science* 237: 70-73.

Sinha R P, Klisch M, Helbling E W, Häder D P (2001). Induction of mycosporine-like amino acids (MAAs) in cyanobacteria by solar ultraviolet-B radiation. *J. Photochem. Photobiol. B: Biol.* 60(2-3): 129-135.

Song T, Mårtensson L, Eriksson T, Zheng W, Rasmussen U (2005). Biodiversity and seasonal variation of the cyanobacterial assemblage in a rice paddy field in Fujian, China. *FEMS Microbiol. Ecol.* 54(1): 131-140.

Sood A, Renuka N, Prasanna R, Ahluwalia A S (2015). Cyanobacteria as potential options for wastewater treatment. In: Ansari A A, Gill S S, Gill R, Lanza G R, Newman, editors. *Phytoremediation: Management of Environmental Contaminants, Volume 2:* Springer International Publishing, Cham, pp 83-93.

Sun Y, Zhang N, Zhou J, Dong S, Zhang X, Guo L, Guo G (2020). Distribution, contents, and types of mycosporine-like amino acids (MAAs) in marine macroalgae and a database for MAAs based on these characteristics. *Mar. Drugs* 18(1): 43.

Tarasuntisuk S, Patipong T, Hibino T, Waditee-Sirisattha R, Kageyama H (2018). Inhibitory effects of mycosporine-2-glycine isolated from a halotolerant cyanobacterium on protein glycation and collagenase activity. *Lett. Appl. Microbiol.* 67: 314-320.

Thakkar K N, Mhatre S S, Parikh R Y (2010). Biological synthesis of metallic nanoparticles. *Nanomedicine* 6(2): 257-262.

Vega J, Bonomi-Barufi J, Gómez-Pinchetti J L, Figueroa F L (2020). Cyanobacteria and red macroalgae as potential sources of antioxidants and UV radiation-absorbing compounds for cosmeceutical applications. *Mar. Drugs* 18(12): 659.

Whitton B A (2002). Soils and rice-fields. In: Whitton BA, Potts M, editors. *The Ecology of Cyanobacteria: Their Diversity in Time and Space:* Springer Netherlands, Dordrecht, pp 233-255.

Woese C R, Fox G E (1977). Phylogenetic structure of the prokaryotic domain: the primary kingdoms. *Proc. Natl. Acad. Sci. USA* 74(11): 5088-5090.

Yamazaki Y, Kitajima M, Arita M, Takayama H, Sudo H, Yamazaki M, Aimi N, Saito K (2004). Biosynthesis of camptothecin. In silico and in vivo tracer study from [1-$^{13}$C] glucose. *Plant Physiol.* 134(1): 161-170.

# Chapter 2

# Cyanobacterial Life History

**Deepak K. Singh[1], Rajeshwar P. Sinha[2] and Jainendra Pathak[3,*]**

[1]Department of Botany, A.N.D. Kisan P.G. College, Babhnan
(Affiliated to R.M.L. Avadh University, Faizabad), Gonda, India
[2]Laboratory of Photobiology and Molecular Microbiology,
Centre of Advanced Study in Botany, Institute of Science, Banaras Hindu University, Varanasi, India
[3]Department of Botany, Pt. Jawaharlal Nehru College
(Affiliated to Bundelkhand University, Jhansi), Banda, India

## Abstract

Cyanobacteria played an important role in the evolution of early Earth and the biosphere. They are responsible for the oxygenation of the atmosphere and oceans since the Great Oxidation Event around 2.4 Ga. They are the major primary producers in past and present oceans, and the ancestors of the chloroplast. Nevertheless, the identification of cyanobacteria in the early fossil record remains ambiguous because the morphological criteria commonly used are not always reliable for microfossil interpretation. In recent times, new biosignatures specific to cyanobacteria were found. This chapter discusses the classical and new cyanobacterial biosignatures, cyanobacteria fossil records and the challenges of molecular approaches on modern cyanobacteria. It is important to understand the possible timing and pattern of the evolution of cyanobacteria and oxygenic photosynthesis.

**Keywords:** biosignatures, cyanobacteria, fossils, mycosporine-like amino acids, scytonemin

[*] Corresponding Author's Email: gpathakvns@gmail.com.

In: Cyanobacteria
Editor: Rajeshwar P. Sinha
ISBN: 979-8-88697-169-9

## Introduction

Modern cyanobacteria constitute an ancient and well-diversified bacterial phylum, with unique complex morphologies and cellular differentiation (Demoulina et al. 2019). They play a key role in food webs as primary producers performing oxygenic photosynthesis. Cyanobacteria also played a major role in early biogeochemical fluxes and in life and Earth evolution (Pathak et al. 2021). They are the only prokaryotic organisms that perform oxygenic photosynthesis, and are thus generally held responsible for the rise of oxygen in the atmosphere and oceans around 2.4 Ga, during the so-called Great Oxidation Event (GOE) (Bekker et al. 2004; Hannah et al. 2004; Demoulina et al. 2019), facilitated by geological processes (Kasting 2013). Oxygenic photosynthesis has enabled the oxygenation of oceanic and terrestrial niches, and the diversification of complex life (Lewis 2017; Demoulina et al. 2019). Modern eukaryotes diversified from a last eukaryotic common ancestor (LECA), an aerobe protist with a mitochondrion (Roger et al. 2017). At least 1.05 Ga ago, oxygenic photosynthesis spread among some eukaryotic clades, giving rise to diverse types of algae and later to plants (Demoulina et al. 2019) (Figure 1).

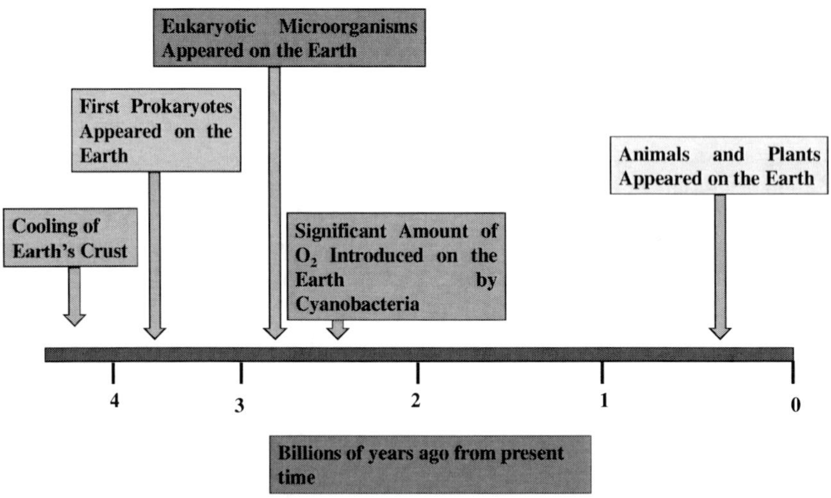


**Figure 1.** Important events which occurred during the evolution of life on the Earth.

This important evolutionary step was due to the primary endosymbiosis of a cyanobacterium within a unicellular eukaryote, and subsequent higher-order endosymbiotic events (Sagan 1967; Delwiche 1999; Schirrmeister et al. 2011), although the identity and habitat of the cyanobacterial ancestor of the chloroplast are still debated (Falcón et al. 2010; Blank 2013; Nakov et al. 2017; Ponce-Toledo et al. 2017; Sánchez-Baracaldo et al. 2017). Despite the importance of cyanobacteria in the early evolution of Earth and life, fundamental questions remain about their origin, the timing and pattern of their diversification, and the origin of oxygenic photosynthesis (Fischer et. al 2016). Several types of evidence are used to reconstruct the fossil record of cyanobacteria, but all have their limitations and challenges (Demoulina et al. 2019).

Stromatolites are usually associated with cyanobacterial activity. However, although conical stromatolites seem to plead for oxygenic photosynthesis (Bosak et al. 2009), other types of stromatolites and microbially induced sedimentary structures (MISS) (Noffke et al. 2001; Heubeck 2009; Homann et al. 2018) may have been produced by non-cyanobacterial lineages, such as anoxygenic phototrophs (Bosak et al. 2013), or by/in association with methanotrophs (Slotznick and Fischer 2016). This suggests that stromatolites and MISS are not necessarily indicative of cyanobacteria activity, and perhaps not even of photosynthesis (Suosaari et al. 2016). Biomarkers (fossil molecules) can indicate the presence of metabolisms such as oxygenic photosynthesis, but they are preserved only in well-preserved unmetamorphosed rocks, and contamination is a challenge (Alleon and Summons 2019). Among those, lipids such as 2-methyl-hopanes are produced by cyanobacteria (Schinteie and Brocks 2017) but not only (Rashby et al. 2007), and pigments such as porphyrins with N isotope composition (Gueneli et al. 2018). Microfossils may provide direct evidence for cyanobacteria, but their identification is often ambiguous. Cyanobacterial identity of only three fossil taxa is not debated: *Eoentophysalis*, *Eohyella* and *Polybessurus*. *Eoentophysalis belcherensis* is the oldest microfossil interpreted with certainty as a cyanobacterium (Hofmann 1976). This microfossil has been described from 1.89-1.84 Ga silicified stromatolites of the Belcher Supergroup, Hudson Bay, Canada (Hofmann 1976). Discrepancies between the geological and fossil records and molecular phylogenies remain, and the origin and evolution of cyanobacteria, oxygenic photosynthesis, and the chloroplast are still debated (Demoulina et al. 2019). This chapter deals with cyanobacterial life history focussing on the classical and new biosignatures of cyanobacteria, their fossil record, molecular phylogenies, and molecular clocks.

## Origin of Cyanobacteria and Oxygenic Photosynthesis

In most cyanobacterial phylogenetic analyses that are using a noncyanobacterial outgroup to root the tree, the reference strain *Gloeobacter violaceus* PCC7421 has a basal position (Seo and Yokota 2003; Hoffmann et al. 2005; Gupta and Mathews 2010). Consequently, the phylogenetic node bearing *Gloeobacter* and the rest of modern cyanobacterial lineages serve as calibration for the origin of cyanobacteria in several studies (Blank and Sanchez-Baracaldo 2010; Schirrmeister et al. 2013; 2015; Demoulina et al. 2019). In these studies, the authors have set different root limit ages, so that the maximum root age may vary between the earliest estimate of abiogenesis around 4.2 Ga (Blank and Sanchez-Baracaldo 2010), the end of the Late Heavy Bombardment at 3.85 Ga (Schirrmeister et al. 2015), or the GOE at 2.4 Ga (Battistuzzi et al. 2004). Recently, two newly discovered lineages were proposed as sister groups of the cyanobacterial phylum, the *Melainabacteria* (Di Rienzi et al. 2013) and the *Sericytochromatia* (Soo et al. 2017). Of note, these lineages (mostly known as metagenomics assemblies) do not contain genes required for photosynthesis nor carbon fixation (Soo et al. 2017). The integration of genetic data from *Melainabacteria* and *Sericytochromatia* as outgroups for molecular clock analyses suggested that cyanobacteria evolved just before the GOE (Shih et al. 2017; Magnabosco et al. 2018). Taken together, this suggests that oxygenic photosynthesis has evolved after the separation of cyanobacteria from *Melainabacteria* (Shih et al. 2017; Magnabosco et al. 2018). However, the loss of photosynthetic capability in the ancestor of the three lineages before or at the time of GOE has been suggested as an alternative hypothesis that cannot be ruled out (Martin et al. 2018). Among bacterial phototrophs (cyanobacteria, green S-bacteria, green non sulfur bacteria, purple bacteria, heliobacteria, some acidobacteria and gemmatimonadetes), cyanobacteria are the only lineage that possesses two photosynthetic reaction centers of the Fe-S type (RCI) and Quinone type (RCII), whereas anoxygenic bacteria possess either the Fe-S or Quinone type (Demoulina et al. 2019). So far, three hypotheses were proposed to explain the presence of both types of RC in modern cyanobacteria. Two of these hypotheses suggest that both RCs would have been present in an anoxygenic phototrophic ancestor (Demoulina et al. 2019). In a first hypothesis, RCs evolved within the common ancestor of (all) bacterial phototrophs. Both of them were kept in the cyanobacterial lineage whereas there was a selective loss of one type of each in the modern anoxygenic lineage ancestors (Olson 1970; Olson and Pierson 1987). In the second hypothesis, the RCI and RCII would have emerged in the protocyanobacterial ancestor by duplication of a unique ancestral RC. This was followed by lateral transfer of a different RC type to the ancestors of the modern anoxygenic phototrophs (Allen and Martin 2007). The existence of an anoxygenic cyanobacterial ancestor may be supported by the occurrence of several genes involved in anoxygenic photosynthesis in modern cyanobacterial genomes (Mulkidjanian et al. 2006), and the cooccurrence of both anoxygenic and oxygenic photosynthesis in several lineages of modern cyanobacteria clades (Cohen et al. 1986; Klatt et al. 2016). A third hypothesis rather suggests the independent evolution of the two RCs in separate lineages of anoxygenic phototrophs and their lateral transfer into a protocyanobacterial ancestor, the so-called fusion hypothesis (Mathis et al. 1990; Blankenship 1992). At least in purple bacteria, the genes for RCII are clustered in an ensemble of operons, the photosynthesis gene cluster (PGC), some organisms even harbor the PGC on large plasmids. This observation makes the transfer of full photosystems highly plausible, and recent events of such kind have been convincingly inferred in Rhodobacteraceae (Brinkmann et al. 2018). In order to test the likelihood of the ancient transfer of photosystems between the bacterial phototrophic lineages, Magnabosco et al. (2018) added horizontal gene transfer events information of two genes (encoding for Mg-chelatase and S-adenosyl-L-homocysteine hydrolase) as additional constraints to their models to estimate the stem age of the bacterial phototrophs (Cyanobacteria, green S-bacteria and green non-sulfur bacteria). It was assumed that such estimates would allow them to investigate the feasibility and timing of the RC transfer events between phototrophic lineages (Demoulina et al. 2019). Their results excluded the possibility of a RC transfer from the green sulfur bacteria to cyanobacteria, and thus, invalidated the fusion hypothesis. However, they were not able to choose between the two hypotheses suggesting that both RCs emerged from a common ancestor (Magnabosco et al. 2018). First, the reaction centers RCI and RCII would operate separately and asynchronously in the same ancestral anoxygenicphototroph organism (Demoulina et al. 2019). The RCI would catalyze $H_2S$ oxidation as in green S-bacteria, while the RCII would act as light-dependent electron transporter as in purple bacteria (Shestakov and Karbysheva 2017). A water-splitting RCII could also have evolved from an ancestral RCII type already capable of photosynthesis and manganese oxidation (Fischer et al. 2015). However, the evolutions of these processes, and the early or late evolution of oxygenic photosynthesis, are still debated (Cardona et al. 2018). Figure 2 shows various processes responsible for evolution of cyanobacteria.

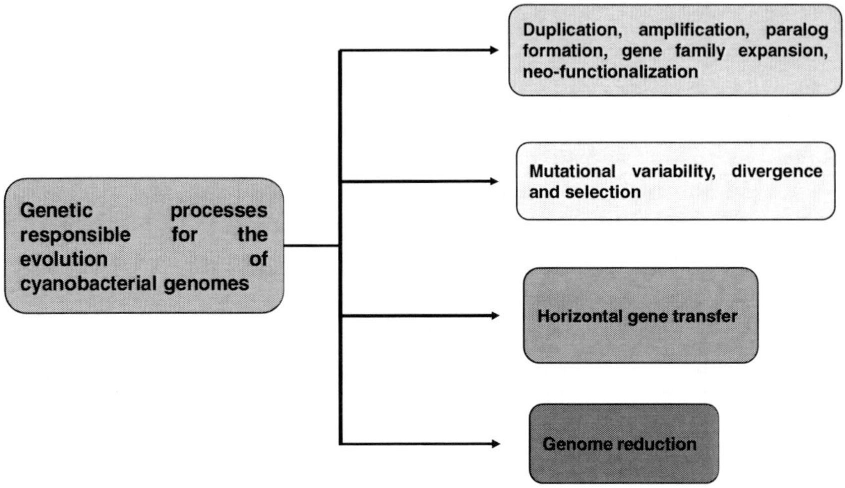


**Figure 2.** Various genetic processes which contributed in the evolution of cyanobacteria.

## Classification, Morphology and Division Pattern of Cyanobacteria

Stanier and colleagues (Stanier et al. 1978) recognized the prokaryotic nature of the cyanobacteria and proposed to follow the International Code of Nomenclature for Bacteria. According to the Bergey's Manual of Systematic Bacteriology, and following the Stanier approach and Rippka et al. (1979) concepts, cyanobacteria were divided into two groups: unicellular and multicellular filamentous cyanobacteria; and five subsections (Figure 3) based on morphological criteria, and corresponding to five former cyanobacterial orders: Chroococcales (section I) and Pleurocapsales (section II) harbor unicellular cells, which divide by binary fission in one or multiple plans and are solitary or arranged in colonies. Pleurocapsales can also produce small easily dispersed cells (baeocytes) after division by multiple fissions (Pathak et al. 2021). Within the multicellular filamentous cyanobacteria, Oscillatoriales (section III) have only vegetative cells arranged in filaments, whereas Nostocales (section IV), and Stigonematales (section V) are capable of producing specialized cells, the heterocytes, which allow N fixation in an anoxic compartment; or can exhibit environmental stress resistant cells called akinetes. Besides, cyanobacteria from section V are also characterized by their ability to divide in more than one plane and form true branched trichomes (Rippka et al. 1979).

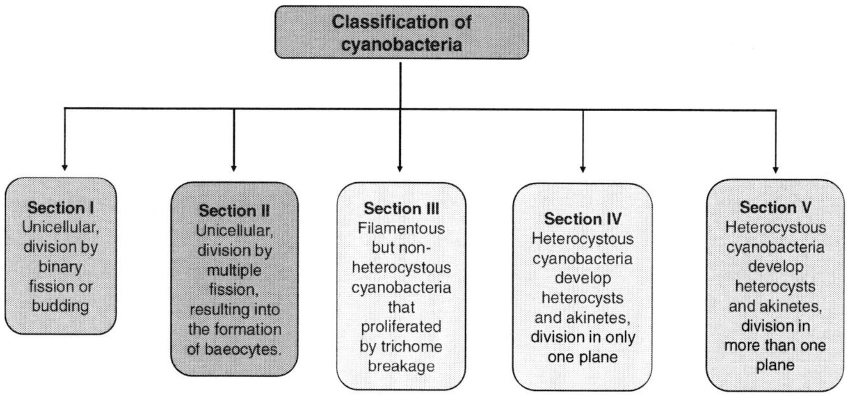


**Figure 3.** Conventional classification of cyanobacteria.

Interpreting microfossils as cyanobacteria is generally based on morphological criteria, and their mode of division (Demoulina et al. 2019). However, the simple shape of many microfossils makes an unambiguous identification very difficult (Schirrmeister et al. 2016). The size of cyanobacteria cells or filaments may be used as a taxonomic criterion for microfossil biological affinity. Moreover, different fossil and modern groups of bacteria, cyanobacteria and eukaryotic algae have overlapping size ranges. Therefore, microfossil size alone does not constitute a reliable criterion for the interpretation of a microfossil as a cyanobacterium (Altermann and Kazmierczak 2003).

Cyanobacteria form spheroidal or rod-shaped cells, filaments or tubes. Some of them occur as spiral filaments (*Spirulina*) but this morphology also occurs in other bacteria (*Leptospira*) (Adler and de la Pena Moctezuma 2010). Cyanobacteria from the Nostocales and Stigonematales orders may present some of the most complex morphologies among prokaryotes, including specialized cells such as larger elongated cells with thicker walls (akinetes) and round cells in or at the end of the filament (heterocytes, fixing nitrogen) (Demoulina et al. 2019). The complex multicellular filamentous forms of nostocalean and (uniseriate or multiseriate) stigonematalean cyanobacteria may also display false or true branching. These more complex cyanobacteria may present unique characters allowing to not only identify their fossils as cyanobacteria, but also as specific cyanobacterial clades, more useful as calibrating points (Demoulina et al. 2019). However, most cyanobacteria have simple morphologies, widespread in the three domains of life. The division pattern of fossil cells, when preserved, indicates their reproduction, and in some cases, may be indicative of particular taxonomic groups (Knoll and Golubic 1979; Golubic and Seong-Joo 1999). Cyanobacteria reproduce asexually by binary or multiple fissions. Multiple fissions may lead to the formation of baeocytes, which are small cells formed within the parental cell (Castenholz et al. 2001). They can also divide in two or three planes of division, such as *Entophysalis* spp., the modern counterpart of *Eoentophysalis*. Moreover, some cyanobacteria occur as colonies either within (*Gloeocapsa* spp., modern counterpart of *Gloeodiniopsis*) or without (*Cyanobium* spp., *Synechococcus* spp., and *Synechocystis* spp.) a thick polysaccharide envelope (Demoulina et al. 2019). They can also present more or less organized and dense aggregates and even cells organized as tablets (*Merismopedia* spp.). Some cyanobacteria can also reproduce with the aid of hormogonia, which are short filaments resulting from break up of longer filaments (Graham and Wilcox 2000). Trichomes are ensheathed individual filaments (Graham and Wilcox 2000).

## Cyanobacterial Ultrastructure

The wall ultrastructure of modern cyanobacteria consists of a peptidoglycan layer of varying thickness in the periplasmic space between a cytoplasmic and an outer membrane, with generally an external layer (Šmarda et al. 2002). In some cases, a transparent or pigmented exopolysaccharidic (EPS) envelope, the so-called sheath, may surround cells, filaments or colonies (Demoulina et al. 2019). Cyanobacterial EPS includes two forms, one attached to the cell wall and one released in the environments. They may be composed by up to twelve different monosaccharides, including pentoses, hexoses, and acid hexoses as well as methyl sugars and/or amino sugars (e.g., N-acetyl glucosamine, 2,3-O-methyl rhamnose, 3-Omethyl rhamnose, 4-O-methyl rhamnose, 3-O-methyl glucose) (Pereira et al. 2009). In the fossil record, sheaths of cyanobacteria are very common given that they are more readily fossilized, or less easily degraded, than the unsheathed ones (Bartley 1996). The association of sheaths with clay minerals is more frequent than with unsheathed cyanobacteria, which helps the preservation of these structures (Lepot et al. 2014; Newman et al. 2017). This association would be due to differences in the chemical composition between the EPS and the sheath, and the rapid coating of sheaths (Newman et al. 2017). As phototrophs, all cyanobacteria but *Gloeobacter* spp (Rippka et al. 1974) possess internal membranes called thylakoids hosting the photosynthetic apparatus, the two photosystems and their pigments. The arrangement of the thylakoid in the cells is well organized and coincides with cyanobacterial lineages taxonomy (Komárek and Kaštovský 2003). Since cyanobacteria are the only oxygenic photosynthetic organisms among prokaryotes, the presence of preserved thylakoids in microfossils would be a reliable criterion to confirm that they were able to make this type of photosynthesis (Pacton et al. 2008). So far, ultralaminae interpreted as thylakoids based on their stacking and thickness, were preserved in microfossils as old as 155 Ma (Pacton et al. 2008). In acid extraction of 600 Ky microbial mats also revealed the resilience of thylakoids displaying their concentric structure over the hosting cell walls (Lepot et al. 2014). Moreover, although eukaryotes also have thylakoids, they are compartmentalized in chloroplasts and often are arranged differently than in cyanobacteria (Demoulina et al. 2019).

## The Cyanobacterial Fossil Record

Microfossils interpreted as cyanobacteria have been reported in rocks as old as the early Archean, but their biogenicity and interpretation is highly debated (Demoulina et al. 2019). The necessity of reliable criteria for the study of microfossils is well illustrated with the famous controversy surrounding the "microfossils" from the 3.45 Ga Apex Chert, Australia (Schopf and Packer 1987). These traces interpreted as fossil cyanobacteria based on their morphology and the geological context (Schopf and Packer 1987) were subsequently reassessed either as pseudofossils (Brasier et al. 2005; Pinti et al. 2009; Bower et al. 2016), contaminants (Altermann and Kazmierczak 2003), mineral artefacts (Javaux and Knoll 2017), or microfossils (Schopf 1993; De Gregorio et al. 2009; Schopf et al. 2017) and the geological context was revised (Brasier et al. 2005). Knoll and Golubic (Knoll and Golubic 1992) determined a confidence range for these microfossils, since most of them are identified based on morphology, sometimes coupled with their occurrence in the photic zone, despite the possibility of morphological convergence. So far, only three taxa are unambiguously identified as cyanobacteria: *Eoentophysalis*, *Polybessurus* and *Eohyella*, because they also present distinctive modes of division (Demoulina et al. 2019).

## Eoentophysalis

Based on morphology, *Entophysalis* belongs to the order Chroococcales (Komárek 1999) and consists of coccoidal unicells forming characteristic pustular palmelloid colonies. *Eoentophysalis belcherensis* was first described in the 1.89-1.84 Ga Belcher Supergroup, Hudson Bay, Canada, where this colonial microorganism formed mats in silicified stromatolites (Golubic and Hofmann 1976; Hofmann 1976). The morphology of *Entophysalis* colonies is due to its mode of cell division by binary fission in three perpendicular planes. Modern *Entophysalis* produce highly hydrated exopolymer envelopes that expand during cell division. Another particularity of *Entophysalis* colonies is the presence in the outer layers of the colonies of a yellow brown UV-protecting pigment, scytonemin that is produced by the most external cells for protection against intense solar radiation (Golubic and Abed 2010). *Entophysalis* cells grow in the form of mats in the intertidal range of shallow marine basins (Golubic and Hofmann 1976; Golubic and Abed 2010). This cyanobacterial genus may also precipitate micrite in stromatolites such as those in the shallow hypersaline Hamelin Pool, Shark Bay (Western Australia), in association with other cyanobacteria but also other bacteria, and halophilic and methanogenic archaea (Papineau et al. 2005; Burns et al. 2009; Golubic and Abed 2010; Suosaari et al. 2016). The microfossils *Eoentophysalis belcherensis* were discovered in silicified stromatolites from the Kasegalik and McLeary formations of the Paleoproterozoic Belcher Supergroup (1.89-1.84 Ga). The geological context of these two formations corresponds to intertidal mudflats as well as shallow subtidal and supratidal zones (Hofmann and Jackson 1969; Hofmann 1976).

## Polybessurus

*Polybessurus* is another Proterozoic microfossil unambiguously identified as a cyanobacterium (Green et al. 1987). *Polybessurus* was formally described from silicified carbonates of the ca. 950-680 Ma Eleonore Bay Group (East Greenland) (Green et al. 1987). Its oldest occurrence is in the ca. 1.35-1.01 GaAvzyan Formation, Russia (Sergeev 1994). *Polybessurus* is a fossil of a coccoidal unicellular microorganism with a very particular morphology unique to cyanobacteria. However, *Polybessurus* reproduced by the formation of baeocytes (smaller cells) which is characteristic of the order Pleurocapsales, but is unlike the modern *Cyanostylon* (Green et al. 1987). The particular morphology of *Polybessurus*, its mode of reproduction and ecology enable its affiliation to cyanobacteria, probably within the Pleurocapsales.

## Eohyella

The Limestone-Dolomite series of the ca. 950-680 Ma Eleonore Bay Group (central East Greenland) also preserve a group of particular microfossils showing distinct endolithic behaviour (Green et al. 1988). Among them, *Eohyella* is a coccoidal microfossil forming pseudofilaments (where juxtaposed cells do not share a common wall), sometimes branching depending on the microfossil species, by the juxtaposition of several cells surrounded by extracellular envelopes. *Eohyella* was qualified as an euendolith cyanobacterium because of its orientation within the substrate: they crosscut the substrate laminae (Green et al. 1988; Tribollet and Golubic 2011). *Eohyella* microfossils occur in oolites and pisolites of shallow peritidal environment. Within the euendolithic cyanobacteria group of the Limestone-Dolomite series, *Eohyella* was the most abundant genus. So far, its oldest occurrence was found in the 1.63 Ga Dahongyu Formation, China (Zhang 1988). The morphology, behaviour and paleoenvironment of *Eohyella* are similar to those of the modern genus *Hyella*, which enabled the identification of this microfossil as a cyanobacterium, of the order of Pleurocapsales.

Other microfossils are also identified but still considered as probable or possible cyanobacteria, such as *Anhuithrix magna* (Pang et al. 2018), *Archaeoellipsoides* (Sergeev et al. 2012), *Eomicrocystis* (Golovenok and Belova 1984), *Eosynechococcus*, *Leiosphaeridia*, *Myxococcoides* (Sergeev et al. 2012), *Gloeodiniopsis* (Schopf 1968), *Obruchevella* (Shi et al. 2017), *Oscillatoriopsis* (Schopf 1968, Mendelson and Schopf 1982; Butterfield et al. 1994), *Palaeolyngbya* (Schopf 1968; Macdonald et al. 2010), *Polysphaeroides* (Timofeev et al. 1976) and *Siphonophycus* (Knoll and Golubic 1992; Javaux et al. 2013).

## Cyanobacterial Molecular Fossils

Cyanobacteria produce lipids (2-methyl-hopanes) (Summons et al. 1999; Brocks et al. 2005) and pigments that can potentially be preserved in the unmetamorphosed geological record (Schinteie and Brocks 2017; Alleon and

Summons 2019). So far, the lipids 2-methyl-hopanes were extracted from bitumen in black shales as old as 1.6 Ga (McArthur basin, Australia) (Summons et al. 1999; Brocks et al. 2005). Their oldest record at 2.7 Ga (Brocks et al. 1999) was reassessed as younger contaminants (Rasmussen et al. 2008; French et al. 2015). These fossilized lipids were first attributed to cyanobacteria (Summons et al. 1999), but it is now acknowledged that they might be signatures of other bacterial lineages (Rashby et al. 2007). Pigments are also used as signatures for (anoxygenic and oxygenic) photosynthesis. More precisely, the presence of ancient chlorophylls can be detected by the preservation of their nitrogen-containing tetrapyrrole (porphyrin) core. Moreover, some fossilized forms of carotenoids, such as okenane and isorenieratane evidence the presence of purple S-bacteria, green S-bacteria and other bacterial lineages including cyanobacteria (Brocks et al. 2005). The oldest porphyrins reported so far are preserved in shales of the 1.1 GaTaoudeni basin, Mauritania (Gueneli et al. 2018), which also preserves exquisite microfossils, including eukaryotes and cyanobacteria (Beghin et al. 2017a). These fossilized pigments exhibit a specific N isotope fractionation indicating a cyanobacterial source and permit to suggest that cyanobacteria were the dominant primary producers in mid-proterozoic oceans (Gueneli et al. 2018). Other UV-protective (sunscreen) pigments may be used as signature for bacterial life, such as the mycosporine-like amino acids (MAAs), and two colored molecules specific of cyanobacteria: scytonemin and gloeocapsin (Demoulina et al. 2019; Singh et al. 2020; Pathak et al. 2021). For instance, the combined analyses of modern cultures and fossil (4500 years BP) microbial mats of cyanobacteria from Antarctica revealed that cells and pigmented filamentous sheaths can withstand acetolysis (used to isolate them from the mineral matrix) and retain their molecular signature identified by FTIR microspectroscopy (Lepot et al. 2014). They are also preserved in siliciclastic sediments by precipitation of nano-aragonite and clay minerals (Lepot et al. 2014). However, the scarce knowledge of the composition of pigments, preservable cells, cysts and other structures produced by modern microorganisms, and of their transformation and alteration through fossilization, limits this approach. Moreover, high temperature and pressure (metamorphism) after burial can alter even more or erase original biological properties (Marshall et al. 2005). Raman microspectroscopy enables estimating the temperature at which the organic material has been submitted (Baludikay et al. 2018), which is necessary to interpret properly the spectra obtained by FTIR microspectroscopy (Marshall et al. 2005). Raman spectroscopy also permits to characterize molecules in modern organisms, including cyanobacteria (Edwards et al. 1999; Jehlièka and Oren 2013; Storme et al. 2015; De Oliveira et al. 2015). Scytonemin is a molecule consisting of phenolic and indolic subunits (Proteau et al. 1993; Rastogi et al. 2013; Pathak et al. 2020). Today it is notably biosynthesized in benthic filaments of *Calothrix* sp. (Lepot et al. 2014), and in the endolithic cyanobacteria *Hyella* sp. and *Solentia* sp., from coastal carbonates (Storme et al. 2015). It may be a promising signature of cyanobacteria given that it can be fossilized (Lepot et al. 2014). In older deposits from Antarctica, derivatives of scytonemin and carotenoids can be extracted from 125000 years BP sediments (Hodgson et al. 1997). However, the preservation potential of scytonemin in older rocks is not known. Artificial taphonomic experiments of decaying cyanobacterial cultures showed the recalcitrance of filamentous polysaccharide sheaths, possibly helped by the presence of pigments (Bartley 1996). However, in lake sediments from Antarctica, both brown (scytonemin-rich) and transparent (scytonemin-poor) filamentous sheaths were well preserved; hence, scytonemin probably was not the factor driving their preservation (Lepot et al. 2014). Gloeocapsin, an enigmatic pigment detected in the thick sheath enveloping colonies of the cyanobacterial genus *Gloeocapsa* growing on carbonate surfaces (Storme et al. 2015) and in lichens (with cyanobacterial symbionts), might also become a useful indicator, but its molecular composition remains to be characterized and its preservation potential is currently unknown (Storme et al. 2015).

## Behavior and Paleoecology of Cyanobacteria

Establishing the paleoecology of fossil assemblages is difficult, although combining micropaleontology and geochemistry at high-resolution along a paleoenvironmental gradient may be informative (Beghin et al. 2017b; Javaux and Knoll 2017). At the microscale, the distribution and the orientation of microfossils within rocks may also provide insights about their ecology (Knoll and Golubic 1992). Moreover, their orientation might help to infer the behavior of microfossils, such as phototropism of filaments erected towards the light or chemotropism, rock-boring by endolithic microbes, mat-building benthic microbes, or plankton settling down with no preferential orientation (Demoulina et al. 2019). Golubic and Seong-Joo (1999) and Green et al. (1988) concluded that *Eohyella* was a euendolithic cyanobacterium owing to its orientation in oolithes, by analogy with its modern counterpart, *Hyella* (Knoll and Golubic 1992). *Euendoliths* are rock-inhabiting microorganisms, which dissolve mineralized substrates to penetrate the rock (Golubic et al. 1981), in contrast to other endoliths, e.g., chasmoendoliths, which are endoliths that colonize existing rock fissures (Golubic et al. 1981).

## Isotopic Fractionation

Analyses of N isotopic fractionation measured on molecular fossils can also indicate metabolism and cyanobacterial affinity (Demoulina et al. 2019). Phototrophic organisms have a specific nitrogen isotopic offset between total biomass and chloropigments (Gueneli et al. 2018). Based on laboratory experiments, this offset is independent of the nitrogen source ($NH_4^+$, $N_2$, or $NO_3^-$) and its isotopic composition and from redox conditions during cell growth. The N isotopic offset remains relatively constant within different phototrophic organisms such as cyanobacteria, bacteria, red or green algae or plants, and thus may help identify the source organisms. This offset was notably measured on in 1.1 Gaporphyrins, permitting to relate them to cyanobacteria (Gueneli et al. 2018).

## Intracellular Biomineralization

In modern oceans and alkaline lakes, some cyanobacteria have the capacity to form beads of intracellular Ca-carbonates (Couradeau et al. 2012; Benzerara et al. 2014), but their preservation potential is unknown. Some extant cyanobacteria also have the capacity to produce intracellular ferric phosphates (Brown et al. 2010). In the 1.88 Ga Gunflint Formation, Canada, specific microfossil taxa preserved in silicified stromatolites contain internal Fe-silicate and Fe-carbonate nanocrystals, absent from the external wall surfaces. This feature and its distribution pattern are consistent with intracellular biomineralization, with subsequent recrystallization, and not with known patterns of diagenesis. Thus, the combination of large size, morphology and intracellular Fe biominerals is consistent with a cyanobacterial affinity, and not with other known Fe-mineralizing microorganisms (Lepot et al. 2017). High-resolution observations (Lepot et al. 2017), might possibly reveal new cyanobacterial signatures at the nano-scale, in the rock record.

## Molecular Dating of Cyanobacteria

Since the late 90's a myriad of phylogenetic studies based on single loci (16S rRNA or some protein) have been carried out (Schirrmeister et al. 2011; Shih et al. 2013). In 2013, the large CyanoGEBA (Genomic Encyclopedia of Bacteria and Archaea) sequencing project led to an improvement in terms of genomic coverage of cyanobacterial taxa, notably by sequencing genomes belonging to sections II and V (Shih et al. 2013). Since then, the number of publicly available cyanobacterial genomes has dramatically increased (Demoulina et al. 2019). Yet, their quality, especially contamination of cyanobacterial assemblies by non-cyanobacterial DNA, has gone worse in parallel, which is a problem for phylogenetic analyses (Cornet et al. 2018a). Moreover, the real biodiversity of cyanobacteria is still under-represented in genomic databases, mainly because of a biased sampling in the sequencing effort (Dvořák et al. 2015; Cornet et al. 2018b). Nevertheless, since Shih et al. (2013), many authors have taken advantage of these new genomes to carry out phylogenomic analyses (Dagan et al. 2013; Li et al. 2014; Ochoa de Alda et al. 2014; Sánchez-Baracaldo 2015; Uyeda et al. 2016; Walter et al. 2017; Magnabosco et al. 2018). Most of these studies focussed on integrating new genomic data to the same set of 100-200 hundred loci (Sánchez-Baracaldo et al. 2017). By doing so, few tried to handle the methodological difficulties associated with the use of largescale data to resolve the phylogeny of old groups such as cyanobacteria, whether during dataset assembly (contamination, horizontal gene transfer, hidden paralogy) or phylogenetic inference (substitutional saturation, compositional bias, heterogeneous evolutionary rate) (Philippe et al. 2011; Irisarri et al. 2017; Philippe et al. 2017; Simion et al. 2017). Consequently, among the ten phylogenomic studies cited above, only three are in agreement on the cyanobacterial backbone (Shih et al. 2013; Sánchez-Baracaldo 2015; Ponce-Toledo et al. 2017).

## Conclusion

Cyanobacterial fossil record starts unambiguously at 1.89-1.84 Ga and the minimum age for the oxygenic photosynthesis starts with the GOE around 2.4 Ga. *Eoentophysalis*, *Polybessurus* and *Eohyella* microfossils present a combination of distinctive morphologies, modes of division and ecology that are representative of the cyanobacterial phylum. Possible geochemical traces of oxygenation and the metabolisms involved in stromatolites and MISS builders in the Archean have been discussed. Still there is debate on the origin and timing of oxygenic photosynthesis although some studies corroborate that the evolution of oxygenic photosynthesis happened right before the GOE which would then be a consequence of this evolution. The origin, timing and environment of the primary endosymbiotic event giving rise to eukaryotic algae are also still debated. Therefore, it is essential to define new biosignatures indicative of cyanobacteria in order to reassess their fossil record and provide new calibration points for molecular clocks. Those biosignatures will combine analyses of the morphology, ultrastructure and ecology of promising microfossils, with their molecular (lipids and pigments), metal and isotopic composition. Identification of these fossils, not only as cyanobacteria, but of specific clades within the cyanobacteria, will help in understanding of their diversification record and provide new calibration points. Coupling these new microfossil calibration points with improved molecular phylogenies and alternative molecular clocks will then enable to date the minimum ages of important biological events such as the origin of oxygenic photosynthesis and the acquisition of chloroplasts among photosynthetic eukaryotic lineages.

## Acknowledgments

Deepak K. Singh and Jainendra Pathak are thankful to Prof. Rajeshwar P. Sinha for providing immense guidance and support during the preparation of manuscript.

## References

Adler B, de la Pena Moctezuma A (2010) *Leptospira* and *Leptospirosis*. *Vet Microbiol* 140: 287-296.

Allen JF, Martin W (2007) Out of thin air. *Nature* 445: 610-612.

Alleon J, Summons RE (2019) Organic geochemical approaches to understanding early life. *Free Radic Biol Med* 140: 103-112.

Altermann W, Kazmierczak J (2003) Archean microfossils: a reappraisal of early life on Earth. *Res Microbiol* 154: 611-617.

Baludikay BK, François C, Sforna MC, Beghin J, Cornet Y, Storme J-Y, Fagel N, Fontaine F, Littke R, Baudet D, Delvaux D, Javaux EJ (2018) Raman microspectroscopy, bitumen reflectance and illite crystallinity scale: comparison of different geothermometry methods on fossiliferous Proterozoic sedimentary basins (DR Congo, Mauritania and Australia). *Int J Coal Geol* 191: 80-94.

Bartley JK (1996) Actualistic taphonomy of cyanobacteria; implications for the Precambrian fossil record. *Palaios* 11: 571-586.

Battistuzzi FU, Feijao A, Hedges SB (2004) A genomic timescale of prokaryote evolution: insights into the origin of methanogenesis, phototrophy, and the colonization of land. *BMC Evol Biol* 4: 1-14.

Beghin J, Storme JY, Blanpied C, Gueneli N, Brocks JJ, Poulton SW, Javaux EJ (2017a) Microfossils from the late Mesoproterozoic- early Neoproterozoicatar/elMreïti group, Taoudeni basin, Mauritania, northwestern Africa. *Precambrian Res* 291: 63-82.

Beghin J, Guilbaud R, Poulton SW, Gueneli N, Brocks JJ, Storme JY, Blanpied C, Javaux EJ (2017b) A palaeoecological model for the late Mesoproterozoic-early Neoproterozoicatar/elMreïti group, Taoudeni basin, Mauritania, northwestern Africa. *Precambrian Res* 299: 1-14.

Bekker A, Holland HD, Wang P, IiiD R, Stein HJ, Hannah JL, Coetzee LL, Beukes NJ (2004) Dating the rise of atmospheric oxygen. *Nature* 427: 117-120.

Benzerara K, Skouri-Panet F, Li J, Ferard C, Gugger M, Laurent T, Couradeau E, Ragon M, Cosmidis J, Menguy N, Margaret-Oliver I, Tavera R, Lopez-Garcia P, Moreira D (2014) Intracellular Ca-carbonate biomineralization is widespread in cyanobacteria. *Proc Natl Acad Sci Unit States Am* 111: 10933-10938.

Blank CE, Sanchez-Baracaldo P (2010) Timing of morphological and ecological innovations in the cyanobacteria-a key to understanding the rise in atmospheric oxygen. *Geobiology* 8: 1-23.

Blank CE (2013) Origin and early evolution of photosynthetic eukaryotes in freshwater environments: reinterpreting Proterozoic paleobiology and biogeochemical processes in light of trait evolution. *J Phycol* 49: 1040-1055.

Blankenship RE (1992) Origin and early evolution of photosynthesis. *Photosynth Res* 33: 91-111.

Bosak T, Liang B, Sim MS, Petroff AP (2009) Morphological record of oxygenic photosynthesis in conical stromatolites. *Proc Natl Acad Sci Unit States Am* 106: 10939-10943.

Bosak T, AH, Petroff AP (2013) The meaning of stromatolites. *Annu Rev Earth Planet Sci* 41: 21-44.

Bower DM, Steele A, Fries MD, Green OR, Lindsay JF (2016) Raman imaging spectroscopy of a putative microfossil from the ~3.46 Ga Apex chert: insights from quartz grain orientation. *Astrobiology* 16: 169-180.

Brasier MD, Green OR, Lindsay JF, McLoughlin N, Steele A, Stoakes C (2005) Critical testing of Earth's oldest putative fossil assemblage from the ~3.5 Ga Apex chert, Chinaman Creek, Western Australia. *Precambrian Res* 140: 55-102.

Brinkmann H, Göker M, Koblížek M, Wagner-Döbler I, Petersen J (2018) Horizontal operon transfer, plasmids, and the evolution of photosynthesis in Rhodobacteraceae. *ISME J* 12: 1994-2010.

Brocks JJ, Logan GA, Buick R, Summons RE (1999) Archean molecular fossils and the early rise of eukaryotes. *Science* 285: 1033-1036.

Brocks JJ, Love GD, Summons RE, Knoll AH, Logan GA, Bowden SA (2005) Biomarker evidence for green and purple sulphur bacteria in a stratified Palaeoproterozoic sea. *Nature* 437: 866-870.

Brown II, Bryant DA, Casamatta D, Thomas-Keprta KL, Sarkisova SA, Shen G, Graham JE, Boyd ES, Peters JW, Garrison DH, McKay DS (2010) Polyphasic characterization of a thermotolerant siderophilic filamentous cyanobacterium that produces intracellular iron deposits. *Appl Environ Microbiol* 76: 6664-6672.

Burns BP, Anitori R, Butterworth P, Henneberger R, Goh F, Allen MA, Ibañez-Peral R, Bergquist PL, Walter MR, Neilan BA (2009) Modern analogues and the early history of microbial life. *Precambrian Res* 173: 10-18.

Butterfield NJ, Knoll AH, Swett K (1994) Paleobiology of the Neoproterozoic svanbergfjellet formation, Spitsbergen. *Foss Strat* 34: 1-84.

Cardona T (2018) Early archean origin of heterodimeric photosystem I. *Heliyon* 4: e00548.

Castenholz RW, Wilmotte A, Herdman M, Rippka R, Waterbury JB, Iteman I, Hoffmann L, Phylum BX (2001) Cyanobacteria. In: Boone DR, Castenholz RW, editors. *Bergey's Manual of Systematic Bacteriology*: Second Edition, Springer, New York. pp. 437: 473-599.

Cohen Y, Jorgensen BB, Revsbech NP, Poplawski R (1986) Adaptation to hydrogen sulfide of oxygenic and anoxygenic photosynthesis among cyanobacteria. *Appl Environ Microbiol* 51: 398-407.

Cornet L, Wilmotte A, Javaux EJ, Baurain D (2018a) A constrained SSU-rRNA phylogeny reveals the unsequenced diversity of photosynthetic cyanobacteria (Oxyphotobacteria). *BMC Res Notes* 11: 435.

Cornet L, Meunier L, Van Vlierberghe M, Léonard RR, Durieu B, Lara Y, Misztak A, Sirjacobs D, Javaux EJ, Philippe H, Wilmotte A, Baurain D (2018b) Consensus assessment of the contamination level of publicly available cyanobacterial genomes. *PLoS One* 13: e0200323

Couradeau E, Benzerara K, Gérard E, Moreira D, Bernard S, Brown GE, López-García P (2012) An early-branching microbialite cyanobacterium forms intracellular carbonates. *Science* 336: 459-462.

Dagan T, Roettger M, Stucken K, Landan G, Koch R, Major P, Gould SB, Goremykin VV, Rippka R, De Marsac NT, Gugger M, Lockhart PJ, Allen JF, Brune I, Maus I, Pühler A, Martin WF (2013) Genomes of stigonematalean cyanobacteria (subsection V) and the evolution of oxygenic photosynthesis from prokaryotes to plastids. *Genome Biol Evol* 5: 31-44.

De Gregorio BT, Sharp TG, Flynn GJ, Wirick S, Hervig RL (2009) Biogenic origin for Earth's oldest putative microfossils. *Geology* 37: 631-634.

De Oliveira VE, Miranda MACN, Soares MCS, Edwards HGM, De Oliveira LFC (2015) Study of carotenoids in cyanobacteria by Raman spectroscopy. *Spectrochim Acta Part A Mol Biomol Spectrosc* 150: 373-380.

Delwiche CF (1999) Tracing the thread of plastid diversity through the tapestry of life. *Am Nat* 154: S164-S177.

Demoulina CF, Laraa YJ, Corneta L, Françoisa C, Baurainb D, Wilmottec A, Javauxa EJ (2019) Cyanobacteria evolution: Insight from the fossil record. *Free Radic Biol Med* 140: 206-223.

Di Rienzi SC, Sharon I, Wrighton KC, Koren O, Hug LA, Thomas BC, Goodrich JK, Bell JT, Spector TD, Banfield JF, Ley RE (2013) The human gut and groundwater harbor non-photosynthetic bacteria belonging to a new candidate phylum sibling to cyanobacteria. *Elife* 2013: 1-25.

Dvořák P, Poulíčková A, Hašler P, Belli M, Casamatta DA, Papini A (2015) Species concepts and speciation factors in cyanobacteria, with connection to the problems of diversity and classification. *Biodivers Conserv* 24: 739-757.

Edwards HGM, Garcia-Pichel F, Newton EM, Wynn-Williams DD (1999) Vibrational Raman spectroscopic study of scytonemin, the UV-protective cyanobacterial pigment. *Spectrochim Acta A Mol Biomol Spectrosc* 56: 193-200.

Falcón LI, Magallón S, Castillo A (2010) Dating the cyanobacterial ancestor of the chloroplast. *ISME J* 4: 777-783.

Fischer WW, Hemp J, Johnson JE (2015) Manganese and the evolution of photosynthesis. *Orig Life Evol Biosph* 45: 351-357.

Fischer WW, Hemp J, Johnson JE (2016) Evolution of oxygenic photosynthesis. *Annu Rev Earth Planet Sci* 44: 647-683.

French KL, Hallmann C, Hope JM, Schoon PL, Zumberge JA, Hoshino Y, Peters CA, George SC, Love GD, Brocks JJ, Buick R, Summons RE (2015) Reappraisal of hydrocarbon biomarkers in Archean rocks. *Proc Natl Acad Sci Unit States Am* 112: 5915-5920.

Golovenok VK, Belova MY (1984) Riphean microbiotas in cherts of the Billyakh group on the anabar uplift. *Paleontol Zh* 4: 20-30.

Golubic S, Hofmann HJ (1976) Comparison of Holocene and mid-Precambrian entophysalidaceae (cyanophyta) in stromatolitic algal mats: cell division and degradation. *J Paleontol* 50: 1074-1082.

Golubic S, Friedman I, Schneider J (1981) The lithobiontic ecological niche, with special reference to microorganisms. *SEPM J Sediment Res* 51: 475-478.

Golubic S, Seong-Joo L (1999) Early cyanobacterial fossil record: preservation, paleoenvironments and identification. *Eur J Phycol* 34: 339-348.

Golubic S, Abed RMM (2010) *Entophysalis* mats as environmental regulators. *Natl Acad Sci Planet Biol* 237-251.

Graham LK, Wilcox LW (2000) The origin of alternation of generations in land plants: a focus on matrotrophy and hexose transport. *Phil Trans R Soc Lond B* 355: 757-767.

Green JW, Knoll AH, Golubic S, Swett K (1987) Paleobiology of distinctive benthic microfossils from the upper Proterozoic Limestone-Dolomite "Series," central East Greenland. *Am J Bot* 74: 928-940.

Green JW, Knoll AH, Swett K (1988) Microfossils from oolites and pisolites of the upper Proterozoic Eleonore Bay group, central East Greenland. *J Paleontol* 62: 835-852.

Gueneli N, Mckenna AM, Ohkouchi N, Boreham CJ, Beghin J, Javaux EJ, Brocks JJ (2018) 1.1-billion-year-old Porphyrins establish a marine ecosystem dominated by bacterial primary producers. *Proc Natl Acad Sci Unit States Am* 115: E6978-E6986.

Gupta RS, Mathews DW (2010) Signature proteins for the major clades of cyanobacteria. *BMC Evol Biol* 10: 1-20.

Hannah JL, Bekker A, Stein HJ, Markey RJ, Holland HD (2004) Primitive Os and 2316 Ma age for marine shale: implications for Paleoproterozoic glacial events and the rise of atmospheric oxygen. *Earth Planet Sci Lett* 225: 43-52.

Heubeck C (2009) An early ecosystem of Archean tidal microbial mats (Moodies Group, South Africa, ca. 3.2 Ga). *Geology* 37: 931-934.

Hodgson DA, Wright SW, Davies N (1997) Mass spectrometry and reverse phase HPLC techniques for the identification of degraded fossil pigments in lake sediments and their application in palaeolimnology. *J Paleolimnol* 18: 335-350.

Hofmann HJ, Jackson GD (1969) Precambrian (aphebian) microfossils from belcher islands, Hudson Bay. *Can J Earth Sci* 6: 1137-1144.

Hofmann HJ (1976) Precambrian microflora, belcher islands, Canada: significance and systematics. *J Paleontol* 50: 1040-1073.

Hoffmann L, Komárek J, Kaštovský J (2005) System of cyanoprokaryotes (cyanobacteria)-state in 2004. *Arch Hydrobiol Suppl Algol Stud* 117: 95-115.

Homann M, Sansjofre P, Van Zuilen M, Heubeck C, Gong J, Killingsworth B, Foster IS, Airo A, Van Kranendonk MJ, Ader M, Lalonde SV (2018) Microbial life and biogeochemical cycling on land 3,220 million years ago. *Nat Geosci* 11: 665-671.

Irisarri I, Baurain D, Brinkmann H, Delsuc F, Sire J-Y, Kupfer A, Petersen J, Jarek M, Meyer A, Vences M (2017) Phylotranscriptomic consolidation of the jawed vertebrate time tree. *Nat Ecol Evol* 1: 1370-1378.

Javaux EJ, Lepot K, Van Zuilen MA, Melezhik VA, Medvedev PV (2013) Palaeoproterozoic microfossils. In: Melezhik VA, editor. *Reading the Archive of Earth's Oxygenation:* Springer, New York, 1352-1370.

Javaux EJ, Knoll AH (2017) Micropaleontology of the lower Mesoproterozoic Roper group, Australia, and implications for early eukaryotic evolution. *J Paleontol* 91: 199-229.

Jehlièka J, Oren A (2013) Raman spectroscopy in halophile research. *Front Microbiol* 4: 1-7.

Kasting JF (2013) What caused the rise of atmospheric $O_2$? *Chem Geol* 362: 13-25.

Klatt JM, de Beer D, Häusler S, Polerecky L (2016) Cyanobacteria in sulfidic spring microbial mats can perform oxygenic and anoxygenic photosynthesis simultaneously during an entire diurnal period. *Front Microbiol* 7: 1-10.

Knoll AH, Golubic S (1979) Anatomy and taphonomy of a Precambrian algal stromatolite. *Precambrian Res* 10: 115-151.

Knoll AH, Golubic S (1992) Proterozoic and living cyanobacteria. In: Schidlowski M, Golubic S, Kimberley MM, McKirdy DM, Trudinger PA, editors. *Early organic evolution*: Springer, Berlin, Heidelberg. pp. 450-462.

Komárek J (1999) Cyanoprokaryota 1. Teil: Chroococcales, Subwasserflora von Mitteleuropa, Spektrum Akademischer Verlag, Heidelberg. 19: 1-548.

Komárek J, Kaštovský J (2003) Coincidences of structural and molecular characters in evolutionary lines of cyanobacteria. *Arch Hydrobiol Suppl Algol Stud* 109: 305-325.

Lepot K, Compère P, Gérard E, Namsaraev Z, Verleyen E, Tavernier I, Hodgson DA, Vyverman W, Gilbert B, Wilmotte A, Javaux EJ (2014) Organic and mineral imprints in fossil photosynthetic mats of an East Antarctic Lake. *Geobiology* 12: 424-450.

Lepot K, Addad A, Knoll AH, Wang J, Troadec D, Béché A, Javaux EJ (2017) Iron minerals within specific microfossil morphospecies of the 1.88 Ga Gunflint Formation. *Nat Commun* 8: 14890.

Lewis LA (2017) Hold the salt: freshwater origin of primary plastids. *Proc Natl Acad Sci Unit States Am* 114: 9759-9760.

Li B, Cox CJ, Lopes JS, Foster PG, Embley TM (2014) Compositional biases among synonymous substitutions cause conflict between gene and protein trees for plastid origins. *Mol Biol Evol* 31: 1697-1709.

Macdonald FA, Schmitz MD, Crowley JL, Roots CF, Jones DS, Maloof AC, Strauss JV, Cohen PA, Johnston DT, Schrag DP (2010) Calibrating the Cryogenian. *Science* 327: 1241-1243.

Magnabosco C, Moore KR, Wolfe JM, Fournier GP (2018) Dating phototrophic microbial lineages with reticulate gene histories. *Geobiology* 16: 179-189.

Marshall CP, Javaux EJ, Knoll AH, Walter MR (2005) Combined micro-Fourier transform infrared (FTIR) spectroscopy and micro-Raman spectroscopy of Proterozoic acritarchs: a new approach to Palaeobiology. *Precambrian Res* 138: 208-224.

Martin WF, Bryant DA, Beatty JT (2018) A physiological perspective on the origin and evolution of photosynthesis. *FEMS Microbiol Rev* 42: 205-231.

Mathis P (1990) Compared structure of plant and bacterial photosynthetic reaction centers. Evolutionary implications. *Biochim Biophys Acta Bioenerg* 1018: 163-167.

Mendelson CV, Schopf JW (1982) Proterozoic microfossils from the sukhayatunguska, shorikha, and yudoma formations of the siberian platform, USSR. *J Paleontol* 56: 42-83.

Mulkidjanian AY, Koonin EV, Makarova KS, Mekhedov SL, Sorokin A, Wolf YI, Dufresne A, Partensky F, Burd H, Kaznadzey D, Haselkorn R, Galperin MY (2006) The cyanobacterial genome core and the origin of photosynthesis. *Proc Natl Acad Sci Unit States Am* 103: 13126-13131.

Nakov T, Boyko JD, Alverson AJ, Beaulieu JM (2017) Models with unequal transition rates favor marine origins of cyanobacteria and photosynthetic eukaryotes. *Proc Natl Acad Sci Unit States Am* 114: E10606-E10607.

Newman SA, Klepac-Ceraj V, Mariotti G, Pruss SB, Watson N, Bosak T (2017) Experimental fossilization of mat-forming cyanobacteria in coarse-grained siliciclastic sediments. *Geobiology* 15: 484-498.

Noffke N, Gerdes G, Klenke T, Krumbein WE (2001) Microbially induced sedimentary structures-a new category within the classification of primary sedimentary structures-discussion. *J Sediment Res* 72: 587-588.

Ochoa de Alda JAG, Esteban R, Diago ML, Houmard J (2014) The plastid ancestor originated among one of the major cyanobacterial lineages. *Nat Commun* 5: 4937.

Olson JM (1970) The Evolution of Photosynthesis: Hypothesis: Photosynthetic bacteria and blue-green algae shared a common photoheterotrophic ancestor. *Science* 168: 438-446.

Olson JM, Pierson BK (1987) Origin and evolution of photosynthetic reaction centers. *Orig Life Evol Biosph* 17: 419-430.

Pacton M, Gorin GE, Fiet N (2008) Unravelling the origin of ultralaminae in sedimentary organic matter: the contribution of bacteria and photosynthetic organisms. *J Sediment Res* 78: 654-667.

Pang K, Tang Q, Chen L, Wan B, Niu C, Yuan X, Xiao S (2018) Nitrogen-fixing heterocystous cyanobacteria in the tonian period. *Curr Biol* 28: 616-622.

Papineau D, Walker JJ, Mojzsis SJ, Pace NR (2005) Composition and structure of microbial communities from stromatolites of Hamelin Pool in Shark Bay, western Australia. *Appl Environ Microbiol* 71: 4822-4832.

Pathak J, Pandey A, Maurya PK, Rajneesh R, Sinha RP, Singh SP (2020) Cyanobacterial secondary metabolite scytonemin: a potential photoprotective and pharmaceutical compound. *Proc Natl Acad Sci India Sect B Biol Sci* 90: 467-481.

Pathak J, Singh PR, Sinha RP, Rastogi RP (2021) Evolution and Distribution of Cyanobacteria. In: Rastogi RP, editor. *Ecophysiology and Biochemistry of Cyanobacteria*: Springer, Singapore. pp. 1.

Pereira S, Zille A, Micheletti E, Moradas-ferreira P, De Philippis R, Tamagnini P (2009) Complexity of cyanobacterial exopolysaccharides: composition, structures, inducing factors and putative genes involved in their biosynthesis and assembly. *FEMS Microbiol Rev* 33: 917-941.

Philippe H, Brinkmann H, Lavrov DV, Littlewood DTJ, Manuel M, Wörheide G, Baurain D (2011) Resolving difficult phylogenetic questions: why more sequences are not enough. *PLoS Biol* 9: e1000602.

Philippe H, de Vienne DM, Ranwez V, Roure B, Baurain D, Delsuc F (2017) Pitfalls in supermatrixphylogenomics. *EurJ Taxon No 283 Pitfalls Supermatrix Phylogenomics* 283: 1-25.

Pinti DL, Mineau R, Clement V (2009) Hydrothermal alteration and microfossil artefacts of the 3,465-million-year-old Apex chert. *Nat Geosci* 2: 640-643.

Ponce-Toledo RI, Deschamps P, López-García P, Zivanovic Y, Benzerara K, Moreira D (2017) An early-branching freshwater cyanobacterium at the origin of plastids. *Curr Biol* 27: 386-391.

Proteau PJ, Gerwick WH, Garcia-Pichel F, Castenholz R (1993) The structure of scytonemin, an ultraviolet sunscreen pigment from the sheaths of cyanobacteria. *Experientia* 49: 825-829.

Rashby SE, Sessions AL, Summons RE, Newman DK (2007) Biosynthesis of 2-methylbacteriohopanepolyols by an anoxygenic phototroph. *Proc Natl Acad Sci Unit States Am* 104: 15099-15104.

Rasmussen B, Fletcher IR, Brocks JJ, Kilburn MR (2008) Reassessing the first appearance of eukaryotes and cyanobacteria. *Nature* 455: 1101-1104.

Rastogi RP, Sinha RP, Incharoensakdi A (2013) Partial characterization, UV-induction and photoprotective function of sunscreen pigment, scytonemin from *Rivularia* sp. HKAR-4. *Chemosphere* 93: 1874-1878.

Rippka R, Waterbury J, Cohen-Bazire G (1974) A cyanobacterium which lacks thylakoids. *Arch Microbiol* 100: 419-436.

Rippka R, Deruelles J, Waterbury JB, Herdman M, Stanier RY (1979) Generic assignments, strain histories and properties of pure cultures of cyanobacteria. *Microbiology* 111: 1-61.

Roger AJ, Muñoz-Gómez SA, Kamikawa R (2017) The origin and diversification of mitochondria. *Curr Biol* 27: R1177-R1192.

Sagan L (1967) On the origin of mitosing cells. *J Theor Biol* 14: 225-274.

Sánchez-Baracaldo P (2015) Origin of marine planktonic cyanobacteria. *Sci Rep* 5: 14-17.

Sánchez-Baracaldo P, Bianchini G, Huelsenbeck JP, Raven JA, Pisani D, Knoll AH (2017) Reply to Nakov et al.: model choice requires biological insight when studying the ancestral habitat of photosynthetic eukaryotes. *Proc Natl Acad Sci Unit States Am* 114: E10608-E10609.

Schinteie R, Brocks JJ (2017) Paleoecology of Neoproterozoic hypersaline environments: Biomarker evidence for haloarchaea, methanogens, and cyanobacteria. *Geobiology* 15: 641-663.

Schirrmeister BE, Anisimova M, Antonelli A, Bagheri HC (2011) Evolution of cyanobacterial morphotypes. Taxa required for improved phylogenomic approaches. *Commun Integr Biol* 11: 45.

Schirrmeister BE, de Vos JM, Antonelli A, Bagheri HC (2013) Evolution of multicellularity coincided with increased diversification of cyanobacteria and the great oxidation event. *Proc Natl Acad Sci Unit States Am* 110: 1791-1796.

Schirrmeister BE, Gugger M, Donoghue PCJ (2015) Cyanobacteria and the Great oxidation event: evidence from genes and fossils. *Palaeontology* 58: 769-785.

Schirrmeister BE, Sanchez-Baracaldo P, Wacey D (2016) Cyanobacterial evolution during the Precambrian. *Int J Astrobiol* 15: 187-204.

Schopf JW (1968) Microflora of the bitter springs formation, late Precambrian, central Australia. *J Paleontol* 42: 651-688.

Schopf JW, Packer BM (1987) Early archean (3.3-3.5-billion-year-old) microfossils from warrawoona group, Australia. *Science* 237: 70-73.

Schopf JW (1993) Microfossils of the early archean Apex chert: new evidence of the antiquity of life. *Science* 260: 640-646.

Schopf JW, Kitajima K, Spicuzza MJ, Kudryavtsev AB, Valley JW (2017) SIMS analyses of the oldest known assemblage of microfossils document their taxon correlated carbon isotope compositions. *Proc Natl Acad Sci Unit States Am* 115: 53-58.

Seo PS, Yokota A (2003) The phylogenetic relationships of cyanobacteria inferred from 16S rRNA, gyrB, rpoC1 and rpoD1 gene sequences. *J Gen Appl Microbiol* 49: 191-203.

Sergeev VN (1994) Microfossils in cherts from the middle riphean (Mesoproterozoic) Avzyan Formation, southern Ural Mountains, Russian federation. *Precambrian Res* 65: 231-254.

Sergeev VN, Sharma M, Shukla Y (2012) Proterozoic fossil cyanobacteria. *Palaeobotanist* 61: 189-358.

Shestakov SV, Karbysheva EA (2017) The origin and evolution of cyanobacteria. *Biol Bull Rev* 7: 259-272.

Shi M, Feng Q, Khan MZ, Zhu S (2017) An eukaryote-bearing microbiota from the early Mesoproterozoic Gaoyuzhuang Formation, Tianjin, China and its significance. *Precambrian Res* 303: 709-726.

Shih PM, Wu D, Latifi A, Axen SD, Fewer DP, Talla E, Calteau A, Cai F, Tandeau de Marsac N, Rippka R, Herdman M, Sivonen K, Coursin T, Laurent T, Goodwin L,

Nolan M, Davenport KW, Han CS, Rubin EM, Eisen JA, Woyke T, Gugger M, Kerfeld CA (2013) Improving the coverage of the cyanobacterial phylum using diversity-driven genome sequencing. *Proc Natl Acad Sci Unit States Am* 110: 1053-1058.

Shih PM, Hemp J, Ward LM, Matzke NJ, Fischer WW (2017) Crown group Oxyphotobacteria postdate the rise of oxygen. *Geobiology* 15: 19-29.

Simion P, Philippe H, Baurain D, Jager M, Richter DJ, Di Franco A, Roure B, Satoh N, Queinnec E, Ereskovsky A (2017) A large and consistent phylogenomic dataset supports sponges as the sister group to all other animals. *Curr Biol* 27: 958-967.

Singh DK, Pathak J, Pandey A, Singh V, Rajneesh HA, Kumar D, Sinha RP (2020) Ultraviolet-screening compound mycosporine-like amino acids in cyanobacteria: Biosynthesis, functions, and applications. In: Singh PK, Kumar A, Singh VK, Shrivastava AK, editors. *Advances in Cyanobacterial Biology*: Academic Press, Cambridge, MA, USA. pp. 219-233.

Slotznick SP, Fischer WW (2016) Examining archean methanotrophy. *Earth Planet Sci Lett* 441: 52-59.

Šmarda J, Šmajs D, Komrska J, Krzyžánek V (2002) S-layers on cell walls of cyanobacteria. *Micron* 33: 257-277.

Soo RM, Hemp J, Parks DH, Fischer WW, Hugenholtz P (2017) On the origins of oxygenic photosynthesis and aerobic respiration in cyanobacteria. *Science* 355: 1436-1440.

Stanier RY, Sistrom WR, Hansen TA, Whitton BA, Castenholz RW, Pfennig N, Gorlenko VN, Kondratieva EN, Eimhjellen KE, Whittenbury R, Gherna RL, Trüper HG (1978) Proposal to place the nomenclature of the cyanobacteria (blue-green algae) under the rules of the International Code of Nomenclature of Bacteria. *Int J Syst Evol Microbiol* 28: 335-336.

Storme JY, Golubic S, Wilmotte A, Kleinteich J, Velázquez D, Javaux EJ (2015) Raman characterization of the UV-protective pigment gloeocapsin and its role in the survival of cyanobacteria. *Astrobiology* 15: 843-857.

Summons RE, Jahnke LL, Hope JM, Logan GA (1999) 2-Methylhopanoids as biomarkers for cyanobacterial oxygenic photosynthesis. *Nature* 400: 554-557.

Suosaari EP, Reid RP, Playford PE, Foster JS, Stolz JF, Casaburi G, Hagan PD, Chirayath V, Macintyre IG, Planavsky NJ, Eberli GP (2016) New multiscale perspectives on the stromatolites of Shark Bay, western Australia. *Sci Rep* 6: 1-13.

Timofeev BV, Hermann TN, Mikhailova NS (1976) Microphytofossils from the Precambrian, Cambrian and Ordovician. *Acad Sci USSR Inst Geol Geochronol Precambrian, Leningrad, Nauka* 106.

Tribollet A, Golubic S (2011) Reef Bioerosion: Agents and Processes. In: Dubinsky Z, Stambler N, editors. *Coral Reefs: An Ecosystem in Transition*: Springer, Dordrecht. pp. 435-449.

Uyeda JC, Harmon LJ, Blank CE (2016) A comprehensive study of cyanobacterial morphological and ecological evolutionary dynamics through deep geologic time. *PLoS One* 11: e0162539.

Walter JM, Coutinho FH, Dutilh BE, Swings J, Thompson FL, Thompson CC (2017) Ecogenomics and taxonomy of cyanobacteria phylum. *Front Microbiol* 8: 1-18.

Zhang Y (1988) Proterozoic stromatolitic micro-organisms from Hebei, North China: cell preservation and cell division. *Precambrian Res* 38: 165-175.

# Chapter 3

# Cyanobacterial Ecology and Systematics: Introduction of a Polyphasic Approach in Cyanobacterial Taxonomy

**Sagarika Pal and Prashant Singh***

Laboratory of Cyanobacterial Systematics and Stress Biology, Institute of Science, Banaras Hindu University, Varanasi, India

## Abstract

Ecological studies have helped to understand the diverse nature of cyanobacteria and to uncover its long history due to morphological plasticity and ecological flexibility. Cyanobacteria are found flourishing all over different habitats on earth, including the extreme environments. They are major players in the biogeochemical cycling of carbon, nitrogen, and oxygen. Cyanobacteria are oxygenic autotrophs and are major nitrogen fixers. The inherent ability to fix atmospheric nitrogen makes them extremely important in maintaining the stability and equilibrium of the ecosystem, thereby providing them a better competitive advantage in environments where nitrogen concentration is comparatively low. The adaptations to different kinds of environmental conditions including extreme pH, salinity, and temperature, make them the most diverse group of organisms on the earth. This feature of adaptability has helped to cultivate cyanobacteria successfully in laboratory conditions as well. Not only ecologically, but cyanobacteria are morphologically diverse too and this morphological complexity makes the taxonomy of this group very difficult and confusing. This complicacy impedes the accurate identification and classification of cyanobacteria, thus making it interesting but challenging. The taxonomic

* Corresponding Author's Email: sps.bhu@gmail.com.

In: Cyanobacteria
Editor: Rajeshwar P. Sinha
ISBN: 979-8-88697-169-9

identity is complicated because of confusions due to morphological plasticity and the role of ecology in classification, thus leading to erroneous identifications. The gradual shift from classical morphological taxonomy to molecular taxonomy has led to the development of a combined "polyphasic approach" which has resulted into the amalgamation of morphological, ecological, genetic, molecular and phylogenetic traits. Thus, cyanobacterial taxonomy has been substantially enhanced by combining different fields and taking into account the analysis of all approaches to establish a united, unambiguous and uncomplicated system of classification.

**Keywords:** classification, cyanobacteria, ecological, polyphasic approach, taxonomy

## Introduction

Cyanobacteria are ancient oxygen-evolving prokaryotes that have a significant role in making earth's anaerobic atmosphere aerobic and hence a diverse group of organisms, found in almost every extreme habitat on earth (Whitton and Potts 2002; Meeks and Elhai 2002; Mataloni and Komárek 2004; Henson et al. 2004; Whitton 2012; Gaysina et al. 2019). The stromatolites from the early Precambrian suggest the existence of cyanobacteria dating back to approximately 2.5 billion years ago and attempts to give a vast idea of how cyanobacteria had luxuriantly spread in virtually every imaginable geographical position on the planet (Schopf 1974). Thus, the study of this group became very important because of its huge biogeographical distribution and long evolutionary history. Cyanobacteria fix both carbon dioxide and atmospheric nitrogen, making them the major primary producers in marine ecosystems, where *Synechococcus* (unicellular cyanobacteria) is the most well distributed and well-studied group of cyanobacteria. Apart from *Synechococcus*, some other oceanic cyanobacteria include *Chroococcus, Prochlorococcus, Synechocystis, and Crocosphaera.* While the former three genera are non-nitrogen fixing, *Crocosphaera* is nitrogen-fixing (Scanlan 2012). Cyanobacteria are also involved in the event of endosymbiosis, which has resulted in the origin of chloroplast in phototrophic eukaryotes and probably helped and structured the development of higher plants (Margulis and Bermudes 1985; Stadnichuk and Kusnetsov 2021). Thus, the correct identification and classification of this group is important, as many changes have happened through the classification system over time, ultimately resulting in a unified methodology, known as the 'Polyphasic Approach'.

## Cyanobacterial Ecology and Diversity

Cyanobacteria constitute a diversified and ecologically important group of phototrophs. They carry out oxygenic photosynthesis and have chlorophyll as a major photosynthetic pigment. They utilize carbon dioxide as the major source of carbon using the pentose phosphate pathway. Their occurrence has been reported from places with extremely harsh conditions ranging from the extreme cold deserts of Arctic and Antarctic zones to the very hot springs and deserts, even in high saline conditions to marshes (Mataloni and Komárek 2004; Whitton and Potts 2012). Cyanobacteria are the dominant photoautotrophs in the Antarctic terrestrial ecosystem as well as freshwater. Their abundance of biomass is generally found in the benthic habitats of the lakes and ponds, where they form microbial mats, which are highly pigmented. They also form structured biofilms that cover the substrate (Vincent 2000).

A better understanding of cyanobacterial diversity will provide more knowledge about their ecological, morphological, and physiological aspects. They are mainly cosmopolitan with their characteristics common to both bacteria and algae (Adams 2000). In commercial use, cyanobacteria are found to be useful in crop productivity in various environments because of their ability to fix nitrogen. Cyanobacteria have proven to be a potential source of biotechnological applications such as food, feed, fuel, medicines, industries, fertilizers, diagnostics, in detecting/combating pollution, etc. (Borowitzka and Borowitzka 1988; De 1939; Gustafson et al. 1989; Kannaiyan et al. 1997; Mitsui et al. 1986; Thajuddin and Subramanian 2005). They have been found to be a good source of metabolites with biological activities like antimalarial, antibacterial, antiviral, antifungal, and cytotoxic properties (Falch et al. 1995). They are also a potential source for industrially important metabolites such as emulsifying agents, surfactants, flocculants, etc. (Bender et al. 1994; De Philippis and Vincenzini 1998). This ability of cyanobacteria to have such variations in their physiological aspects is due to their high "adaptability". Their adaptability to various habitats can be attributed to the long evolutionary history that they hold. These are ultimately responsible for oxygenation of the earth's atmosphere and are important footsteps of the evolutionary ancestry of the photosynthesizing plastids of algae and higher plants. To understand the diversity of these oxygenic photosynthetic microorganisms in-depth, many polyphasic studies with light microscopy, cultures and the comparative analysis of *16S rRNA* gene have been done (Garcia-Pichel et al. 2001; Lôpez-Cortés et al. 2001).

## Different Habitats of Cyanobacteria

### Aquatic Cyanobacteria

Cyanobacteria are found in diverse habitats, such as in marine ecosystems, including coastal shores and in open ocean. Planktonic and benthic are the two major ecological categories of aquatic cyanobacteria, where they are found in sediments, rocks, stones, algae, and aquatic plants (Fogg 1973). A large number of cyanobacterial communities are also observed in littoral zones, as they form prominent intertidal and intralittoral mats and create symbiotic partnerships, particularly with sponges and ascidians. Nitrogen fixation is the primary function of cyanobacteria in tropical marine environments as well as terrestrial ecosystems. (Hoffman 1999). Nitrogen-fixing, non-heterocytous *Trichodesmium* sp. can be found in the three major oceans (Atlantic, Pacific, and Indian) in mid and low latitude zones. Nitrogen-fixing cyanobacteria, *Richelia* and other cyanobacteria in the North Pacific, Caribbean, and North Atlantic establish distinct symbioses with the centric diatoms *Rhizosolenia* and *Hemiaulus*. *Nodularia spumigena*, a heterocytous and highly poisonous, diazotrophic cyanobacteria, is confined to mainly Baltic Sea's brackish waters and a coastal estuary in southern Australia, where it flourished by increased phosphorus intake from anthropogenic activities (Sellner 1997).

Planktic cyanobacteria adapt buoyancy regulation as one of the major functions in vertical migrations. The development of aerotopes, which includes gas vesicles, are another adaptive mechanism (Brookes and Ganf 2001; Reynolds 1987). *Prochlorococcus*, a cyanobacterial genus that was discovered about 20 years ago, is probably the most common photosynthetic organism in the world (Scanlan et al. 2009). *Prochlorococcus* dominates phytoplankton communities up to a significant depth in the central ocean. It has been reported that the population size of *Prochlorococcus* is limited by latitude, therefore it is understandable that it was absent at temperatures below 15°C (Partensky et al. 1999; Patterson et al. 1994; Zubkov et al. 2003). *Synechococcus* has a greater geographical spread than *Prochlorococcus*, and it is found in both polar and high-nutrient water. In marine planktonic communities, the genera *Cyanobium* and *Synechocystis* are also common. (Guedes et al. 2014). A filamentous cyanobacteria, *Trichodesmium* has a pan-global distribution in oligotrophic settings in tropical and subtropical oceans, making it one of the most luxuriant bloom-forming groups in the pelagic zone of seas and oceans (Capone et al. 1997; LaRoche and Breitbarth 2005). *Trichodesmium* has the potential to thrive at temperatures above 20°C, which means it could expand its range in the face of global warming (Hutchins et al. 2007; Stal and Zehr 2008). *Crocosphaera*, a genus of nitrogen-fixing coccoids, has also been discovered in tropical oceans (Zehr et al. 2007). *Crocosphaera watsonii,* is well-known for its benefaction to marine nitrogen fixation and, can also deplete its iron-metalloenzyme reserves thus making it possible to survive in iron deficient conditions (Zehr et al. 2001; Montoya et al. 2004; Moisander et al. 2010; Saito et al. 2011). The Baltic Sea's most important bloom-forming cyanobacteria include *Nodularia*, *Aphanizomenon*, and *Anabaena* (Knelman et al. 2012) and each year, the cyanobacteria *Nodularia spumigena* produces poisonous blooms in this habitat (Teikari et al. 2018). There are both planktonic and benthic *Nodularia* species: *Nodularia baltica*, *N. litorea,* and *N. spumigena* are planktonic species, whereas *N. harveyana* and *N. sphaerocapsa* are benthic species (Komárek 2003). Benthic filamentous cyanobacteria typically have a distinct gliding movement and can create mats, which are dense multi-layered communities of several species that grow on sediments from various settings (Hoiczyk 2000; Stal 1995). These blooms are more luxuriant, extensive, and tenacious in several tropical and subtropical ecosystems, estuaries, and reef habitats across the world, with many detrimental environmental and human health consequences (Paerl and Fulton 2006). A filamentous cyanobacteria *Lyngbya*, can be found in benthic communities all over the world, where it creates dominating cover and blooms in tropical and subtropical coral reef and seagrass habitats (Paul et al. 2005; Jones et al. 2011; Knelman et al. 2012). *Moorea*, a filamentous cyanobacterial genus, is part of a worldwide pantropical biological group that is prevalent in the marine benthos. Members of this genus are photosynthetic, non-diazotrophic organisms that frequently form mats or blooms in shallow waters. Recent research using phylogenetic inference of small-subunit (SSU) rRNA genes has revealed a considerable amount of unique diversification of benthic cyanobacteria from tropical and subtropical marine settings (Engene et al. 2013).

## Cyanobacteria in Mangrove Habitats

Several surveys of the cyanobacterial diversity of mangroves from various tropical and subtropical regions across the world have been published (Branco et al. 1996; Silambarasan et al. 2012; Silva et al. 2014; Alvarenga et al. 2015). The major cyanobacterial mangrove dwellers are *Aphanocapsa*, *Oscillatoria*, *Chroococcus*, *Schizothrix*, *Lyngbya*, *Calothrix,* and *Coleofas-ciculus*.

Woitchik et al. (1997) in their research discovered that around 13 to 21% of biological nitrogen is being fixed among the total biological nitrogen-fixation from decaying mangrove leaves of the coastal lagoon in Kenya. Furthermore, it was discovered that light and temperature are the major environmental factors that influence nitrogen fixation on pneumatophores. Thus, the cyanobacterial diversity and the nitrogen fixation process may help to promote primary productivity in the mangrove environment (Kyaruzi et al. 2004). Temporal differences in distribution were identified, with pre-monsoon and monsoon seasons having higher numerical density and less diversity in benthic cyanophytes, and post-monsoon (POM) seasons having lower abundance and substantially higher diversity in the mangrove habitats. The physio-chemical parameters of sediments and pores revealed that sediment temperature was positively connected with cyanobacterial abundance, but dissolved oxygen and pore water dissolved nitrate were negatively correlated. In considerably fewer saline conditions, pore water salinity also altered the community makeup of cyanobacteria, with heterocytous cyanobacteria dominating (Benny et al. 2021).

## Cyanobacteria in Freshwater Conditions

The cyanobacteria are well-known to flourish in freshwater ecosystems but are also found in brackish, salty, cold, hot springs, and places of extreme environmental conditions where almost no microalgae can survive (De Nobel et al. 1998). During bloom formation, many secondary metabolites and various bioactive compounds with varying chemical structures are produced by cyanobacteria. Observations were made in different cases where in this situation some autotrophic cyanobacteria produce chemicals, that possess allelopathic effects, which impact the biological processes of other aquatic plants or phytoplankton (Leão et al. 2009).

One of the major dominating species of cyanobacteria found in oligotrophic lakes are phycoerythrin-rich freshwater cyanobacteria, *Synechococcus.* Some other species found along with *Synechococcus,* include *Cyanobium* and *Synechocystis*, which are also significant in freshwater ecosystems (Albertano et al. 1997; Stockner 1988; Komárek 2003). Upon analysis of 16S ribosomal DNA in *Synechococcus* species, the polyphyletic nature of the group has been clearly observed (Robertson et al. 2001; Urbach et al. 1998). The genus *Aphanothece,* is majorly found to thrive in freshwater bodies while other cyanobacterial taxa include *Chroococcus, Gomphosphaeria, Cyanodictyon*, *Merismopedia*, *Rhabdoderma, Coelosphaerium, Snowella*, and *Coelomoron* (Komárek 2003).

Some of the cyanobacteria present at the bottom sediment of water are called benthic cyanobacteria and they form cohesive mats (Mur 1999). The most common epilithic cyanobacterial genera found in freshwater streams in Assam's Kakoijana Reserve Forest are *Leptolyngbya, Nostoc, Chroococcus, Aphanocapsa,* and *Aphanothece* (Saha et al. 2007). Benthic cyanobacteria are well-known for producing toxins. Their recent involvement, well known now in animal poisonings has expanded significantly during the previous decade. Cyanobacterial blooms prevail in benthic zones due to increased anthropogenic activity on waterways and climate change (Catherine et al. 2013).

## Cyanobacteria in Symbiotic Associations

Cyanobacteria are different from all prokaryotes because of their ability to establish symbiotic partnerships with a diverse spectrum of hosts. Their ability of symbiosis is mainly due to their very efficient photosynthetic mechanism, which allows them to sustain this activity even under very low light levels, as usual in most symbiotic relationships (Usher 2008). In symbiosis they function as 'chloroplasts' with different types of non-photosynthetic partners, such as marine invertebrates, where cyanobionts feed carbon to the host partners. Cyanobacteria and marine organisms are a good example of a type of symbiotic interaction that contributes significantly to the global ocean's oligotrophic water, where cyanobionts help a wide range of marine organisms to live in sometimes nutrient-depleted waters. These interactions are important constituents of the ecology and overall life cycle pattern of many cyanobacterial lineages. Interactions with plants, fungi, mammals, and eukaryotic algae are some common examples (Janson 2002; Murakami et al.

2004; Rai et al. 2000; Wilkinson 1987). In the colonies of the Caribbean coral *Montastraea cavernosa* there is a endosymbiotic cyanobacteria that expresses nitrogenase and fixes nitrogen for the host. *Zooxanthellae* show mutualistic symbiotic relationship with corals, using the products of nitrogen fixation, as revealed by studies of stable isotopes (Lesser et al. 2007).

Some intriguing species that exist in a symbiotic relationship with numerous commonly found diatoms such as *Hemiaulus*, *Rhizosolenia*, and *Chaetoceros* cyanobacteria include heterocytous cyanobacteria, *Richelia intracellularis*, and *Calothrix rhizosoleniae*. These cyanobionts are important in this scenario because they offer fixed nitrogen to their nitrogen-limited hosts. In this way, both the host and the ecosystem would be benefitted (Yeung et al. 2012). Blooms of the phytoplankton *Hemiaulus hauckii* and its cyanobacterial parasite *Richelia intracellularis* can reach extraordinarily high densities, having significant ecological consequences for the ecosystem. These blooms dominate carbon deposition in the offshore tropical river plumes and so significantly subsidize additional nitrogen inputs.

Dominance of cyanobacteria in epiphytic communities has been recorded in depth using light microscopy, *16S rRNA* and *nif*H gene analysis. The *nif*H gene expression patterns also show that the heterocytous phylotypes may be the dominant diazotrophs during the day, whereas the non-heterocytous types may be dominant at night (Hamisi et al. 2013). The filamentous heterocytous genus *Nostoc* is generally known for being a productive symbiotic partner with a variety of eukaryotes namely protists, mammals, fungi, and plants (Nilsson et al. 2002). Numerous phylogenetic analyses utilizing various molecular markers (*16S rRNA*, *rpo*B, *rbc*L, and *nif*D or *nif*H) indicated that the nearest relatives of the genus *Nostoc* are members of the family Nostocaceae including the genera *Anabaena, Aphanizomenon,* and *Trichormus* (Henson et al. 2002; Lohtander et al. 2003; Svenning et al. 2004; Rajaniemi et al. 2005; Singh et al. 2012). *Nostoc* has a complex life cycle with differentiation into heterocytes (nitrogen-fixing cells), akinetes (resting spores), and hormogonia (motile filaments), which function as dispersal and infective units in symbiosis (Lazaroff and Vishniac 1961; Stanier et al. 1979; Mollenhauer et al. 1999). In an imperative study, the phylogenetic analysis revealed that two distinct patterns of symbiotic behavior evolution might exist for the symbiotic Nostocacean cyanobacteria, one that leads to the symbiotic association of *Nostoc* species with a variety of plants and the other leading to the association of a unique cyanobacterial type with the water fern *Azolla* (Papaefthimiou et al. 2008).

## Terrestrial Cyanobacteria

Soil cyanobacteria are a well-researched group of cyanobacteria known for being ubiquitous in different kinds of terrestrial ecosystems. Cyanobacteria can be found in some of the most extreme terrestrial settings, including scorching and cold desert rocks, Antarctic soils, and highly acidic post-mining sites. Environmental conditions such as humidity, soil temperature, soil texture, pH, salt concentration, nutrition and biotic interactions influence soil cyanobacteria. Higher plants have a considerable impact on the cyanobacterial flora as well (Metting 1981; Friedmann and Ocampo-Friedmann 1984; Broady 1996; Lukešová 2001; Bakieva et al. 2012). There are number of articles on presence of cyanobacteria in arid and semiarid soils, particularly cyanobacteria in biological crust communities. BSCs (macrobiotic crusts, cryptogamic crusts) are found throughout the world's arid regions, including the polar regions (Büdel et al. 2008). These crusts are composed of water-stable surface soil that has been occupied by algae, cyanobacteria, fungi, lichens, and mosses (Johansen 1993). It should be emphasized that these BSCs serve an important role in dry ecosystems, reducing soil erosion and also in the restoration of disturbed soil. Cyanobacteria can be found in surface cryptogamic crusts, as free-living organisms in water bodies, and within or on rocks in arid environments. Cyanobacteria have some primary responsibilities in arid environments such as soil improvement and protection, nitrogen addition to desert ecosystems via nitrogen-fixing cyanobacteria and lichens, and primary biomass production for use as food and other secondary output. Physical enhancement and protection of arid soils can control desertification and rehabilitate arid places. Cultivating algae and cyanobacteria for biomass production, which is already being used in non-arid conditions in agriculture, aquaculture, and now in the biochemical sector, has promising possibilities in dry locations due to the abundance of sunlight. The organism's primary production can also be employed for direct human and cattle consumption, as well as in urban garbage treatment. Thus, biomass production can serve as a form of resource diversification, relieving pressure on fragile arid lands (Isichei 2009). While studying the biological crusts in different parts of India, genera such as *Lyngbya, Plectonema* and *Scytonema* which have a definite sheath, and representatives of the genera *Oscillatoria*, *Phormidium*, *Microcoleus*. *Nostoc*, *Calothrix*, *Aulosira*, *Fischerella*, *Westiellopsis*, and *Hapalosiphon* have been found in majority (Tirkey and Adhikary 2006). The dominating taxa found out in Mexico's Baja California Desert were *Schizothrix calcicola* and *Desmonostoc muscorum* (Flechtner 1998). When presence of cyanobacteria comes to temperate areas, there is insufficient data on biodiversity of cyanobacteria present there. It was discovered that cyanobacteria are quite uncommon in forest soils. However, abundance of cyanobacteria is majorly dependent on the forest type. Cyanobacteria were found in floodplain soils under oak and pine woods. *Desmonostoc muscorum* (as *Nostoc muscorum), Nostoc punctiforme*, *Microcoleus autumnalis* (as *Phormidium autumnale)*, and *Leptolyngbya foveolarum* (as *Phormidium foveolarum)* were the main cyanobacterial species found in forest soils (Aleksakhina and Shtina 1984).

**Figure 1.** Cyanobacteria in diverse habitats: (A) Cyanobacteria in a freshwater pond, Tripura, India; (B) Roots of *Areca catechu* surrounded by algal and cyanobacterial mats, Tripura, India; (C) Cyanobacterial mats in a wooden slap just above a small drainage system, Meghalaya, India; (D) Freshwater dwelling cyanobacteria in a rock shelter, Assam, India; (E) Epilithic cyanobacteria in Arctic streams; (F) Freshwater dwelling cyanobacteria in a stream found attached with the rock, Tripura, India; (G) Algal and cyanobacterial mats on the bark of *Areca catechu* plant, Tripura, India; (H) Cyanobacterial growth beneath ice; (I) Cyanobacterial mats attached to the rock surface in a hilly area, Assam, India; (J) Algal growth on small rock surfaces; (K) Cyanobacterial balls, attached to the roots of *Oryza sativa* collected from a paddy field in Tripura, India; (L) Algal growth in glacial streams; (M) Algal growth in shallow running streams; (N) Symbiotic association of cyanobacteria with Fungi – Crustose lichens strongly adhering to the surface of *Areca catechu* plant, Tripura, India.

## Reasons for Cyanobacterial Diversity

Cyanobacteria exhibit themselves in diverse habitats and can prosper in saline, brackish, and freshwater, and surprisingly even in frigid regions, hot springs, and deserts, where no other microalgae can exist. Nitrogen serves as the necessary macronutrient, with ammonia being the preferred supply (Herrero et al. 2001). They can, however, assimilate nitrate and urea. Though all cyanobacteria exhibit oxygenic photosynthesis, some species can changeover to anoxygenic photosynthesis by utilizing sulfide (Cohen et al. 1986; Garcia-Pichel and Castenholz 1990) and ferrous ion (Widdel et al. 1993) as electron suppliers.

Some cyanobacteria perform fermentation process in dark under anoxic conditions (Stal and Moezelaar 1997; Steunou et al. 2006). The metabolism of intracellularly stored carbohydrate occurs in dark by aerobic respiration, where oxygen acts as a final electron acceptor. This enzyme-based catalysis performs the fermentation process which is expressed constitutively, and hence transition to fermentation can be quick in cyanobacterial metabolism. Cyanobacteria have also been reported to show mixotrophy and heterotrophy in nature (Eiler 2006; Fay 1965). Some cyanobacteria may act as heterotrophs on organic matters in both dark and light conditions. *Cyanothece*, *Aphanocapsa*, *Cholorogloea*, and *Agmenellum quadruplicaturm* have been shown to exhibit photoheterotrophy and chemoheterotrophy (Rippka et al. 1974; Lambert and Stevens 1986; Feng et al. 2010). Cyanobacteria can utilise and store essential nutrients and metabolites for later consumption and during unfavourable conditions. Akinetes are resting cells holding reserve substances that allow them to survive under unfavourable conditions. Furthermore, cytoplasmic inclusions such as glycogen granules, lipid globules, cyanophycin granules, polyphosphate bodies, and carboxysomes have also been identified (Fay and Van Baalen 1987). Cyanophycin (non-ribosomal polypeptide), a nitrogen-storage body (Simon and Weathers 1976), can also be used as an energy storage material (Smith and Russell 1982). Phycobiliproteins have been proposed as the foundation of a nitrogen storage system and they start decaying when nitrogen (or sulphur) are limited (Kana et al. 1992; Wyman et al. 1985). Furthermore, many cyanobacteria accumulate phosphate in the form of polyphosphates, which can be used as insecticides, detergent additives, and antibiotics (Martin et al. 2005; Singh et al. 2009). Many cyanobacteria develop extracellular polysaccharides (EPS) sheaths, which act as a barrier between the cell and the environment and aid in desiccation tolerance. This desiccation resistance property is aided further by the development of sheath pigments such as gloeocapsin, fuscorhodin, and fuscochlorin (Bewley 1979; Shirkey et al. 2000). Quorum sensing has been observed in cyanobacteria in response to cell-population density variations, and this may enable cooperative behaviour to promote fitness. Several studies have shown that production of quorum sensing acyl- homoserine lactones influences biofilm development and nitrogen fixation (Sharif et al. 2008; Romero et al. 2011). Under natural conditions, cyanobacteria inhabiting microbial mats in extreme hypersaline systems were found to have quorum sensing inhibitory actions (Dobretsov et al. 2010). The biochemical substances (allelochemicals) are released in surrounding when there is any allelopathic interaction occurring between cyanobacteria and other species in the community. Some species are responsible for the production of bioactive chemicals with broad range of biological functions. Cyanobacteria are also well known to produce compounds that target other cyanobacteria and microalgae, and these can also be utilized to reduce algal blooms (Dahms et al. 2007). A number of these allelochemicals have been discovered, for example, *Scytonema hofmanni* produces cyanobacterin, *Fischerella muscicola* produces Fischerellin A, *Hapalosiphon* and *Fischerella* sp. produces hapalindoles. Furthermore, cyanotoxins have been reported to be allelochemicals involved in interspecific interactions (Bar-Yosef et al. 2010; Berry et al. 2008; Hogmann and Juttner 1996; Mason et al. 1982; Thanh Doan et al. 2000). Secondary cyanobacterial metabolites having economic applications may be found in marine habitats with changeable environmental conditions (Bhadury and Wright 2004; Burja et al. 2001). Some marine cyanobacteria, possess the potential of producing vitamins, such as vitamin B and E (Plavšić et al. 2002) and other chemicals may operate as medicines, pesticides, and antifouling agents (Abed et al. 2009; Berry et al. 2008; Burja et al. 2001; Shimizu 2003; Boopathy and Kathiresan 2010). Cyanobacteria are the primary nitrogen fixing organisms in nitrogen limited photic environments and the nitrogen fixation process is catalysed by nitrogenase enzyme (Zehr 2011). It consists of two proteins: dinitrogenase reductase encoded by the gene *nif*H, and dinitrogenase encoded by the genes *nif*D and *nif*K. Oxygen interferes in nitrogen fixation as it irreversibly inactivates nitrogenase activity and hence, nitrogen fixation occurs under anoxic or low oxygen conditions only. To protect their nitrogenase from oxygen, diazotrophic cyanobacteria have evolved a variety of techniques with varying degrees of sophistication. The most sophisticated and effective nitrogenase protection approach is geographical separation through the differentiation and formation of specialized cells for nitrogen-fixing known as "heterocyst" (now termed heterocyte). Heterocytes have strong cell walls to prevent oxygen diffusion and do not produce oxygen via photosystem II (Adams and Duggan 1999).

## Cyanobacteria Living in High-Temperature Habitats

Filamentous cyanobacteria such as *Mastigocladus laminosus*, *Phormidium* sp. and several *Oscillatoria* species have an upper-temperature limit for growth that ranges between 55 and 62°C. Unicellular cyanobacteria of the genus *Synechococcus* (*Thermosynechococcus*) tend to be better adapted in both planktonic and benthic habitats at the greatest temperatures (>65°C and up to 74°C) (Steunou et al. 2006; Ward et al. 1998). Filamentous heterocytous cyanobacteria or heterotrophic bacteria are responsible for nitrogen fixation in hot springs (Erdmann and Hagemann 2001; Thomsen and Cox 1993). Since most eukaryotes cannot survive in hot springs, however, bacteria and archaea are the dominant species here, with cyanobacteria being particularly prevalent. Cyanobacteria are also frequently found in alkaline hot springs (Miller et al. 1998). Nitrogen fixation in hot springs has been attributed primarily to filamentous heterocytous cyanobacteria or heterotrophic bacteria (Thomsen and Cox 1993; Erdmann and Hagemann 2001). Since cyanobacteria are frequently found in alkaline hot springs, they become particularly prevalent and can survive in such extreme conditions. Most eukaryotes cannot survive in such extreme conditions but bacteria and archaea have been observed to thrive in such conditions (Miller et al. 1998).

Over the last decade, advancement in nucleic acid sequencing technologies has resulted in more comprehensive surveys of cyanobacterial diversity in geothermal systems around the world, thus providing better knowledge of well-studied hot spring cyanobacterial ecosystems. As a result, a more detailed understanding of biogeographical and local distributions of cyanobacteria in these settings can now be built on top of long-term studies of these systems based on morphological assessments, pure cultures and early *16S rRNA* gene data. Causes of evolution and current ecology of thermophilic cyanobacteria have been characterized by population genetics studies with higher molecular, geographic and temporal resolution. Metagenomics research has begun to uncover the functional gene repertoires of the dominant cyanobacteria and associated members of the communities in which they live and interact. Gene expression studies such as meta transcriptomic investigations, have begun to show in-situ gene expression patterns. The generation of thermostable enzymes (polymerases, proteinases, amylases and xylanases) by cyanobacteria has discussed the successful creation of D-amino acid dipeptides in a high-temperature ATP regeneration system utilizing a thermostable polyphosphate kinase from the thermophilic cyanobacterium *Thermosynechococcus elongatus* BP-1 (Piechula et al. 2001; Sato et al. 2007).

## Cyanobacteria Living in Low-Temperature Zones

Previously, it was widely assumed that Antarctic habitats were devoid of life as it has been observed that Antarctic habitats are among the world's driest and coldest ecosystems. Recent studies have revealed the presence of microbes in the snow near the South Pole (Carpenter et al. 2000), 3.5 km deep in Vostok ice (Karl et al. 1999; Priscu et al. 1999), exposed soils, sandstones (Friedmann et al. 1993), meltwater ponds (Vincent 2004), and liquid water column which are permanently frozen (Moorhead and Priscu 1998; Psenner et al. 1999). The majority of bacteria discovered in these environments are prokaryotic and photosynthetically active. Cyanobacteria are one of the most essential components of these microbial communities as they have the ability to produce an adequate amount of fixed carbon through photosynthesis that helps in maintaining the ecosystem (Gordon et al. 2000; Vincent 2004). Cyanobacteria in Antarctic habitats are adapted and acclimated to their environment in terms of temperature, freeze/thaw survival, photoprotection and light acquisition for photosynthesis (Nadeau et al. 1999; Tang and Vincent 1999). Some cyanobacterial groups, like *Leptolyngbya, Phormidium, Oscillatoria* and *Nostoc*, are global and can be found in quite diverse habitats (Singh and Elster 2007). The persistent freezing temperatures in these severe ecosystems are frequently accompanied by freeze-thaw cycles, substantial swings in irradiance (including UV radiation), and large variations in nutrient availability and salinity. As a result of these limits, polar and alpine habitats have lower biodiversity, with prokaryotes accounting for a large portion of total ecosystem biomass and species richness. Cyanobacteria are particularly intriguing because they are frequently the major phototrophs in such ecosystems. A variety of cyanobacteria can be found in polar and alpine habitats as they can regulate their lipid content to modify their membrane fluidity and function at low temperature and therefore, they have the ability to withstand the abiotic stressors that prevail in these frigid conditions (Řeháková et al. 2007). Cyanobacteria frequently form visible crusts, mats, and biofilms on substrates such as snow, ice, soil and rocks, where they are primary producers. Very recently, signals of the *nif*H gene have been detected in Arctic saltwater and hypersaline locales in the ice, in numerous groups of cyanobacteria which have the potential to fix nitrogen (Díez et al. 2012).

## Cyanobacteria Inhabiting High Saline Environment

In salt lakes, hypersaline lagoons and some solar salterns, halophilic cyanobacteria contribute immensely to primary production (Golubic and Seong-Joo 1999; Oren 2009). The filamentous *Microcoleus chthonoplastes* and *Halospirulina tapeticola*, as well as the unicellular *Aphanothece halophytica*, are the most widely dispersed species. Halophilic and halotolerant cyanobacteria maintain osmotic balance with the environment by reducing intracellular ionic concentration or accumulating suitable low molecular weight solutes. The build-up of disaccharides such as sucrose and trehalose in less tolerant strains such as *Chroococcidiopsis* (Hershkovitz et al. 1991), glucosyl glycerol in *Synechocystis sp.* PCC 6803 (Erdmann and Hagemnn 2001), and glycine betaine and L-glutamate betaine in strains evolved to the highest salt concentrations (Mackay et al. 1984). The later halotolerant cyanobacteria also use a $Na^+$ – ATPase to remove excess $Na^+$ from their cells to maintain the cytoplasmic $Na^+$ content (Oren 2007; Wiangnon et al. 2007). And also, there are certain halophilic cyanobacteria which accumulate particularly high quantities of MAA, which has been proposed to play a function as a compatible solute.

Researchers have studied the salt tolerance mechanisms in cyanobacteria from all around the world. Cyanobacteria developed around 3,000 million years ago and are thought to be the primary colonists of hostile habitats. The physiological components of cyanobacterial adaptation to high salinities include (a) the production and storage of osmo-protective chemicals, (b) preservation of low internal inorganic ion concentrations, and (c) the expression of a collection of salt-stress proteins. The exposure of abiotic stimulus in cyanobacterial cells resulted in the fast production of multiple stress-regulated proteins and also plays vital role in targeting major changes in protein synthetic machinery. Organic solutes such as disaccharides like sucrose, trehalose, and glucosyl glycerol, quaternary amines, like glycine betaine, and free amino acids such as glutamine have all been synthesized to regulate the abiotic stress. The formation of certain fatty acids, sucrose- and osmotic-stress-induced proteins has also been recorded to protect cyanobacteria from alkaline environments. The accumulation of internal osmoticum in the form of inorganic ions, as well as the avoidance of intracellular $Na^+$ accumulation by the inhibition of $Na^+$ inflow and effective active efflux mechanisms or metabolic modifications has been well studied. In cyanobacteria, $Na^+$ extrusion is mainly mediated by a $Na^+/H^+$ antiporter that is activated by increased cytochrome oxidase activity. The suppression of sodium ion inflow appears to be a crucial mechanism for cyanobacterial survival under salt stress, and the creation of salt-stress proteins has been discovered in cyanobacteria (Singh and Dhar 2010).

## Cyanobacteria Living under High Radiation

The intense radiation from the sun in harsh environments is relatively stressful for cyanobacteria. The presence of an exceptionally thick exterior covering known as the EPS (extracellular polymeric substances) helps these organisms to have a better survival opportunity from ultraviolet light as well as desiccation (Shirkey et al. 2000). In some cyanobacterial taxa such as *Nostoc commune*, the EPS layer is well known as a prominent UVB protectant along with providing desiccation tolerance and protection against freeze and thaw in order to maintain the integrity of the cell membrane (Hill et al. 1997; Tamaru et al. 2005). The *de novo* synthesis of less-sensitive isoforms of D1 and D2 polypeptides in the core of photosystem II reaction centre is another adaptation for UVB protection (Gao and Garcia-Pichel 2011). Some filamentous cyanobacteria from alkaline hot springs are found to be wrapped in amorphous silica matrix which helps them to limit the UV-B radiation but allowing limited PAR levels to penetrate (Phoenix et al. 2006).The generation and accumulation of non-enzymatic antioxidants like ascorbic acid, carotenoids, reduced glutathione, and tocopherols helps to scavenge UVB-induced ROS production. Another important compound scytonemin, which is found in the sheath, surrounding the cells of special filamentous cyanobacteria such as *Scytonema*, *Nostoc*, *Lyngbya*, and *Calothrix* and unicellular cyanobacteria *Chlorogloeopsis* sp. and *Chroococcidiopsis* sp. is one of the most effective radiation-protective natural compound found in cyanobacteria (Garcia-Pichel et al. 1992; Dillon et al. 2003).

## Development of Cyanobacterial Systematics and Taxonomy

Taxonomy helps to understand and unfold the extent of biodiversity in an ecosystem. Cyanobacteria are both ecologically and evolutionary diverse group of organisms and were nomenclated into different groups time to time such as Cyanophyta, Myxophyceae, Cyanophyceae, and Cyanochloronta (Smith 1938; Stitzenberger 1860; Sachs 1874; Bold and Wynne 1978; Fay 1983). Despite having a wide adaptive nature and evolutionary significance, a large gap persists within the taxonomy of this group. The problems are excavated due to limited and sporadic sampling, difficulty in culturing in certain groups of cyanobacteria, and phenotypic variability as well as ecological flexibility among the various groups (Castenholz 1992; Komárek and Anagnostidis 1998, 2005; Johansen and Casamatta, 2005; Komárek et al. 2014; Dvořák et al. 2018). The earliest classifications were mainly based on the botanical code of nomenclature where cyanobacteria were referred as blue-green algae, until Stanier (1977) first proposed to classify them according to the bacteriological code of nomenclature and hence gave the name cyanobacteria, for the first time. The prokaryotic nature of the cell was the main reason that Stanier proposed the name cyanobacteria (Stanier et al. 1978). The classification system has changed over time, as many taxonomists classified the groups to form a stable and uniformed system of classification. Traditionally cyanobacteria were classified based on morphology of filaments, cells, branching patters, characteristics of sheath, reproductive structures such as akinetes, hormogones, etc. (Bornet and Flahault 1888; Geitler 1932; 1942; Fritsch 1945; Drouet and Daily 1956; Drouet 1981; Desikachary 1959). One of the earliest classifications was proposed by Thuret (1875) in which cyanobacteria were divided into two tribes- Coccogoneae and Hormogoneae. The tribe Chroococcaceae, representing single-celled reproductive structures (Nägeli 1849; Hansgirg 1892) were studied individually and not divided further, whereas Hormogoneae, which reproduced through hormogones were further subdivided into 2 subtribes and 5 families (Rabenhorst 1865; Thuret 1875; Bornet and Flahault 1885; 1888). Geitler (1932) laid a comprehensive taxonomical classification where he presented 4 orders- Chroococcales, Pleurocapsales, Dermocarpales and Hormogonales, 20 families and 145 genera. He was persuasive about the possible similarity between *Stigonema* and *Siphononema* in his study. His ideas of classification have provided a better view of taxonomy to many other classifications that were proposed later (Fritsch 1945; Drouet and Daily 1956; Desikachary 1959; Bourelly 1970). The branching pattern in cyanobacteria has played an important role in classification, where filamentous forms can be unbranched or branched. Again, the branching pattern can be true or false. The order Stigonematales and Nostocales are examples where the branching pattern has been taken into consideration (Geitler 1932; 1942; Fritsch 1945). At the same time, Elenkin

(1936, 1938 and 1949) laid out different ideas on the evolution of the order Hormogonales, where he suggested that the order may have developed from the family Chroococcaceae, first through the genus *Siphononema* or second via the genus *Tubiella.* The classification by Fritsch (1942, 1944 and 1945) mainly takes into account the morphological characters where the main focus was on heterocytous cyanobacteria. He proposed 5 orders, where Chroococcales (unicellular, palmelloid forms, reproduction through endospores and cell division), Chamaesiphonales (unicellular, shows polarity, reproduction by endospores as well as exospores) and Pleurocapsales (filamentous, non-heterocytous, heterotrichous type, reproduction through endospores), are placed together as they lack hormogone formation, whereas Nostocales (filamentous, both non-heterocytous and heterocytous, some false-branched, multiplication by hormocyte, hormogones and akinetes) and Stigonematales (heterocytous, true-branched, pit-connections between cells, and reproduction through hormogones and hormocytes, but rarely through akinetes) are placed together because they form hormogones. He postulated Pleurocapsales to be the most evolved among the three non hormogone producing orders, along with also indicating towards a sort of link between Pleurocapsales and Stigonematales. Similarly, the Stigonematales was placed as the most evolved order among the latter two heterocytous orders. Desikachary (1959) in his work overall accepted Fritsch's classification and kept the five orders intact – Chroococcales, Chamaesiphonales, Pleurocapsales, Nostocales and Stigonematales. He too considered Stigonematales to be the most evolved group. Desikachary also postulated the possibility of evolution of heterocytous cyanobacteria from some simple non-heterocytous groups by secondary reduction. Desikachary also transferred the genus *Camptylonema* under Stigonemataceae but did not agree for its placement under Fischerellaceae, as *Camptylonema* possessed both true as well as false branching. He also established a new genus *Camptylonemopsis* (Desikachary 1959), where he included all the heterocytous cyanobacteria, having crescent shaped filaments that do not show any type of true branching. The placement of *Plectonema,* a false branching genus was done in Scytonemataceae whereas closely related genera which are non-false branching, such as *Phormidium* and *Lyngbya*, were placed in Oscillatoriaceae (Desikachary 1973). Along with time, the modern age of cyanobacterial classification started to emerge. This age started with Prescott (1962), who classified cyanobacteria in 3 orders, Chroococcales, Chamaesiphonales, and Hormogonales, almost in agreement with Frémy's (1929) classification. The incorporation of cyanobacteria in the 8$^{th}$ edition (II$^{nd}$ division) of Bergey's

Manual of Determinative Bacteriology, under phototropic bacteria by International Committee on Systematic Bacteriology (ICSB) of the International Association of Microbiological Societies (IAMS), has brought a new perspective in cyanobacterial taxonomy. Stanier proposed to classify cyanobacteria under the Bacteriological code of nomenclature (ICBN) as there were many developments that had been made to study pure cyanobacterial cultures according to morphological, physiological, ecological and genetic methods (Stanier 1977; 1978). It was eventually Rippka et al. (1979) who first classified cyanobacteria according to ICBN and divided them into 5 sections by designating roman numbers. In Section I, unicellular cyanobacteria which reproduce by binary fission were kept and were further divided into 3 groups on the basic of baeocyte formation. Section II comprised again of unicellular cyanobacteria and further divided into 3 groups - the first group representing members with immotile baeocytes, second with motile baeocytes and third group with motile baeocytes and pseudofilamentous structure which are formed through cellular aggregation via binary fission. Section III consisted of filamentous, non-heterocytous cyanobacteria. Section IV comprised of filamentous, heterocytous cyanobacteria which can divide in one plane. At the end, section V included heterocytous, filamentous and branched forms that divided in more than one plane. This classification revolutionized the systematics of cyanobacteria that had been going through decades purely on morphological basis and governed under the provisions of the Botanical Code.

## Introduction to Modern Taxonomy- Beginning of a "Polyphasic Approach"

At the start of 1980's, a new school of thought started to emerge, where a combined total evidence approach began to spread it roots in cyanobacterial taxonomy. Initially it was only the morphological characters along with some ecological and biochemical traits, which were used in this system of classification. Subsequently, with more refinements and clarity, regarding the usage of the *16S rRNA* gene as a molecular marker, cyanobacterial taxonomy improved further. The universal nature and high information content help to unravel and understand major evolutionary changes in cyanobacteria (Woese and Fox 1977). At that time, major reforms were in queue as the botanical as well as bacteriological code was taken into account, for classification of cyanobacteria. Major revisions were also carried out at that point of time

(Komárek and Anagnostidis 1989). Thus, reconciliation efforts were designed to unite both the codes of nomenclature with an ultimate aim to bring up a taxonomic system which could help to understand and study cyanobacteria in a new perspective. But it was not easy as more than a century's work on cyanobacterial work was based on botanical nomenclature and thus many proposals were subsequently made aiming to bring up a unified system of cyanobacterial taxonomy (Oren 2004, 2011; Hoffmann 2005; Oren and Tindall 2005; Oren et al. 2009; Oren and Garrity 2014; Pinevich 2015; Oren 2020; Oren et al. 2021). Meanwhile, the term "Polyphasic" was first given by Hoffmann (2005), where he designated cyanobacteria as Cyanoprokaryota and subdivided the class into three sub-classes- Synechococcineae, Oscillatorineae and Nostocineae. From that moment, a lot of new knowledge and improvements have been added to this field of study by many workers all across the globe. Notably, the use of 16S-23S ITS region in cyanobacterial taxonomy has been another step that has further helped to differentiate closely related genera using the polyphasic approach (Iteman et al. 2000; Boyer et al. 2021). Eventually, it was Komárek et al. (2014) who proposed a new system of classification based solely on the polyphasic approach which as of now constitutes the spirit of the present-day taxonomy of cyanobacteria. Many changes were done in the new system of classification by using the methodological polyphasic approach. Notably, the two orders, Synechococcales and Chroococcales were separated from each other. Major revisions were done in the family Oscillatoriaceae on the basis of cytological, morphological, ecological, and molecular evidences. Further, the entire heterocytous cyanobacteria were also grouped together in a single order, Nostocales.

In spite of the universal consensus and agreement on usage of the polyphasic approach, there are still several problematic cases too which need to be addressed. The lack of correct naming and improper typification of several taxa has brought about a lot of confusing scenarios in cyanobacterial taxonomy. Another important aspect is the polyphyletic nature of several taxa such as *Nostoc, Leptolyngbya, Calothrix, Trichormus, Anabaena* and many others which needs to be solved, so as to attain monophyly in these large groups (Komárek 2018).

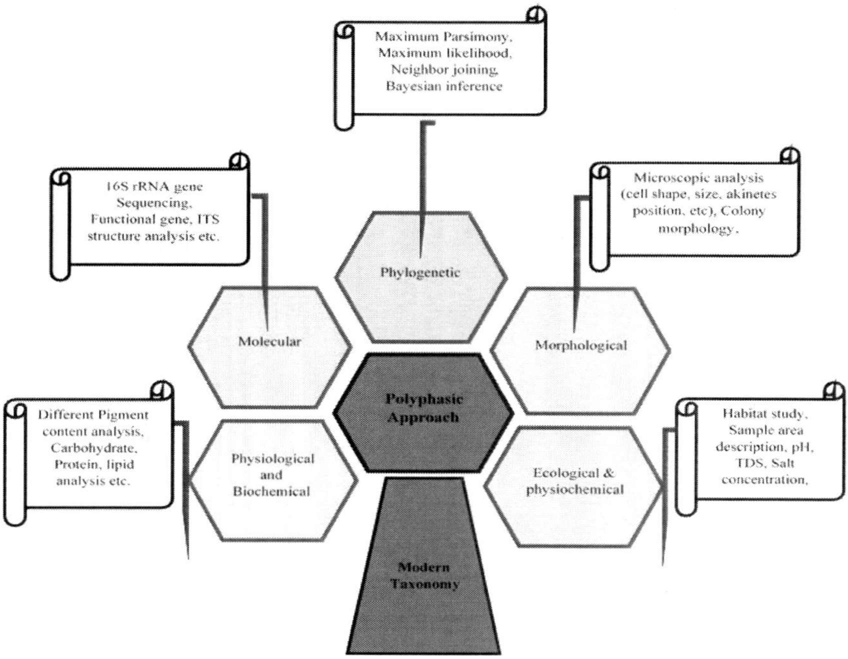


**Figure 2.** Various components of a polyphasic approach in case of cyanobacteria.

Moving on, in spite of many challenges that still remain unaddressed, the polyphasic approach has definitely changed the taxonomy of cyanobacteria and has provided an immense scope to describe new families, genera and species. After the classification system by Komárek et al. (2014), at least 10 new families have been described along with 80 new genera. Alone in India, many new genera have been described by using polyphasic approach and some of them include - *Desertifilum* (Dadheech et al. 2012), *Aliinostoc* (Bagchi et al. 2017), *Leptoelongatus, Euryhalinema* (Chakraborty et al. 2019), *Desikacharya, Dulcicalothrix,* (Saraf et al. 2019a, 2019b), *Constrictifilum* (Chavadar et al. 2021) etc. Family level revision was also done in Calotrichaceae (Saraf et al. 2019b), which now includes 3 genera- *Dulcicalothrix, Calothrix* and *Macrochaete.* We bring herein, some of the major taxonomic studies performed in India, in which new taxa were described using a polyphasic approach.

**Table 1.** Some of the cyanobacterial genera and species described from India by using polyphasic approach in tabulated form

| Name of the Strain | Habitat | Place | Order | Family | References |
|---|---|---|---|---|---|
| *Desertifilum tharense* gen. et sp. nov. | Desert crusts | Thar Desert | Oscillatoriales | Desertifilaceae | Dadheech et al. 2012 |
| *Schmidleinema santiniketanensis* sp. nov. | Lime washed concrete building | Department of Biotechnology, Santiniketan | Nostocales | Hapalosiphonaceae | Keshari et al. 2016 |
| *Scytonema bilaspurensis* sp. nov. | Freshwater body | Bilaspur, Chhattisgarh | Nostocales | Scytonemataceae | Singh et al. 2016a |
| *Desmonostoc punense* sp. nov. | Freshwater body | Panchvati locality of Pune | Nostocales | Nostocaceae | (Singh et. al. 2016b) Saraf et al. 2018b |
| *Scytonema singhii* sp. nov. | Freshwater body | Bilaspur, Chhattisgarh | Nostocales | Scytonemataceae | Singh et al. 2017a |
| *Westiellopsis ramosa* sp. nov. | Freshwater body | Jabalpur, Madhya Pradesh | Nostocales | Hapalosiphonaceae | Singh et al. 2017b |
| *Aliinostoc morphoplasticum* gen. et. sp. nov. | Eutrophic water body | Sihora, Jabalpur | Nostocales | Nostocaceae | Bagchi et al. 2017 |
| *Desikacharya thermotolerans* sp. nov. | Soil-dwelling | Mandsaur, Madhya Pradesh | Nostocales | Nostocaceae | Saraf et al. 2019a |
| *Leptolyngbya indica* sp. nov. | Alluvial plain rice field | Lower Gangetic Plains | Synechococcales | Leptolyngbyaceae | Debnath et al. 2017 |
| *Scytonema pachmarhiense* sp. nov. | Freshwater body | Pachmarhi Biosphere Reserve, Madhya Pradesh | Nostocales | Scytonemataceae | Saraf et al. 2018a |
| *Aliinostoc tiwarii* sp. nov. | Freshwater for hilly area. | Pachmarhi Biosphere Reserve, Madhya Pradesh | Nostocales | Nostocaceae | Saraf et al. 2018b |
| *Aliinostoc soli* sp. nov. | Soil isolate from hilly area. | Mumbra, Mumbai. | Nostocales | Nostocaceae | Saraf et al. 2018b |

| Name of the Strain | Habitat | Place | Order | Family | References |
|---|---|---|---|---|---|
| *Desmonostoc magnisporum* sp. nov. | Freshwater body | Shrirampur, Maharashtra | Nostocales | Nostocaceae | Saraf et al. 2018b |
| *Oxynema aestuarii* sp. nov. | soil biofilms | Sundarbans | Oscillatoriales | Microcoleaceae | Chakraborty et al. 2018 |
| *Leptoelongatus litoralis* gen. et. sp. nov. | Soil surface | Sundarbans (mangrove forest) | Synechococcales | Leptolyngbyaceae | Chakraborty et al. 2019 |
| *Euryhalinema mangrovii* gen. et. sp. nov. | Soil surface | Sundarbans (mangrove forest) | Synechococcales | Leptolyngbyaceae | Chakraborty et al. 2019 |
| *Desikacharya nostocoides* gen. et. sp. nov. | Soil surface and seepage | Mandsaur, Madhya Pradesh | Nostocales | Nostocaceae | Saraf et al. 2019a |
| *Desikacharya soli* sp. nov. | Soil surface and seepage | Mandsaur, Madhya Pradesh | Nostocales | Nostocaceae | Saraf et al. 2019a |
| *Dulcicalothrix necridiiformans* gen. et. sp. nov. | Oligotrophic pond | Ahmednagar, Maharashtra | Nostocales | Calotrichaceae | Saraf et al. 2019b |
| *Westiellopsis akinetica* sp. nov. | Soil-dwelling | Banaras Hindu University, Varanasi, Uttar Pradesh | Nostocales | Hapalosiphonaceae | Mishra et al. 2020 |
| *Iningainema sahyadrensis* sp. nov. | Subaerial outcrops near to a waterfall | Sahyadri mountain range, Western Ghats | Nostocales | Scytonemataceae | Maltsev et al. 2021 |
| *Mastigocladus ambikapurensis* sp. nov. | Wall of hot spring | Tatapani hot spring, Balarampur, Chhattisgarh | Nostocales | Hapalosiphonaceae | Jaiswal et al. 2021 |
| *Constrictifilum karadense* gen. et. sp. nov. | Water logged sugarcane field | Malkapur, Karad, Maharashtra | Nostocales | Nostocaceae | Chavadar et al. 2021 |
| *Aerofilum fasciculatum* gen. et. sp. nov. | Biofilms present on the soil surface | Sagar islands of the Indian Sundarban | Synechococcales | Oculatellaceae | Chakraborty et al. 2021. |
| *Euryhalinema pallustris* sp. nov. | Biofilms present on the soil surface | Lothian islands of the Indian Sundarban | Synechococcales | Prochlorotrichaceae | Chakraborty et al. 2021 |

## Conclusion

Study of cyanobacterial ecology and diversity has allowed us to uncover the long history of cyanobacterial evolution and to understand the widespread distribution of this group. The entire group possesses well differentiated characters and a lot of morphological variations and this possibly explains the reasons why the taxonomy of cyanobacteria still remains confusing and complicated. More than a century-long debate has come to a point where a unified system of classification has eventually been shaped up in the form of a polyphasic approach. This approach based on many evidences has provided us a way to solve various taxonomic problems and to understand in a better way, the complex classification system of cyanobacteria. One important step would be, to honour the work of all the taxonomists who helped us on the way to find a unified, unambiguous, and uncomplicated system of classification. It must also be remembered that continuous revisions and corrections are necessary for the continued and sustained evolution of a system of classification (Komárek 2018). One should not be hesitant to bring new ideas to improvise the present system, as continued discoveries and adaptations are a part of this field of science.

## Acknowledgements

We thank the Head, Department of Botany, Banaras Hindu University (BHU), for providing the necessary facilities and encouragement. SP thanks DST for DST-INSPIRE Junior Research Fellowship (JRF). Department of Science and Technology (DST-SERB), India is acknowledged for the Core Research Grant Project CRG/2018/004111. The Banaras Hindu University is also acknowledged for providing Seed Grant under the Institution of Eminence (IoE) Scheme.

## References

Abed RM, Dobretsov S, Sudesh K (2009) Applications of cyanobacteria in biotechnology. *J Appl Microbiol* 106(1): 1-2.

Adams DG (2000) Heterocyst formation in cyanobacteria. *Curr Opin Microbiol* 3(6): 618-24.

Adams DG, Duggan PS, Tansley (1999) Heterocyst and akinete differentiation in cyanobacteria. *New Phytol* 144(1): 3-3.

Albertano P, Somma DD, Capucci E (1997) Cyanobacterial picoplankton from the Central Baltic Sea: cell size classification by image-analyzed fluorescence microscopy. *J plankton res* 119(10): 1405-16.

Aleksakhina TI, Shtina EA (1984) Soil algae in the rhizosphere of the dominant plants of forest biogeocenoses. *PubMed.*

Alvarenga DO, Rigonato J, Branco LH, Fiore MF (2015) Cyanobacteria in mangrove ecosystems. *Biodivers Conserv* 24(4): 799-817.

Bagchi SN, Dubey N, Singh P (2017) Phylogenetically distant clade of *Nostoc*-like taxa with the description of *Aliinostoc* gen. nov. and *Aliinostoc morphoplasticum* sp. Nov. *Int J Syst Evol Microbiol* 67(9): 3329-38.

Bakieva GR, Khaibullina LS, Gaisina LA, Kabirov RR (2012) Ecological-floristic analysis of soil algae and cyanobacteria on the Tra-Tau and Yurak-Tau Mounts, Bashkiria. *Eurasian soil sci* 45(9): 873-81.

Bar-Yosef Y, Sukenik A, Hadas O, Viner-Mozzini Y, Kaplan A (2010) Enslavement in the Water Body by Toxic *Aphanizomenon ovalisporum*, Inducing Alkaline Phosphatase in Phytoplanktons. *Curr Biol* 20(17): 1557-1561.

Bender J, Gould JP, Vatcharapijarn Y, Young JS, Phillips P (1994) Removal of zinc and manganese from contaminated water with cyanobacteria mats. *Water Environ Res* 66(5): 679-83.

Benny N, Thomas LC, Padmakumar KB (2021) Environmental influences on the cyanobacterial mat formation in the mangrove ecosystems along the southwest coast of India. *Mar Ecol* e12685.

Berry JP, Gantar M, Perez MH, Berry G, Noriega FG (2008) Cyanobacterial toxins as allelochemicals with potential applications as algaecides, herbicides and insecticides. *Mar drugs* 6(2): 117-46.

Bewley JD (1979) Physiological aspects of desiccation tolerance. *Annu Rev Plant Physiol* 30(1): 195-238.

Bhadury P, Wright PC (2004) Exploitation of marine algae: biogenic compounds for potential antifouling applications. *Planta* 219(4): 561-78.

Bold HC, Wynne MJ (1978) *Introduction to Algal Structure and Reproduction.* Prentice Hall, Englewood Cliff, NJ.

Boone DR, Castenholz RW, Garrity GM (2001) Vol. 1: *The Archaea and the deeply branching and phototrophic bacteria.* Springer New York.

Bornet É, Flahault C (1885) Tableau synoptique des Nostochacées filamenteuses hétérocystées. *Mémoires de la Soc nat des Sci nat de Cherboug* 25: 195-223.

Bornet É, Flahault C (1888) Note sur deux nouveaux genres d'algues perforantes/Note on two new genera of borer algae. *J Bot* 2: 161-165.

Borowitzka MA, Borowitzka LJ (1988) Micro-algal biotechnology. Cambridge University Press. Bourrelly P (1970) *Les algues d'eau douce III/ Freshwater Algae III.* N. Boubée & Cie, Paris.

Boyer SL, Flechtner VR Johansen, J.R. (2001). Is the 16S–23S rRNA internal transcribed spacer region a good tool for use in molecular systematics and population genetics? A case study in cyanobacteria. *Mol Biol Evol* 18: 1057-1069.

Branco LHZ, Sant'Anna CL, Azevedo MTP, Sormus L (1996) Cyanophyte flora from Cardoso Island mangroves, Sao Paulo state, Brazil. 1. Chroococcales. *Arch Hydrobiol Suppl Algol Stud* 80:101–113.

Broady PA (1996) Diversity, distribution and dispersal of Antarctic terrestrial algae. *Biodivers Conserve* 5(11): 1307-35.

Brookes JD, Ganf GG (2001) Variations in the buoyancy response of *Microcystis aeruginosa* to nitrogen, phosphorus and light. *J Plankton Res* 23(12): 1399-1411.

Büdel B, Darienko T, Deutschewitz K, Dojani S, Friedl T, Mohr KI, Salisch M, Reisser W, Weber B (2008). Southern African Biological Soil Crusts are Ubiquitous and Highly Diverse in Drylands, Being Restricted by Rainfall Frequency. *Microbial Ecol* 57(2): 229-247.

Burja AM, Tamagnini P, Bustard MT, Wright PC (2001) Identification of the green alga, Chlorella vulgaris (SDC1) using cyanobacteria derived 16S rDNA primers: targeting the chloroplast. *FEMS Microbiol Lett* 202(2): 195-203.

Callieri C, Stockner JG (2002) Freshwater autotrophic picoplankton: a review. *J Limnol* 61(1): 1-14.

Capone DG, Zehr JP, Paerl HW, Bergman B, Carpenter EJ (1997) *Trichodesmium*, a Globally Significant Marine Cyanobacterium. *Sci* 276(5316): 1221-1229.

Carpenter EJ, Lin S, Capone DG (2000) Bacterial activity in South Pole snow. *Appl Environ Microbiol* 66(10): 4514-4517.

Castenholz RW (1992) Species usage, concept, and evolution in the cyanobacteria (blue-green algae). *J Phycol* 28: 737-745.

Castenholz RW, Wilmotte A, Herdman M, Rippka R, Waterbury JB, Iteman I, Hoffmann L (2001) Phylum BX. Cyanobacteria. *Bergey's Manual® of Systematic Bacteriology* 473-599.

Catherine Q, Susanna W, Isidora ES, Mark H, Aurélie V, Jean-François H (2013) A review of current knowledge on toxic benthic freshwater cyanobacteria – Ecology, toxin production and risk management. *Water Res* 47(15): 5464-5479.

Chakraborty S, Maruthanayagam V, Achari A, Mahansaria R, Pramanik A, Jaisankar P, Mukherjee J (2018) *Oxynema aestuarii* sp. nov. (Microcoleaceae) isolated from an Indian mangrove forest. *Phytotaxa* 374: 24-40.

Chakraborty S, Maruthanayagam V, Achari A, Pramanik A, Jaisankar P, Mukherjee J (2019) *Euryhalinema mangrovii* gen. nov., sp. nov. and *Leptoelongatus litoralis* gen. nov., sp. nov. (Leptolyngbyaceae) isolated from an Indian mangrove forest. *Phytotaxa* 422(1): 58-74.

Chakraborty S, Maruthanayagam V, Achari A, Pramanik A, Jaisankar P, Mukherjee J (2021) *Aerofilum fasciculatum* gen. nov., sp. nov. (Oculatellaceae) and *Euryhalinema pallustris* sp. nov (Prochlorotrichaceae) isolated from an Indian mangrove forest. *Phytotaxa* 11: 522(3):165-86.

Chavadar M, Saraf A, Suradkar A, Mishra D, Kumar N, Singh P (2021) *Constrictifilum karadense* gen. et sp. nov., a new Nostocalean genus from Maharashtra, India. *FEMS Microbiol Lett* 368(11): fnab057.

Cohen Y, Jorgensen BB, Revsbech NP, Poplawski R (1986) Adaptation to Hydrogen Sulfide of Oxygenic and Anoxygenic Photosynthesis among Cyanobacteria. *Appl Environ Microbiol* 51(2): 398-407.

Dadheech PK, Abed RM, Mahmoud H, Mohan MK, Krienitz L (2012) Polyphasic characterization of cyanobacteria isolated from desert crusts, and the description of *Desertifilum tharense* gen. et sp. nov (Oscillatoriales). *Phycologia* 51(3): 260-70.

Dahms HU, Ying X, Pfeiffer C (2006) Antifouling potential of cyanobacteria: a mini-review. *Biofouling* 22(5): 317-27.

De Nobel, WT, Matthijs HCP, Von Elert E, Mur LR (1998) Comparison of the light-limited growth of the nitrogen-fixing cyanobacteria *Anabaena* and *Aphanizomenon*. *New Phytol* 138(4): 579-587.

De Philippis R, Vincenzini M (1998) Exocellular polysaccharides from cyanobacteria and their possible applications. *FEMS Microbiol Rev 22*(3): 151-175.

De PK (1939) The role of blue-green algae in nitrogen fixation in rice-fields. *Proc R Soc B Biol Sci* 127(846): 121-139.

Debnath M, Singh T, Bhadury P (2017) New records of Cyanobacterial morphotypes with Leptolyngbya indica sp. nov. from terrestrial biofilms of the Lower Gangetic Plain, India. *Phytotaxa* 316(2): 101-20.

Desikachary TV (1959) Cyanophyta, ICAR. *Monograph of algae* ICAR, New Delhi.

Desikachary TV (1973) Status of classical taxonomy. *Botany monographs*.

Díez B, Bergman B, Pedrós-Alió C, Antó M, Snoeijs P (2012) High cyanobacterial nifH gene diversity in Arctic seawater and sea ice brine. *Environmental Microbiol Rep* 4(3): 360-366.

Dillon JG, Miller SR, Castenholz RW (2003) UV-acclimation responses in natural populations of cyanobacteria (*Calothrix* sp.). *Environ Microbiol* 5(6): 473-483.

Dobretsov S, Abed RMM, Al Maskari SMS, Al Sabahi JN, Victor R (2010) Cyanobacterial mats from hot springs produce antimicrobial compounds and quorum-sensing inhibitors under natural conditions. *J Appl Phycol 23:6 23*(6): 983-993.

Drouet F, Daily WA (1956) Revision of the coccoid Myxophyceae. *Butler University Botl Studies* 12: 1-218.

Drouet, F (1981) Revision of the Stigonemataceae with a summary of the classification of the blue-green algae. *Nova Hedwigia* 66: 1-221.

Dvořák P, Jahodářová E, Casamatta DA, Hašler P, Poulíčková A (2018) Difference without distinction? Gaps in cyanobacterial systematics; when more is just too much. *Fottea* 18(1): 130-6.

Eiler A (2006) Evidence for the Ubiquity of Mixotrophic Bacteria in the Upper Ocean: Implications and Consequences. *Appl Environ Microbiol* 72(12): 7431.

Elenkin AA (1936) Monographie algarum Cyanophycearum aquidulcium et terrestrium infinibus URSS inventatrum, vol 1. *Acad Nauk URSS Moscow*, Leningrad 675.

Elenkin AA (1938) Monographia algarum cyanophycearum aquidulcium et terrestrium in finibus URSS inventarum. *Sinezelenje vodorosli SSSR* 1: 1-908.

Elenkin AA (1949) Monographia algarum cyanophycearum aquidulcium et terrestrium in finibus URSS inventarum [Sinezelenye vodorosli SSSR]. *Pars Specialis (Systematica)* 2: 1908.

Engene N, Paul VJ, Byrum T, Gerwick WH, Thor A, Ellisman MH (2013) Five chemically rich species of tropical marine cyanobacteria of the genus Okeania gen. nov. (Oscillatoriales, Cyanoprokaryota). *J Phycol* 49(6): 1095-1106.

Erdmann N, Hagemann M (2001) Salt Acclimation of Algae and Cyanobacteria: A Comparison. *Algal Adaptation to Environmental Stresses* 323-361.

Falch BS, Konig GM, Wright AD, Sticher O, Angerhofer CK, Pezzuto JM Bachmann H (1995) Biological activities of cyanobacteria: Evaluation of extracts and pure compounds. *Planta Medica* 61(4): 321-328.

Fay P (1965) Heterotrophy and nitrogen fixation in *Chlorogloea fritschii*. *Microbiol* 39(1): 11-20.

Fay P (1983) *The Blue-Greens.* Edward Arnold Publisher, London.

Fay P, Van Baalen C (1987) *The cyanobacteria.* Elsevier.

Feng X, Bandyopadhyay A, Berla B, Page L, Wu B, Pakrasi HB, Tang YJ (2010) Mixotrophic and photoheterotrophic metabolism in *Cyanothece* sp. ATCC 51142 under continuous light. *Microbiol* 156(8): 2566-2574.

Flechtner VR, Johansen JR, Clark WH (1998) Algal composition of microbiotic crusts from the central desert of Baja California, Mexico. *The Great Basin Naturalist* 1: 295-311.

Fogg GE (1973) Phosphorus in primary aquatic plants. *Phosphorus in Fresh Water and the Marine Environment* 77-91.

Frémy P (1929) Les Nostocacées de la Normandie. Not Mem Doc Soc Agric Archeol Hist nat Manche 41: 197-228.

Friedmann EI, Ocampo-Friedmann R (1984). The antarctic cryptoendolithic ecosystem: Relevance to exobiology. *Ori Life* 14(1): 771-776.

Friedmann EI, Kappen L, Meyer MA, Nienow JA (1993) Long-term productivity in the cryptoendolithic microbial community of the Ross Desert, Antarctica. *Microbial Ecol* 25(1): 51-69.

Fritsch FE (1942) The interrelations and classification of the Myxophyceae (Cyanophyceae). *New Phytol* 41(2): 134-148.

Fritsch FE (1944) Present-day classification of algae. *Bot Rev* 10: 233-277.

Fritsch FE (1945) *The structure and reproduction of the algae, Foreword, Phaeophyceae, Rhodophyceae,* Myxophyceae, vol II. Cambridge University Press, Cambridge 1-939.

Gao Q, Garcia-Pichel F (2011) Microbial ultraviolet sunscreens. *Nat Rev Microbiol* 9(11): 791-802.

Garcia-Pichel F, Castenholz RW (1990) Comparative anoxygenic photosynthetic capacity in 7 strains of a thermophilic cyanobacterium. *Archives Microbiol* 153(4): 344-351.

Garcia-Pichel F, López-Cortés A, Nübel U (2001) Phylogenetic and morphological diversity of cyanobacteria in soil desert crusts from the Colorado plateau. *Appl Environ Microbiol* 67(4): 1902-1910.

Garcia-Pichel F, Sherry ND, Castenholz RW (1992) Evidence for an ultraviolet sunscreen role of the extracellular pigment scytonemin in the terrestrial cyanobacterium *Chiorogloeopsis* sp. *Photochem photobiol* 56(1): 17-23.

Gaysina L, Saraf A, Singh P (2019) Cyanobacteria in diverse habitats. In: Mishra AK, Tiwari DN, Rai AN (eds) *Cyanobacteria.* From basic science to applications. Academic, London 1-28.

Geitler L (1932) Cyanophyceae. In: *Rabenhorst's Kryptogamen Flora von Deutschland,* vol 14. Akad Verlagsges, Leipzig 1196.

Geitler L (1942) Schizophyta (Klasse Schizophyceae). In: Engler A, Prantl K (eds) *Natürliche Pflanzenfamilien* 1b. Duncker & Humblot, Berlin 1-232.

GM (1938) *Cryptogamic Botany,* Vol. 1, McGraw Hill, New York.

Golubic S, Seong-Joo L (1999). Early cyanobacterial fossil record: preservation, palaeoenvironments and identification. *Eur J Phycol* 34(4): 339-348.

Gordon DA, Priscu JC, Giovannoni S (2000) Distribution and phylogeny of bacterial communities associated with mineral particles in Antarctic Lake ice. *Microb Ecol* 39: 197-202.

Guedes IA, da Costa Leite DM, Manhães LA, Bisch PM, Azevedo SMFO, Pacheco ABF (2014) Fluctuations in microcystin concentrations, potentially toxic *Microcystis* and genotype diversity in a cyanobacterial community from a tropical reservoir. *Harmful Algae* 39: 303-309.

Gustafson KR, Cardellina JH, Fuller RW, Weislow OS, Kiser RF, Snader KM, Patterson GML, Boyd MR (1989) AIDS-antiviral sulfolipids from cyanobacteria (blue-green algae). *J Natl Cancer Inst* 81(16): 1254-1258.

Hagmann L, Jüttner F (1996) Fischerellin A, a novel photosystem-II-inhibiting allelochemical of the cyanobacterium *Fischerella muscicola* with antifungal and herbicidal activity. *Tetrahedron Lett* 37(36): 6539-6542.

Hamisi M, Díez B, Lyimo T, Ininbergs K, Bergman B (2013) Epiphytic cyanobacteria of the seagrass Cymodocea rotundata: diversity, diel nifH expression and nitrogenase activity. *Environ Microbiol Rep* 5(3): 367-376.

Hansgirg AN (1892) Prodromus der Algenflora von Böhmen. *2 Arch Naturwiss Landesdurchforsch Böhmen* [Prodromus of the algal flora of Bohemia. *2 Arch Naturwiss country search Bohemia*] 8: 1-268.

Henson BJ, Hesselbrock SM, Watson LE, Barnum SR (2004) Molecular phylogeny of the heterocystous cyanobacteria (subsections IV and V) based on nifD. *Int J Syst Evol Microbiol* 54(2): 493-497.

Henson BJ, Watson LE, Barnum SR (2002) Molecular Differentiation of the Heterocystous Cyanobacteria, *Nostoc* and *Anabaena*, Based on Complete NifD Sequences. *Curr Microbiol* 45(3): 161-164.

Herrero A, Muro-Pastor AM, Flores E (2001) Nitrogen control in cyanobacteria. *J Bacteriol* 183(2): 411-425.

Hershkovitz N, Oren A, Cohen Y (1991) Accumulation of Trehalose and Sucrose in Cyanobacteria Exposed to Matric Water Stress. *Appl Environ Microbiol* 57(3): 645-648.

Hill DR, Keenan TW, Helm RF, Potts M, Crowe LM, Crowe JH (1997) Extracellular polysaccharide of *Nostoc commune* (Cyanobacteria) inhibits fusion of membrane vesicles during desiccation. *J Appl Phycol* 9(3): 237-248.

Hoffman L (1999) Marine cyanobacteria in tropical regions: diversity and ecology. *Eur J Phycol* 34(4): 371-379.

Hoiczyk E (2000) Gliding motility in cyanobacterial: observations and possible explanations. *Archives of Microbiol* 174(1–2): 11-17.

Hutchins DA, Fu FX, Zhang Y, Warner ME, Feng Y, Portune K, Bernhardt PW, Mulholland MR (2007) CO2 control of *Trichodesmium* N2 fixation, photosynthesis, growth rates, and elemental ratios: Implications for past, present, and future ocean biogeochemistry. *Limnol Oceanogr* 52(4): 1293-1304.

Isichei AO (1990) The role of algae and cyanobacteria in arid lands. A review. *Arid Land Res Manag* 4(1): 1-7.

Iteman I, Rippka R, de Marsac NT, Herdman M (2000) Comparison of conserved structural and regulatory domains within divergent 16S rRNA–23S rRNA spacer sequences of cyanobacteria. The GenBank accession numbers for the sequences reported in this paper are AF180968 and AF180969 for ITS-L and ITS-S, respectively. *Microbiol* 146(6): 1275-1286.

Jaiswal TP, Chakraborty S, Singh P, Mishra AK, Singh SS (2021) Description of hot spring dwelling *Mastigocladus ambikapurensis* sp. nov., using a polyphasic approach. *Plant Syst Evol* 307(3): 1-2.

Janson S (2002) Cyanobacteria in Symbiosis with Diatoms. *Cyanobacteria in Symbiosis* 1-10.

Johansen JR (1993) Cryptogamic crusts of semiarid and arid lands of North America. *J phycol* 29(2): 140-7.

Johansen JR, Casamatta DA (2005) Recognizing cyanobacterial diversity through adoption of a new species paradigm. *Algological studies* 117(1): 71-93.

Jones AC, Monroe EA, Podell S, Hess WR, Klages S, Esquenazi E, Niessen S, Hoover H, Rothmann M, Lasken RS, Yates JR, Reinhardt R, Kube M, Burkart MD, Allen EE, Dorrestein PC, Gerwick WH, Gerwick L (2011) Genomic insights into the physiology and ecology of the marine filamentous cyanobacterium *Lyngbya majuscula*. *Proc Natl Acad Sci USA* 108(21): 8815-8820.

Kana TM, Feiwel NL, Flynn LC (1992) Nitrogen starvation in marine *Synechococcus* strains: clonal differences in phycobiliprotein breakdown and energy coupling. *Mar Ecol Prog Ser* 88: 75-82.

Kannaiyan S, Aruna SJ, Merina Prem Kumari S, Hall DO (1997) Immobilized cyanobacteria as a biofertilizer for rice crops; Intl. Conference on Applied Algology, Knysna, South Africa, April 1996. *J Appl Phycol* 9(2): 167-174.

Karl DM, Bird DF, Björkman K, Houlihan T, Shackelford R, Tupas L (1999) Microorganisms in the Accreted Ice of Lake Vostok, Antarctica. *Sci* 286(5447): 2144-2147.

Keshari N, Das SK, Adhikary SP (2016) *Schmidleinema santiniketanense* sp. nov. (Cyanobacteria/Cyanoprokaryota) from a building facade in Santiniketan, India. *Phytotaxa* 3: 283(2): 181-7.

Knelman JE, Legg TM, O'Neill SP, Washenberger CL, González A, Cleveland CC, Nemergut DR (2012) Bacterial community structure and function change in association with colonizer plants during early primary succession in a glacier forefield. *Soil Biol Biochem* 46: 172-180.

Komárek J (2003) Problem of the taxonomic category. *Algological Studies/Archiv für Hydrobiologie, Supplement Volumes* 1: 281-97.

Komárek J (2006) Cyanobacterial taxonomy: current problems and prospects for the integration of traditional and molecular approaches. *Algae* 21(4): 349-75.

Komárek J (2018) Several problems of the polyphasic approach in the modern cyanobacterial system. *Hydrobiologia* 811(1):7-17.

Komárek J, Anagnostidis K (1989) Modern approach to the classification system of Cyanophytes 4-Nostocales. Archiv für Hydrobiologie. Supplement band. *Monographische Beiträge* 82(3): 247-345.

Komárek J, Anagnostidis K (1998) Freshwater Flora of Central Europe, Vol 19(1): Cyanoprokaryota: Chroococcales (Süßwasserflora von Mitteleuropa, Bd. 19(1): Cyanoprokaryota: Chroococcales) (H. Ettl, J. Gerloff, H. Heynig & D. Mollenhauer, Eds.). Süßwasserflora von Mitteleuropa.

Komárek J, Anagnostidis K (2005) Cyanoprokaryota -2. Teil/2nd Part: Oscillatoriales. In Büdel, B., L. Krienitz, G. Gärtner & M. Schagerl (eds), *Süßwasserflora von Mitteleuropa* 19(2) Elsevier/Spektrum Heidelberg.

Komárek J, Kaštovský J, Mareš J, Johansen JR (2014) Taxonomic classification of cyanoprokaryota (cyanobacterial genera), using a polyphasic approach. *Preslia* 86: 295-335.

Komárek J, Kaštovský J, Ventura S, Turicchia SŠ (2009) The cyanobacterial genus *Phormidesmis. Algological studies* 1: 41-59.

Kyaruzi JJ, Kyewalyanga MS, Muruke MHS (2004) Cyanobacteria Composition and Impact of Seasonality on their *In Situ* Nitrogen Fixation Rate in a Mangrove Ecosystem Adjacent to Zanzibar Town. *Western Indian Ocean J Mar Sci 2*(1): 35-44.

Lambert DH, Stevens SE (1986) Photoheterotrophic growth of Agmenellum quadruplicatum PR-6. *J Bacteriol* 165(2): 654-656.

LaRoche J, Breitbarth E (2005) Importance of the diazotrophs as a source of new nitrogen in the ocean. *J Sea Res* 53(1-2): 67-91.

Lazaroff N, Vishniac W (1961) The effect of light on the developmental cycle of *Nostoc muscorum*, a filamentous blue-green alga. *J Gen Microbiol* 25(3): 365-74.

Leao PN, Vasconcelos MT, Vasconcelos VM (2009) Allelopathy in freshwater cyanobacteria. *Crit rev microbiol* 35(4): 271-82.

Lesser MP, Falcón LI, Rodríguez-Román A, Enríquez S, Hoegh-Guldberg O, Iglesias-Prieto R (2007) Nitrogen fixation by symbiotic cyanobacteria provides a source of nitrogen for the scleractinian coral *Montastraea cavernosa*. *Mar Ecol Prog Ser* 346: 143-152.

Lohtander K, Oksanen I, Rikkinen J (2003) Genetic diversity of green algal and cyanobacterial photobionts in Nephroma (Peltigerales). *The Lichenologist* 35(4): 325-339.

Lôpez-Cortés A, Garcia-Pichel F, Nübel U, Vâzquez-Juârez R (2001) Cyanobacterial diversity in extreme environments in Baja California, Mexico: a polyphasic study. *Int Microbiol* 4(4): 227-236.

Lu W (1999) *Application of molecular biological techniques to the identification of cyanobacteria.* Master's thesis, Liverpool John Moores University, Liverpool, England.

Ludwig W, Klenk H (2001) Overview: a phylogenetic backbone and taxonomic framework for procaryotic systematics, In: Boone DR, Castenholz, RW Eds. *Bergey's Manual of Systematic Bacteriology*, vol. 1. Springer-Verlag, New York, NY, USA 49-65.

Lukešová A (2001) Soil Algae in Brown Coal and Lignite Post-Mining Areas in Central Europe (Czech Republic and Germany). *Restoration Ecol 9*(4): 341-350.

Mackay MA, Norton RS, Borowitzka LJ (1984) Organic osmoregulatory solutes in cyanobacteria. *J Gen Microbiol* 130(9): 2177-2191.

Maltsev Y, Kezlya E, Maltseva S, Karthick B, Dvořák P, Kociolek JP, Kulikovskiy M (2021) A new species of the previously monotypic genus Iningainema (Cyanobacteria, Scytonemataceae) from the Western Ghats, India. *Eur J Phycol* 16: 1-1.

Margulis L, Bermudes D (1985) Symbiosis as a mechanism of evolution: status of cell symbiosis theory. *Symbiosis*.

Marin B, Nowack EC, Melkonian M (2005) A plastid in the making: evidence for a second primary endosymbiosis. *Protist* 156(4): 425-32.

Mason CP, Edwards KR, Carlson RE, Pignatello J, Gleason FK, Wood JM (1982) Isolation of Chlorine-Containing Antibiotic From the Freshwater Cyanobacterium *Scytonema hofmanni*. *Sci* 215(4531): 400-402.

Mataloni G, Komárek J (2004) Gloeocapsopsis aurea, a new subaerophytic cyanobacterium from maritime Antarctica. *Polar Biol* 27(10): 623-8.

Meeks JC, Elhai J (2002) Regulation of cellular differentiation in filamentous cyanobacteria in free-living and plant-associated symbiotic growth states. *Microbiol Rev* 66(1): 94-121.

Metting B (1981) The systematics and ecology of soil algae. *Bot Rev* 47(2): 195-312.

Miller SR, Wingard CE, Castenholz RW (1998) Effects of visible light and UV radiation on photosynthesis in a population of a hot spring cyanobacterium, a *Synechococcus* sp., subjected to high-temperature stress. *Appl Environ Microbiol* 64(10): 3893-3899.

Mishra D, Suradkar A, Saraf A, Singh P (2020) Phylogenetic evaluation of the true-branched heterocytous cyanobacteria and description of soil dwelling *Westiellopsis akinetica* sp. nov. *FEMS microbiol let* 367(5): fnaa046.

Mitsui A, Kumazawa S, Takahashi A, Ikemoto H, Cao S, Arai T (1986) Strategy by which nitrogen-fixing unicellular cyanobacteria grow photoautotrophically. *Nature* 323(6090): 720-722.

Moisander PH, Beinart RA, Hewson I, White AE, Johnson KS, Carlson CA, Montoya JP, Zehr JP (2010) Unicellular cyanobacterial distributions broaden the oceanic N2 fixation domain. *Sci (New York, N.Y.)* 327(5972): 1512-1514.

Mollenhauer D, Bengtsson R, Lindstrøm EA (1999) Macroscopic cyanobacteria of the genus *Nostoc*: a neglected and endangered constituent of European inland aquatic biodiversity. *Eur J Phycol* 34(4): 349-360.

Montoya JP, Holl CM, Zehr JP, Hansen A, Villareal TA, Capone DG (2004) High rates of $N_2$ fixation by unicellular diazotrophs in the oligotrophic Pacific Ocean. *Nature 430*(7003): 1027-1031.

Moorhead DL, Priscu JC (1998) Linkages among ecosystem components within the McMurdo Dry valleys: A synthesis. Ecosystem Dynamics in a Polar Desert: TheMcMurdo Dry Valleys. *Antarctic Res Ser, American Geophysical Union* 72: 351-64.

Mur (1999) *Toxic Cyanobacteria in Water: A guide to their public health consequences, monitoring and management*. Chapter 2.Cyanobacteria in the environment.

Murakami A, Miyashita H, Iseki M, Adachi K, Mimuro M (2004) Chlorophyll d in an epiphytic cyanobacterium of red algae. *Sci* 303(5664): 1633-1634.

Nadeau TL, Howard-Williams C, Castenholz RW (1999) Effects of solar UV and visible irradiance on photosynthesis and vertical migration of Oscillatoria sp. (Cyanobacteria) in an Antarctic microbial mat. *Aquatic Microbial Ecol 20*(3): 231-243.

Nägeli C (1849) *Gattungen einzelliger Algen: physiologisch und systematisch bearbeitet. Friedrich Schulthess.*

Nilsson M, Bhattacharya J, Rai AN, Bergman B (2002) Colonization of roots of rice (*Oryza sativa*) by symbiotic *Nostoc* strains. *The New Phytologist* 156(3): 517-525.

Oren A (2004) A proposal for further integration of the Cyanobacteria under the Bacteriological Code. *Int J Syst Evol Microbiol* 54: 1895-1902.

Oren A (2007) *Diversity of Organic Osmotic Compounds and Osmotic Adaptation in Cyanobacteria and Algae* 639-655.

Oren A (2009) *Mycosporine-like amino acids as osmotic solutes in a community of halophilic cyanobacteria* 14(3): 231-240.

Oren A (2011) Naming Cyanophyta/Cyanobacteria – a bacteriologist's view. *Fottea* 11(1): 9-16.

Oren A (2020). Three alternative proposals to emend the Rules of the International Code of Nomenclature of Prokaryotes to resolve the status of the Cyanobacteria in the prokaryotic nomenclature *Int J Syst Evol Microbiol* 70(7): 4406-4408.

Oren A, Arahal DR, Rosselló-Móra R, Sutcliffe IC, Moore ER (2021). Emendation of General Consideration 5 and Rules 18a, 24a and 30 of the International Code of Nomenclature of Prokaryotes to resolve the status of the Cyanobacteria in the prokaryotic nomenclature. *Int J Syst Evol Microbiol* 71(8): 004939.

Oren A, Garrity GM (2014) Proposal to change General Consideration 5 and Principle 2 of the International Code of Nomenclature of Prokaryotes. *Int J Syst Evol Microbiol* 64: 309-310.

Oren A, Komárek J, Hoffmann L (2009) Nomenclature of the Cyanophyta/Cyanobacteria/ Cyanoprokaryotes: what has happened since IAC Luxembourg? *Algol Stud* 130: 17-26.

Oren A, Tindall BJ (2005) Nomenclature of the cyanophyta/cyanobacteria/ cyanoprokaryotes under the International Code of Nomenclature of Prokaryotes. *Algol Stud* 117(6): 39-52.

Paerl HW, Fulton RS (2006) Ecology of Harmful Cyanobacteria. *Undefined* 95-109.

Papaefthimiou D, Hrouzek P, Mugnai MA, Lukesova A, Turicchia S, Rasmussen U, Ventura S (2008) Differential patterns of evolution and distribution of the symbiotic behaviour in nostocacean cyanobacteria. *Int J Syst Evol Microbiol* 58(3): 553-564.

Partensky F, Hess W R, Vaulot D (1999) *Prochlorococcus*, a Marine Photosynthetic Prokaryote of Global Significance . *Microbiol Mol Biol Rev* 63(1): 106-127.

Patterson GML, Larsen LK., & Moore RE (1994) Bioactive natural products from blue-green algae. *J Appl Phycol* 6(2): 151-157.

Paul VJ, Thacker RW, Banks K, Golubic S (2005) Benthic cyanobacterial bloom impacts the reefs of South Florida (Broward County, USA). *Coral Reefs* 24(4): 693-697.

Phoenix VR, Bennett PC, Engel AS, Tyler SW, Ferris FG (2006) Chilean high-altitude hot-spring sinters: a model system for UV screening mechanisms by early Precambrian cyanobacteria. *Geobiol* 4(1): 15-28.

Piechula S, Waleron K, è wia L tek WS, Biedrzycka I, Podhajska AJ (2001) Mesophilic cyanobacteria producing thermophilic restriction endonucleases. *FEMS Microbiol Lett* 198(2): 135-140.

Pinevich AV (2015). Proposal to consistently apply the International Code of Nomenclature of Prokaryotes (ICNP) to names of the oxygenic photosynthetic bacteria (cyanobacteria), including those validly published under the International Code of Botanical Nomenclature (ICBN)/International Code of Nomenclature for algae, fungi and plants (ICN), and proposal to change Principle 2 of the ICNP. *Int J Syst Evol Microbiol* 65(3): 1070-1074.

Plavšić M, Terzic S, Ahel M, Van Den Berg CMG (2002) Folic acid in coastal waters of the Adriatic Sea. *Mar Freshwater Res* 53(8): 1245-1252.

Prescott GW (1962) Algae of the western Great Lakes area. W.C. Brown Co, Dubuque, IA.

Priscu JC, Adams EE, Lyons WB, Voytek MA, Mogk DW, Brown RL, McKay CP, Takacs CD, Welch KA, Wolf CF, Kirshtein JD, Avci R (1999) Geomicrobiology of Subglacial Ice Above Lake Vostok, Antarctica. *Sci* 286(5447): 2141-2144.

Psenner R, Sattler B, Wille A, Fritsen CH, Priscu JC, Felip M, Catalan J (1999) Lake ice microbial communities in alpine and antarctic lakes. *Cold-Adapted Organisms* 17-31.

Rabenhorst L (1865) *Flora europaea algarum acquae dulcis et submarinae auctore Ludovico Rabenhorst: Algas phycochromaceas complectens.* E. Kummerum.

Rai AN, Söderbäck E, Bergman B (2000) Tansley Review No. 116 Cyanobacterium–plant symbioses. *The New Phytologist* 147(3): 449-481.

Rajaniemi P, Hrouzek P, Kaštovská K, Willame R, Rantala A, Hoffmann L, Komárek J, Sivonen K (2005) Phylogenetic and morphological evaluation of the genera *Anabaena, Aphanizomenon, Trichormus and Nostoc* (Nostacales, cyanobacteria). *Int J Syst Evol Microbiol* 55(1): 11-26.

Řeháková K, Johansen JR, Casamatta DA, Xuesong L, Vincent J (2007) Morphological and molecular characterization of selected desert soil cyanobacteria: Three species new to science including *Mojavia* pulchra gen. et sp. nov. *Phycologia* 46(5): 481-502.

Reynolds CS (1987) Cyanobacterial Water-Blooms. *Adv Bot Res 13*(C): 67-143.

Rippka R, Stanier RY, Deruelles J, Herdman M, Waterbury JB (1979) Generic assignments, strain histories and properties of pure cultures of Cyanobacteria. *Microbiol* 111: 1-61.

Rippka R, Waterbury J, Cohen-Bazire G (1974) A cyanobacterium which lacks thylakoids. *Arch Microbiol* 100(1): 419-436.

Robertson BR, Tezuka N, Watanabe MM (2001) Phylogenetic analyses of *Synechococcus* strains (cyanobacteria) using sequences of 16S rDNA and part of the phycocyanin operon reveal multiple evolutionary lines and reflect phycobilin content. *Int J Syst Evol Microbiol* 51(3): 861-871.

Romero M, Muro-Pastor AM, Otero A (2011) Quorum sensing N-acylhomoserine lactone signals affect nitrogen fixation in the cyanobacterium *Anabaena* sp. PCC7120. *FEMS Microbiol Lett* 315(2): 101-108.

Sachs J (1874) *Traité de botanique conforme à l'état présent de la science.* F Savy.

Saha SK, Das R, Bora KN, Uma L (2007) Biodiversity of epilithic cyanobacteria from freshwater streams of Kakoijana reserve forest, Assam, India. *Ind J Microbiol* 47(3): 219-232.

Saito MA, Bertrand EM, Dutkiewicz S, Bulygin VV, Moran DM, Monteiro FM, Follows MJ, Valois FW, Waterbury JB (2011) Iron conservation by reduction of metalloenzyme inventories in the marine diazotroph *Crocosphaera watsonii*. *Proceedings of the National Academy of Sciences of the United States of America*, 108(6): 2184-2189.

Saraf A, Dawda HG, Suradkar A, Batule P, Behere I, Kotulkar M, Kumat A, Singh P (2018a) Insights into the phylogeny of false-branching heterocytous cyanobacteria with the description of Scytonema pachmarhiense sp. nov. isolated from Pachmarhi Biosphere Reserve, India. *FEMS Microbiol Let* 365(15): fny160.

Saraf A, Dawda HG, Suradkar A, Behere I, Kotulkar M, Shaikh ZM, Kumat A, Batule P, Mishra D, Singh P (2018b) Description of two new species of *Aliinostoc* and one new species of *Desmonostoc* from India based on the polyphasic approach and reclassification of *Nostoc punensis* to *Desmonostoc punense* comb. nov. *FEMS Microbiol Let* 365(24): fny272.

Saraf A, Suradkar A, Dawda HG, Gaysina LA, Gabidullin Y, Kumat A, Behere I, Kotulkar M, Batule P, Singh P (2019b) Phylogenetic complexities of the members of Rivulariaceae with the re-creation of the family Calotrichaceae and description of *Dulcicalothrix necridiiformans* gen nov., sp nov., and reclassification of *Calothrix desertica*. *FEMS Microbiol Let* 366(17): fnz219.

Saraf AG, Dawda HG, Singh P (2019a) *Desikacharya* gen. nov., a phylogenetically distinct genus of Cyanobacteria along with the description of two new species, *Desikacharya nostocoides* sp. nov. and *Desikacharya soli* sp. nov., and reclassification of *Nostoc thermotolerans* to *Desikacharya thermotolerans* comb. nov. *Int J Syst Evol Microbiol* 1: 69(2): 307-15.

Sato M, Nimura-Matsune K, Watanabe S, Chibazakura T, Yoshikawa H (2007) Expression analysis of multiple dnaK genes in the cyanobacterium *Synechococcus elongatus* PCC 7942. *J Bacteriol* 189(10): 3751-3758.

Scanlan DJ (2012) Marine picocyanobacteria. In *Ecology of Cyanobacteria II*. Springer, Dordrecht 503-533.

Scanlan DJ, Ostrowski M, Mazard S, Dufresne A, Garczarek L, Hess WR, Post AF, Hagemann M, Paulsen I, Partensky F (2009) Ecological Genomics of Marine Picocyanobacteria. *Microbiol Mol Biol Rev* 73(2): 249-299.

Schopf JW (1974) Paleobiology of the Precambrian: the age of blue-green algae. In *Evolutionary biology*. Springer Boston MA 1-43.

Sellner KG (1997) Physiology, ecology, and toxic properties of marine cyanobacteria blooms. *Limnol Oceanogr 42*(5part2): 1089-1104.

Sharif DI, Gallon J, Smith CJ, Dudley E (2008) Quorum sensing in Cyanobacteria: N-octanoyl-homoserine lactone release and response, by the epilithic colonial cyanobacterium *Gloeothece* PCC6909. *ISME J* 2(12): 1171-1182.

Shimizu Y (2003) Microalgal metabolites. *Curr Opin Microbiol* 6(3): 236-243.

Shirkey B, Kovarcik DP, Wright DJ, Wilmoth G, Prickett TF, Helm RF, Gregory EM, Potts M (2000) Active Fe-containing superoxide dismutase and abundant sodF mRNA in *Nostoc commune* (cyanobacteria) after years of desiccation. *J Bacteriol* 182(1): 189-197.

Silambarasan G, Ramanathan T, Kathiresan K (2012) Diversity of marine cyanobacteria from three mangrove environment in Tamil Nadu Coast, south east coast of India. *Curr Res J Biol Sci* 4:235–238.

Silva CSP, Genuario DB, Vaz MGMV, Fiore MF (2014) Phylogeny of culturable cyanobacteria from Brazilian mangroves. *Syst Appl Microbiol* 37(2):100-112.

Simon RD, Weathers P (1976) Determination of the structure of the novel polypeptide containing aspartic acid and arginine which is found in cyanobacteria. *Biochimica et Biophysica Acta (BBA) - Protein Structure* 420(1): 165-176.

Singh AK, Bhattacharyya-Pakrasi M, Elvitigala T, Ghosh B, Aurora R, Pakrasi HB (2009) A systems-level analysis of the effects of light quality on the metabolism of a cyanobacterium. *Plant Physiol* 151(3): 1596-1608.

Singh NK, Dhar DW (2010) Cyanobacterial Reclamation of Salt-Affected Soil 243-275.

Singh P, Dubey N, Bagchi SN (2017b) *Westiellopsis ramosa* sp. nov., intensely branched species of *Westiellopsis* (cyanobacteria) from a freshwater habitat of Jabalpur, Madhya Pradesh, India. *Plant Syst Evol* 303(9): 1239-49.

Singh P, Minj RA, Kunui K, Shaikh ZM, Suradkar A, Shouche YS, Mishra AK, Singh SS (2016a) A new species of *Scytonema* isolated from Bilaspur, Chhattisgarh, India. *J Syst Evol* 54(5): 519-27.

Singh P, Minz RA, Kunui K, Shaikh ZM, Suradkar A, Shouche YS, Mishra AK, Singh SS (2017a) A new species of *Scytonema* isolated from Bilaspur, Chhattisgarh, India using the polyphasic approach. *Plant Syst Evol* 1: 303(2): 249-58.

Singh P, Shaikh ZM, Gaysina LA, Suradkar A, Samanta U (2016b) New species of *Nostoc* (cyanobacteria) isolated from Pune, India, using morphological, ecological and molecular attributes. *Plant Syst Evol* 302(10): 1381-94.

Singh P, Singh SS, Elster J, Mishra AK (2012) Molecular phylogeny, population genetics, and evolution of heterocystous cyanobacteria using nifH gene sequences. *Protoplasma* 250(3): 751-764.

Smith GM (1938) *Cryptogamic Botany.* vol. I. McGraw Hill Book Company Inc, New York.

Smith Singh SM, Elster J (2007) *Cyanobacteria in Antarctic Lake Environments* 303-320.

Smith VR, Russell S (1982) Acetylene reduction by bryophyte-cyanobacteria associations on a Subantarctic island. *Polar Biol* 1(3): 153-157.

Stadnichuk IN, Kusnetsov VV (2021) Endosymbiotic Origin of Chloroplasts in Plant Cells' Evolution. *Russ J Plant Physiol* 68(1): 1-6.

Stal LJ (1995) Physiological ecology of cyanobacteria in microbial mats and other communities. *New Phytologist* 131(1): 1-32.

Stal LJ, Zehr JP (2008) Cyanobacterial nitrogen fixation in the ocean : diversity, regulation and ecology.

Stal Lucas J, Moezelaar R (1997) Fermentation in cyanobacteria. *FEMS Microbiol Rev* 21(2): 179-211.

Stanier R, Deruelles J, Rippka R, Herdman M, Waterbury JB (1979) Generic Assignments, Strain Histories and Properties of Pure Cultures of Cyanobacteria. *Microbiol* 9(3): 801-810.

Stanier RY (1977) The position of cyanobacteria in the world of phototrophs. *Carlsberg Res Commun* 42: 77.

Stanier RY, Sistrom WR, Hansen TA, Whitton BA, Castenholz RW, Pfennig N, Gorlenko VN, Kondratieva EN, Eimhjellen KE, Whittenburry R, Gherna RL, Trüper HG (1978) Proposal to place the Nomenclature of the Cyanobacteria (Blue-Green Algae) under the rules of the International Code of Nomenclature of Bacteria. *Int J Syst Evol Microbiol* 28: 335-336.

Steunou AS, Bhaya D, Bateson MM, Melendrez MC, Ward DM, Brecht E, Peters JW, Kühl M, Grossman AR (2006) In situ analysis of nitrogen fixation and metabolic switching in unicellular thermophilic cyanobacteria inhabiting hot spring microbial mats. *Proc Natl Acad Sci* 103(7): 2398-2403.

Stitzenberger E (1860) Dr. L Rabenhoosts Algew Sachens, *manual of phycology*. Chronica Botanica, Waltham, MA 243-262.

Stockner JG (1988) Phototrophic picoplankton: An overview from marine and freshwater ecosystems. *Limnol Oceanogr* 33(4part2): 765-775.

Svenning MM, Eriksson T, Rasmussen U (2004) Phylogeny of symbiotic cyanobacteria within the genus *Nostoc* based on 16S rDNA sequence analyses. *Arch Microbiol* 183(1): 19-26.

Tamaru Y, Takani Y, Yoshida T, Sakamoto T (2005) Crucial role of extracellular polysaccharides in desiccation and freezing tolerance in the terrestrial cyanobacterium *Nostoc commune*. *Appl Environ Microbiol* 71(11): 7327-7333.

Tang EPY, Vincent WF (1999) Strategies of thermal adaptation by high-latitude cyanobacteria. *The New Phytologist* 142(2): 315-323.

Thajuddin N, Subramanian G (2005) *Cyanobacterial biodiversity and potential applications in biotechnology*. Undefined.

Thanh Doan N, Rickards RW, Rothschild JM, Smith GD (2000) Allelopathic actions of the alkaloid 12-epi-hapalindole E isonitrile and calothrixin A from cyanobacteria of the genera *Fischerella* and *Calothrix*. *J Appl Phycol* 12(3): 409-416.

Thomsen JK, Cox RP (1993) Upper temperature limits for growth and diazotrophy in the thermophilic cyanobacterium HTF *Chlorogloeopsis*. *Arch Microbiol* 159(5): 423-427.

Thu NK, Tanabe Y, Yoshida M, Matsuura H, Watanabe MM (2012) *Aerosakkonema funiforme* gen. et sp. nov. (Oscillatoriales), a new gas-vacuolated oscillatorioid cyanobacterium isolated from a mesotrophic reservoir. *Phycologia* 1: 51(6): 672-83.

Thuret G (1875) Essai de classification des Nostochinées. *Ann Sci Nat Bot* 6(1): 372-382.

Tirkey J, Adhikary SP (2006) Blue green algae in the biological soil crusts of different regions of India. *Feddes Repertorium* 117(3–4): 280-306.

Urbach E, Scanlan DJ, Distel DL, Waterbury JB, Chisholm SW (1998) Rapid Diversification of Marine Picophytoplankton with Dissimilar Light-Harvesting Structures Inferred from Sequences of *Prochlorococcus* and *Synechococcus* (Cyanobacteria). *J Mol Evol* 46(2): 188-201.

Usher KM (2008) The ecology and phylogeny of cyanobacterial symbionts in sponges. *Mar Ecol* 29(2): 178-192.

Vincent WF (2000) Cyanobacterial dominance in the polar regions. In *The ecology of cyanobacteria*. Springer, Dordrecht 321-340.

Vincent WF (2004) *Microbial ecosystems of Antarctica*. Cambridge University Press.

Ward DM, Ferris MJ, Nold SC, Bateson MM (1998) A Natural View of Microbial Biodiversity within Hot Spring Cyanobacterial Mat Communities. *Microbiol Mol Biol Rev* 62(4): 1353-1370.

Whitton BA ed. (2012) *The ecology of cyanobacteria II. Their diversity in time and space.* Springer, Netherlands.

Whitton BA, Potts M (2012) Introduction to the Cyanobacteria. *Ecology of Cyanobacteria II: Their Diversity in Space and Time* 1-13.

Whitton BA, Potts M eds. (2002) *The ecology of cyanobacteria. Their diversity in time and space.* Springer, Netherlands.

Wiangnon K, Raksajit W, Incharoensakdi A (2007) Presence of a Na+-stimulated P-type ATPase in the plasma membrane of the alkaliphilic halotolerant cyanobacterium *Aphanothece halophytica*. *FEMS Microbiol Lett* 270(1): 139-145.

Widdel F, Schnell S, Heising S, Ehrenreich A, Assmus B, Schink B (1993) Ferrous iron oxidation by anoxygenic phototrophic bacteria. *Nature* 362(6423): 834-836.

Wilkinson CR (1987) Significance of Microbial Symbionts Evolution and Ecology in Sponge. *Symbiosis 4*: 135–146.

Woese CR, Fox GE (1977) Phylogenetic structure of the prokaryotic domain: the primary kingdoms. *Proc Natl Acad Sci USA* 74(11): 5088-5090.

Woitchik AF, Ohowa B, Kazungu JM, Rao RG, Goeyens L, Dehairs F (1997) Nitrogen enrichment during decomposition of mangrove leaf litter in an east African coastal lagoon (Kenya): Relative importance of biological nitrogen fixation. *Biogeochem* 39(1): 15-35.

Wyman M, Gregory RPF, Carr NG (1985) Novel Role for Phycoerythrin in a Marine Cyanobacterium, *Synechococcus* Strain DC2. *Sci* 230(4727): 818-820.

Yeung LY, Berelson WM, Young ED, Prokopenko MG, Rollins N, Coles VJ, Montoya JP, Carpenter EJ, Steinberg DK, Foster RA, Capone DG, Yager PL (2012) Impact of diatom-diazotroph associations on carbon export in the Amazon River plume. *Geophys Res Lett* 39(18): 18609.

Zehr JP (2011) Nitrogen fixation by marine cyanobacteria. *Trends Microbiol* 19(4): 162-173.

Zehr JP, Bench SR, Mondragon EA, McCarren J, DeLong EF (2007) Low genomic diversity in tropical oceanic N2-fixing cyanobacteria. *Proc Natl Acad Sci* 104(45): 17807-17812.

Zehr JP, Waterbury JB, Turner PJ, Montoya JP, Omoregie E, Steward GF, Hansen A, Karl DM (2001) Unicellular cyanobacteria fix N2 in the subtropical North Pacific Ocean. *Nature* 412(6847): 635-638.

Zubkov MV, Fuchs BM, Tarran GA, Burkill PH, Amann R (2003) High rate of uptake of organic nitrogen compounds by *Prochlorococcus* cyanobacteria as a key to their dominance in oligotrophic oceanic waters. *Appl Environ Microbiol* 69(2): 1299-1304.

## Chapter 4

# Cyanobacteria: Role and Utilities in Agriculture, Ecosystems and Environmental Sustainability

**Varsha K. Singh, Neha Kumari, Jyoti Jaiswal and Rajeshwar P. Sinha***

Laboratory of Photobiology and Molecular Microbiology, Centre of Advanced Study in Botany, Institute of Science, Banaras Hindu University, Varanasi, India

## Abstract

Cyanobacteria are the largest primitive group of Gram-negative, cosmopolitan, photoautotrophic prokaryotic microflora, which can thrive in marginal areas with a bare minimum requirement of light and nutrients. These are some of the potential organisms that are responsible for a major global evolutionary transformation ultimately resulting in the origin of aerobic metabolism.

In regards to the present challenges concerning agricultural-ecosystem and depleting environmental standards, molecular and biotechnological advancements offer a more definitive approach to address food security for later generations and also to resolve the complex problems of the environment. Certain specific features of cyanobacteria such as their ubiquitous presence, oxygenic photosynthesis, short generation time, rapid growth rate, high biomass yield, production of useful by-products, and biofuels, enhancement of soil fertility and reduction of greenhouse gas emissions, collectively offered these bio-agents as a gem for sustainable development. Their aptitude to fix atmospheric $N_2$ has rapidly increased their demands as

* Corresponding Author's Email: rpsinhabhu@gmail.com.

In: Cyanobacteria
Editor: Rajeshwar P. Sinha
ISBN: 979-8-88697-169-9

bioinoculants for increasing soil fertility and improving environmental standards.

Genetically manipulated cyanobacteria can be designed with desired genes for the production of several targeted chemicals and bio-fuels such as bio-gas, bio-diesel, etc. resulting in a new pavement for the generation of biofuels that are economically befitted. This chapter is an attempt to assess the importance of ecological information for rational management and exploitation of these organisms and also to highlight the informative data about the utilities of cyanobacteria and their magnificent roles implied in resolving the agro-ecological problems for the future welfare of human and the planet.

**Keywords:** agro-ecosystem, bio-diesels, bio-fuels, biotechnological advancements, environment, sustainability

## Introduction

Cyanobacteria are Gram-negative, cosmopolitan, photolysis mediated oxygen-evolving, nitrogen-fixing prokaryotes which can survive and flourish in almost every plausible habitat starting from hot springs to the Arctic and Antarctic regions likewise as within the sort of symbionts in plants, lichens and several other protists. Their ability to conduct oxygenic photosynthesis represents a primary mechanism of utilizing the earth's alternative energy and therefore the cycling of atmospheric carbon and oxygen. These photosynthetic prokaryotes possess some specific features like extremely high tolerance level of low oxygen and free sulfide, the flexibility to utilize $H_2S$ as a photoreductant rather than $H_2O$, and the power to mitigate the highly energetic UV-B and C radiations, which have strongly supported their evolutionary history on Earth (Whitton and Potts 2000).

The present world population records to 7.9 billion which is increasing at a rate of around 1.05% per annum and is estimated to achieve 10 billion till the year 2057 in step with the foremost recent UN estimates elaborated by Worldometer. With the increasing population increases the food requirements, and to satisfy such high demands of the rising population the cereals production is needed to be increased by 50%. Fulfilling such high demand in food production may be a challenging task and it may be achieved in two ways, either by increasing the area under cultivation or by enhancing the productivity of cultivable land already available. Moreover, the largest challenge for today's agriculture is to boost productivity to meet the wants of both present and future generations within the limited resources available without disrupting environmental harmony (Singh and Strong 2016). The possible answer to the current problem can be the adaptability of sustainable agricultural practices which may be proved as an efficient step in achieving the growing needs of food moreover as for the upkeep of environmental harmony and standards (Mason 2003).

The word "sustainable agriculture" simply means the adaptation of low price effective, eco-friendly agricultural practices during which the use of microorganisms is involved. It mainly relies on the presence of conserved resources such as water and soil, amidst reducing the price of agricultural practices and waste generation thereby maintaining environmental peace and harmony.

Reduction and depletion of cultivable areas are the results of anthropogenic socio-economic activities, drawing our main focus towards the preservation of productive land, which is possible by ensuring the coverage of the extended cultivable area. Specializing in such areas is often proved efficient in supplying sufficient food for meeting the rising demands of society (Pathak et al. 2017). Sustainable technologies have implicated the utilization of biofertilizers consisting of microbes like cyanobacteria, *Azotobacter*, *Rhizobia*, *Azospirillum*, arbuscular mycorrhizal fungi, *Azolla*, phosphate-solubilizing microorganisms and plant growth-promoting rhizobacteria (PGPR). This step might be effective in improving and restoring soil fertility, ultimately promoting agricultural sustainability. Moreover, these microorganisms are significant for reducing the energy demand within the sort of manures and compost and have the aptitude to mitigate stressed agricultural systems and wastelands.

The use of cyanobacterial biomass in the style of fields such as for biofuels, bio-diesel, bioremediators, bioenergy resource, food supplements, bio-fertilizers and their utilization for sustainable agriculture has witnessed a surge within the recent researches (Benson et al. 2014) (Figure 1). Cyanobacteria are taken into account as a beneficial bio-agent because it suppresses the expansion of pathogenic microorganisms and is additionally involved in regulating plant productivity.

Nowadays, sustainable agriculture practices are utilizing these microorganisms in achieving food security thereby meeting the rising food demands without depleting the resources available. Being, a beneficial microorganism, cyanobacteria play an important role in the enhancement of agricultural productivity, improving and restoring soil fertility, maintaining environmental standards and mitigation of GHGs emissions (Singh 2011;

Singh et al. 2011). Also, cyanobacteria are implicated as an important bio-agent in ecological restoration in degraded lands (Singh 2014).

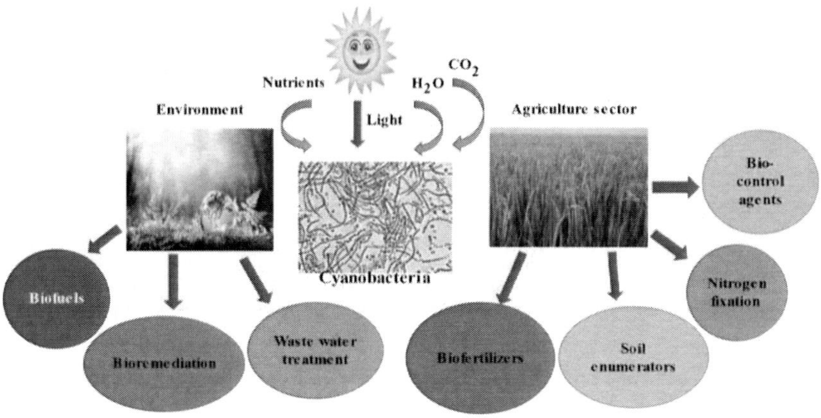


**Figure 1.** Role of cyanobacteria in sustainable agriculture and the environment.

In an environmental aspect, cyanobacteria are useful in wastewater treatment and also, they have the power to detoxify various toxic compounds including pesticides. A study was presented where wastewater treatment was dispensed in column treatment plants using several cyanobacterial species such as *Nostoc* sp., *Phormidum* sp. and *Spirulina* sp. (Sood et al. 2015). Amongst these strains, $PO_4$-P, $NO_3$-N and $NH_4$-N were removed in the proportion ranging from 47 to 99%. This chapter mainly aims to spotlight the role of cyanobacteria as a bioenergy resource, ecological restoration, agricultural and environmental sustainability.

## Utilities of Cyanobacteria in the Agricultural Sector

Around 1950s, the term "algalisation" was described as the application of dried cyanobacterial biomass to the soil. The method of algalisation is obliging in escorting the changes pertaining to improved soil fertility and enhanced crop productivity by almost 15-20% in field experiments. Moreover, the tactic implied is natural and economically significant (Mishra and Pabbi 2004). Cyanobacteria are beneficial microorganisms, having diversified applications in the agricultural and environmental sector (Figure 2) (Whitton 2002; Deviram et al. 2013). These organisms are found inhabiting various soil types mostly rice fields, mines, and drylands, and are chargeable for bringing direct or indirect positive alteration within the physical and chemical properties of soil by nourishing them with carbon, nitrogen, iron, available phosphorus and other minerals and subsequently facilitating increased use of these minerals by plants which ultimately leads to plant growth promotion for increased crop production. Cyanobacterial species have a significant role in humus formation as they are involved in two major biological processes i.e., fixation of carbon through photosynthesis and fixation of atmospheric nitrogen. Many cyanobacteria produce extracellular polymeric substances (EPS) which are mainly of polysaccharidic nature contributing to around twenty-five percent of the whole biomass of cyanobacteria (Nisha et al. 2007). The benefits to the agricultural ecosystem offered through the use of cyanobacteria are as mentioned below:

- Increased solubilization and mobility of nutrients that are of limited supply.
- Complexing of heavy metals and xenobiotics so that their mobility and transport in plants are limited.
- Increased assimilation of amino acids through their mineralization.
- Suppressed growth of pathogenic insects and diseases thereby acting as bio-control agents.
- Stimulation of the plant growth due to their attributes as plant growth promoters.
- Improving the physio-chemical and socio-economic conditions of the soil.

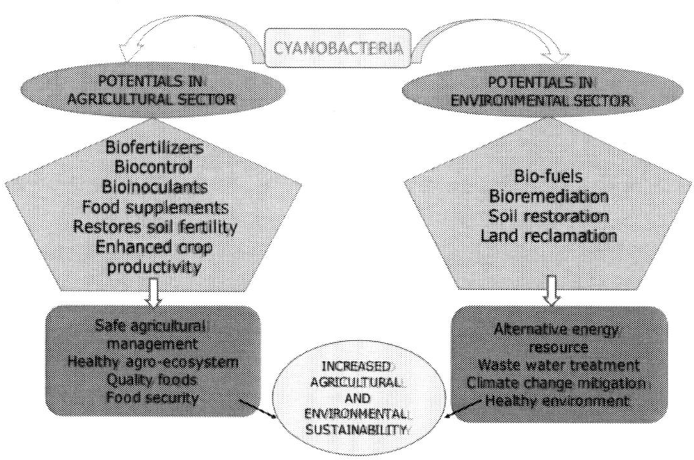


**Figure 2.** Role of cyanobacteria in enhancing agricultural and environmental sustainability.

## Cyanobacteria as $N_2$–Fixers

Nitrogen is the critical limiting macronutrient needed for plant growth and development. Nitrogen fixation is vital to life, as "fixed nitrogen" compounds are required for the biosynthesis of all "nitrogen-containing organic compounds" such as amino acids, proteins and nucleic acids like DNA. Biological nitrogen fixation (BNF) is the most important source of soil nitrogen, which normally is taken care of and maintained by the associated soil microbes (Mahanty et al. 2017). Cyanobacteria that are involved in nitrogen fixation can either be free-living or symbiotically associated and since they are phototrophic, they avoid competition for carbon and energy with crop plants and heterotrophic soil microflora. Cyanobacteria are a significant component of assorted soils. These tiny surface crusts are involved in the biological process of nitrogen fixation i.e., they have the flexibility to convert atmospheric dinitrogen ($N_2$) to nitrites or nitrates (Tharifkhan et al. 2018). Additionally, cyanobacteria supply various plant growth regulators (phytohormones) to crop plants which are mainly gibberellic acid, cytokinins, auxin IAA, etc., also they can supply several beneficiary substances such as cobalamin, free amino acids like serine, glycine, aspartic acid, extra and intracellular polysaccharides like xylose, vitamins like vitamin E, H, and vitamin B-complex which are involved in positive regulation of growth and development of crop plants (El-Enany and Issa 2000).

The permanent establishment of cyanobacteria within the field can be achieved if cyanobacterial biomass is consecutively applied for three or more crop seasons. Cyanobacteria that are involved in nitrogen fixation can either be free-living or symbiotically associated or can make use of specialized thick-walled cells commonly known as 'heterocysts'. These cells are considered as key sites involved in the regulation of nitrogen fixation because of the presence of an enzyme complex, nitrogenase. The enzyme complex nitrogenase (E.C.1.18.6.1) is responsible for the reduction of molecular nitrogen to ammonia. It is a dimeric complex consisting of two protein components, the homodimeric Fe-protein (the dinitrogen reductase), the structural subunits of which are encoded by the nifH gene, and the heterotetrameric MoFe-protein (the dinitrogenase) encoded by structural genes of nifD and nifK (Singh et al. 2011). Both of these enzymes are sensitive to $O_2$, which irreversibly inhibits the enzyme activity. Filamentous heterocystous BGA significantly performs extended respiratory activity by creating a diffusion barrier for gases and thus provides the required anaerobic environment as it lacks oxygenic photosystem II (Scherer et al. 1988).

Photosynthates transported from vegetative cells to the heterocysts provide a reductant supply of nitrogenase via ferredoxin and ATP generation is catalyzed by oxidative phosphorylation activity in the heterocysts (Stewart and Rowell 1986). Microbial degradation after the cell death is responsible for the release of fixed nitrogen within the sort of ammonia, vitamins, polypeptides, free amino acids, and auxin-like substances (Subramanian and Sundaram 1986). Along with heterocystous cyanobacteria, several non-heterocystous unicellular and filamentous genera also show the flexibility of nitrogen fixation. Cyanobacteria can be taken into consideration as a replacement to costly chemical nitrogen fertilizer because of its significant contribution of organic matter to the soil and around 20-30 kg N $ha^{-1}$ as well (Issa et al. 2014). The foremost efficient nitrogen-fixing cyanobacterial species are *Nostoc linkia*, *Anabaena variabilis*, *Aulosira fertilisimia*, *Rivularia* sp., *Calothrix* sp., *Tolypothrix* sp., and *Scytonema* sp. and are mostly present within the paddy field area. *Anabaena* and *Nostoc* can thrive on the marginal areas of soil, stones and rocks, and fixes around 20-25 kg/ha of atmospheric nitrogen. *Anabaena* alone can fix around 60 kg/ha/season of nitrogen and also enhance soil organic matter. The organic process rates for a symbiotic association of cyanobacteria with bryophytes are several-fold on top of the identical free-living cyanobacteria. This increase is anticipated greatly due to the high frequency of heterocyst (Meeks and Elhai 2002). GS-GOGAT (glutamine synthetase-glutamate synthase) pathway is the first step involved in ammonia assimilation in cyanobacteria. In general, two types of GS (GSI and GSIII) and two types of GOGAT (ferredoxin-GOGAT and NADH-GOGAT) are identified and characterized in cyanobacteria. Regarding maintaining amino acids pool homeostasis and C-N proportion, ammonia assimilation is tightly regulated. The extent of GS protein is the key regulatory point and presumably, it is a plant-regulated, transcriptional and post-translational modification of the enzyme.

## Cyanobacteria as Biofertilizers

Cyanobacteria are emerging as one of the foremost microorganisms to be used for sustainable agricultural development (Singh et al. 2016; Singh et al. 2017a; 2017b). Cyanobacteria are the diazotrophs that are useful for the generation of eco-friendly biofertilizers to be made available at a less expensive cost. They mainly function to regulate the nitrogen deficiency in plants, improve the aeration of the soil, increase water holding capacity of soil and add B vitamins to the plants (Song et al. 2005). Cyanobacteria are independent within the sense that it does not require a host for their growth, nourishment, development, and production of beneficial organic products. The Association of *Azolla-Anabaena* is an example of symbiosis for organic process and nutrient enrichment within the field. They lyse lignin and pectin of cyto-membrane and release phenolic compounds which promote sporulation of the organism (Malliga et al. 1996). Applications of these biofertilizers are reported in barley, oats, tomato, radish, cotton, sugarcane, maize, chilli and lettuce (Thajuddin and Subramanian 2005).

Cyanobacteria are found to boost the bioavailability of phosphorus to the plants with the assistance of enzymes like phosphatase which are known to enhance the solubility as well the mobility of insoluble organic phosphates present within the soil (Bose et al. 1971; Wolf et al. 1985; Cameron and Julian 1988). Two hypotheses are being proposed regarding the mechanism of solubilization of phosphate by cyanobacteria and these are mentioned below:

a. Cyanobacteria synthesize $Ca^{2+}$ which acts as a chelator and is answerable for driving the forward reaction without altering the pH of the growth medium (Cameron and Julian 1988; Roychoudhury and Kaushik 1989) as mentioned below:

   $$Ca_{10}(OH)_2(PO_4)_6 \rightarrow 10Ca^{2+} + 2OH^- + 6PO_4^{3-}$$

b. The opposite hypothesis is that the secretions of cyanobacteria such as organic acids are responsible for solubilizing inorganic phosphates in the soil through the following reaction (Bose et al. 1971) as given below:

   $$Ca_3(PO_4)_2 + 2H_2CO_3 \rightarrow 2CaHPO_4 + Ca(HCO_3)_2$$

However, amidst the 2 above-mentioned mechanisms, there could also be a third possibility. The $PO_4^{3-}$ released on the solubilization of an orthophosphate can be taken up by the cyanobacteria themselves and consumed by them for the sake of their own nutrition needs and further when dead locked $PO_4^{3-}$ can be released within the soils, now being available to plants and soils following mineralization (Saha and Mandal 1979; Mandal et al. 1992; 1999).

Cyanobacteria are considered as a natural biofertilizer because they play a significant role in the maintenance and build-up of soil fertility (Sood et al. 2015) (Table 1). The key functions of cyanobacteria as a biofertilizer includes:

a. Increase soil porosity and may be utilized for the production of adhesive substances.
b. Release of phytohormones like auxins and gibberellins, vitamins, growth-promoting substances and amino acids (Roger and Reynaud 1982; Rodriguez et al. 2006).
c. Their characteristic jelly structure enhances the moisture-holding capacity of the soil (Roger and Reynaud 1982).
d. Death and decomposition can result in an increment in soil biomass (Saadatnia and Riahi 2009).
e. Deduction in soil saline conditions (Saadatnia and Riahi 2009).
f. Suppression of weeds and pathogenic microorganism expansion (Saadatnia and Riahi 2009).
g. The release of organic acids can increase the availability of soil phosphate (Wilson 2006).
h. Efficient absorption of heavy metals and removal of contaminants, pollutants and toxins from the microbial surface (bioremediation) (Ibraheem 2007).

## Cyanobacteria as Soil Enumerators

Cyanobacteria are emerging as useful organisms within the field of soil restoration and are considered as soil modifiers. Cyanobacteria when applied to the soil will amend the soil stabilization, mineralization, fertility, nourishment and moisture-holding capacity (Rossi et al. 2017). Few investigations based on cyanobacteria and soil estimated that these microscopic organisms have immense positive effects on crust formation and also help in uplifting soil health and environmental standards. A study on cyanobacterial succession was performed in usar soil comprising of quite 12 forms namely, *Aphanothece* sp., *Aphanothece parietina, Nostoc calcicola, Aulosira fertilissima, Calothrix brevissima, Oscillatoria* sp., *Cylindrospermum licheniforme, Cylindrospermum fertilissima, Phormidium* sp., *Microcoleus* sp., *Nostoc punctiforme,* and among all *N. calcicola* was the dominant form, which was seen regulating soil pH (Pandey et al. 2005). After 6 weeks of inoculation by the cyanobacterial filaments in combination with EPS resulted in improved physio-chemical and morphological properties of the soil thereby signifying the functions of cyanobacteria in the fields of soil aggregation (Malam Issa et al. 2007). The cyanobacterium *Phormidium ambiguum* is reported to enhance the nitrate content of soil by 15%.

**Table 1.** Role of cyanobacteria in soil enumeration and biofertilizers

| Cyanobacteria | Crop | Type | References |
|---|---|---|---|
| *Anabaena* sp. | Rice | Act as biofertilizer | Prasanna et al. 2015a |
| *Spirulina platensis* | Wheat | Weight of grain increased | Michalak et al. 2016 |
| *Anabaena* sp. | Maize | Promote growth | Prasanna et al. 2015b |
| *Spirulina platensis* | Cucumber | Yield and fruit quality enhanced | El Eslamboly et al. 2019 |
| *Anabaena azollae* | Rice, Sugarcane, Maize | Add nitrogen | Reddy and Triveni 2016 |
| *Nostoc carneum* TUBT04 | Rice | Act as biofertilizer | Chittapun et al. 2018 |
| *Phormidium ambiguum* | Increased soil stability | Soil crust | Chamizo et al. 2018 |
| *Anabaena, Phormidium* | Increased grain yield and nutritional value | Wheat | Renuka et al. 2017 |
| *Azolla-Anabaena* | Nitogen fixation | Rice | Chittora et al. 2020 |
| *Phormidium kutzing* | Soil enrichment | Biological soil crust | Park et al. 2017a |
| *Scytonema arcangeli, Nostoc* sp. | Soil restoration | Soil crust | Park et al. 2017b |
| *Nostoc commune, Scytonema arcangeli* | Soil restoration | Soil crust | Román et al. 2018 |
| *Oscillatoria princeps, Oscillatoria nigra* | Promotes growth | Maize, Barley | Mohan and Kumar 2017 |

## Cyanobacteria as Functional Foods

Functional foods are the ones that are the source of ample amount of essential nutritive substances, such as proteins, carbohydrates, lipids, antioxidants, vitamins, calcium, fats, and minerals (El-Sohaimy 2012). These may contain bioactive compounds which are natural chemicals that are obtained from different sources like plants, animals or micro-organisms and are effective in positively regulating the growth and development of living organisms. Cyanobacteria being high in protein content could even be used as potential food supplements and are found to supply nutritional, pharmaceutical, therapeutic and other beneficial values. There are several attributes of cyanobacteria that promote them as a beautiful substitute for food security and sustainability: cosmopolitan distributions, short generation time, rapid growth rate, high nutrient content, the requirement of minimal light and nutrients for growth and development, survival in extreme conditions, higher digestibility rate, product constancy over an in-depth pH and temperature range, etc. (Meena et al. 2018a; 2018b; Panjiar et al. 2018).

Presently, cyanobacteria are accessible from the market in a variety of forms like instance capsules, supplements, tablets and drinks (Singh et al. 2016). These are considered to extend the nutritive values of packed food items like chips, cookies, candies, gums and beverages and are serving as the source of natural food coloring agents. In recent times, the foremost frequent cyanobacterial strain used for nutritive sources is *Spirulina* (*Arthrospria*), because of its extremely high protein content and magnificent nutritive properties (Soletto et al. 2005; Meena et al. 2015).

Cyanobacteria secrete bioactive compounds that are rich in protein, minerals and vitamins which are useful in serving the purpose of maintenance and fixing of bones and teeth. They contain axerophthol, thiamine, riboflavin, niacin, pyridoxine, vitamin E, H, folacin, pantothenic acid, inositol and are abundant in B-complex Vitamin (Shetty et al. 2006). They also produce secondary metabolites which are the source of bioactive molecules including cytotoxic (41%), antitumor (13%), antiviral (4%), and antimicrobial (12%), and other compounds (18%) such as antimalarials, antifungal, antimycotics, multidrug resistance reversers, pesticides, herbicides, insecticides, algaecides, and immunosuppressive agents (Cardellina et al. 2010). All the above-mentioned characteristics make cyanobacteria an opportunist to be considered as an emerging and attractive alternative source for food sustainability.

## Cyanobacteria as Bio-Control Agents

Cyanobacteria are taken under consideration as a beneficial bio-agent because it suppresses the expansion of pathogenic microorganisms and is additionally involved in regulating plant productivity. Cyanobacteria produce a range of bioactive metabolites having antibacterial, antiviral, antifungal and anticancerous properties that are eco-friendly and should be suitably applied to the control of phytopathogens (Dahms et al. 2006). These bioactive compounds are among the assorted categories of polyketides, amides, amines, alkaloids, fatty acids, indoles, and lipopeptides. Additionally, cyanobacteria produce a broad spectrum of anti-algal metabolites which inhibit the expansion of pathogenic organisms by interfering with their metabolic and physio-chemical activities (Dahms et al. 2006). Cell constituents of cyanobacteria are known to cut back the incidence of *Erysiphe polygoni* and *Botrytis cinerea* which are pathogenic and are responsible for causing diseases in plants. These also reduce the expansion of saprophytes like *Chaetomium globosum*, *Aspergillus oryzae,* and *Cunninghamella blakesleeana*, as well phytopathogens such as *Sclerotinia sclerotiorum* and *Rhizoctonia solani* (Kulik 1995).

Previous researchers reported that compounds like Fischerellin obtained from *Fischerella muscicola*, have antifungal activity against some phytopathogenic fungi such as *Uromyces appendiculatus* causing brown rust, *Pyricularia oryzae* causing rice blast, and *Phytophthora infestans* causing late blight of potato (Hagmann and Juttner 1996). *Nostoc muscorum* exhibits antifungal properties against soil fungi and in particular to those producing "damping off" diseases in plants (De Caire et al. 1990). The fungus *Sclerotinia sclerotiorum* causes "white mold," one in every of the foremost polytheist plant pathogens, mostly affecting lettuce (*Lactuca sativa L.*) of Compositae and other species of rosette plants (Tassara et al. 2008). It is found that cell extracts of *N. muscorum* can inhibit the in vitro growth of several fungal phytopathogens such as *Rhizoctonia solani* which causes root and stem rots (Kulik 1995). It is estimated that *Nostoc* sp., produces a potential source of natural pesticides known as cryptophycin, which is pathogenic against certain fungi, insects, and nematodes (Biondi et al. 2004). Nowadays, cyanobacterial strains are effectively used as bio-control agents enhancing agriculture yield thereby ensuring food security. Recent advancements are implying cyanobacterial metabolites as a useful commercial product for promoting agricultural sustainability. However, the idea regarding cyanobacteria as bio-control agents is mostly implemented and conducted under lab conditions and very few within the natural field conditions. Therefore, there's a need for in-depth research to seek out the feasibility of cyanobacteria to be used as the potential bio-control agents against various pathogenic organisms.

## Cyanobacterial Applications in Ecosystem Restoration

Cyanobacteria are photosynthetic prokaryotes that represent amongst the most important and most versatile groups of microorganisms found in a sort of habitats (Gaysina et al. 2019). Their ability to thrive in extreme environments underlines the likelihood of using cyanobacteria as nature-based solutions for ecosystem restoration (Singh et al. 2016). Cyanobacteria are distributed over a wide range of aquatic and terrestrial ecosystems and climatic zones and have affected major geochemical cycles (carbon, nitrogen, and oxygen) on Earth for billions of years (Singh et al. 2018; Gaysina et al. 2019). Irrespective of their color, all algae possess chlorophyll a as the major pigment, consecutively by their photosynthetic action they take up the $CO_2$ and aerate their respective habitats. Cyanobacteria being cosmopolitan are involved in the aeration of habitats like fish ponds.

Several species of cyanobacteria are known to secrete antibiotics resulting in the death of pathogens like *Nostoc* is thought to secrete an antibiotic that is referred to as bacteriocin that can kill related strains of alga. Bacteriocin is a proteinaceous antibiotic active against prokaryotic strains that are closely associated with the organism that produces the antibiotic. Also, *Scytonema* sp. is thought to produce cyanobacterin, chlorine-containing gamma lactone (Pignantello et al. 1983). These antibiotics are significant within the sense that they inhibit the expansion of competing organisms as *Microcystis* has been observed to inhibit the growth of *Staphylococcus* and *Closterium* (Lee 1989).

Cyanobacteria are helpful in land reclamation as these are the primary colonizers of marshlands. These are significant in ecological succession. Soils with increased salinity and alkaline conditions are generally unsuitable for raising crops but cyanobacteria are helpful in the reclamation of such soils. Cyanobacteria are involved in the remediation of salt-affected soil (Figure 3). This is often because BGA is seen playing role in preferential adsorption of sodium. The expansion of these BGA is found efficient in reducing salinity, pH, exchangeable sodium and also electrical conductivity when grown in saline-alkaline habitats. Although, the mechanism by which the BGA scavenge sodium is not yet reported upto the mark still they are chargeable for increased aggregation, hydraulic conductivity, moisture content, soil organic matter and permeability. Some cyanobacterial species such as *Cylindrospermum* sp., *Calothrix* sp., *Pseudoanabaena* sp., *Scytonema* sp. and *Thricomus* sp. are extremely sensitive to disturbance (Zuncan et al. 2006) and subsequently, their absence indicates highly disturbed soils. They are often found growing symbiotically with plants, fungi and microorganisms, benefitting them with shelter from abiotic stresses and serving as an input of nutrients into soils (Foster et al. 2019). Cyanobacteria being the first dominant colonizers of biocrusts act as primary producers which enable further colonization of biocrusts by lichens or mosses. Their ability to manuver throughout the soil profile allows the subsequent colonization by less mobile bacteria (Chamizo et al. 2019a). Biocrusts, a community of interacting organisms, including cyanobacteria, algae, lichens, and bryophytes, promote ecosystem functions by playing an important role in nutrient cycling, soil hydrological, chemical and physical properties and thermal energy balance (Bowker et al. 2018).

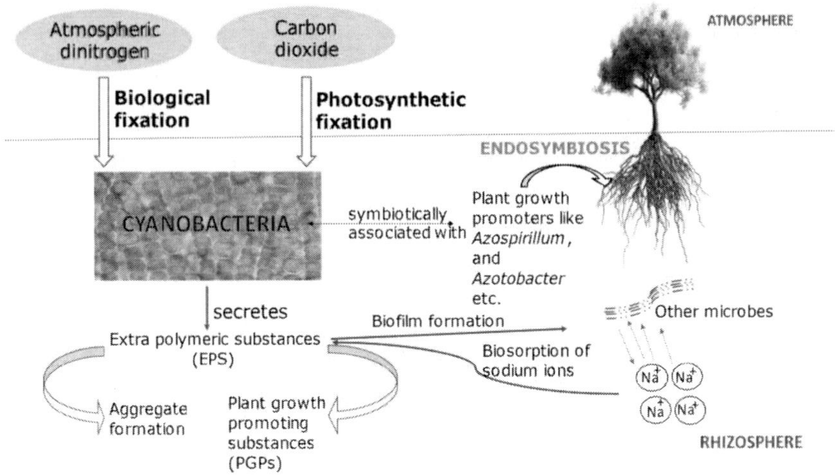


**Figure 3.** General mechanism involved behind remediation of salt-affected soil mediated by cyanobacteria. Cyanobacteria can fix both atmospheric dinitrogen through biological fixation and $CO_2$ through photosynthesis and can either be free-living or symbiotically associated. They release EPS which are chargeable for biofilm formation and sodium biosorption. EPS are further involved in soil aggregation and release of PGPs. Plant growth promoting bacteria (PGPBs) show endosymbiotic, endophytic and rhizospheric associations with plants.

Cyanobacteria possess characteristic features that assist in soil restoration via multiple processes which may include nitrogen fixation, production of extracellular polymeric substance (EPS) and increment in organic carbon content. Moreover, cyanobacteria form a crucial superficial matrix of EPS, water, lipids, proteins, and other compounds that promotes soil colonization by retaining moisture and enhancing soil particle aggregation (Roncero-Ramos et al. 2019; Singh et al. 2019).

Several species of cyanobacteria such as *Scytonema*, *Oscillatoria* have the power to bind sand and soil particles thereby preventing erosion. Improved carbon and nitrogen fixation, runoff inhibition and aggregate formation through secreted EPS are the major processes involved in erosion-prone soil restoration. This ability extends to desert ecosystems where cyanobacteria from biocrusts effectively improve nutrient accumulation and promote edaphic conditions (He et al. 2019). Their growth significantly increases the water-stable aggregates, soil aggregation and arrangement, influences infiltration rate, aeration and soil temperature parallelly improving the physical environment of the crop (Kaushik 1997). They are seen effectively maintaining the water content of the soil via reducing water loss through direct evaporation from open lands e.g., *Oscillatoria.*

Cyanobacteria short wet cycles make them excellent candidates for desert soils restoration (Perera et al. 2018). Recently, the aptitude of cyanobacteria inoculation for the recovery of soils in burned areas has shown promising results (Chamizo et al. 2019b). Restoring post-mining sites requires a holistic approach through which the disturbance of topsoil incorporating soil microbial communities may end up in an exceedingly modification of ecosystem functions (Williams et al. 2019). Muñoz-Rojas et al. (2018) highlighted the potential of cyanobacteria to extend soil functions in mine-site restoration by promoting soil carbon sequestration. Similarly, Chua et al. (2019) demonstrated that bio-priming native plant seeds with indigenous cyanobacteria promote seedling growth in soils commonly employed in mine restoration. Moreover, these organisms can retain pollutants with high efficiency because of diverse proteins and polysaccharide receptors on their surface, acting as prominent agents within the remediation of heavy metal and hydrocarbon polluted soils.

## Role of Cyanobacteria in Bioremediation

Cyanobacteria can remove environmental contaminants like phenol, crude oil, pesticides, textile dyes, xenobiotics and heavy metals (Table 2). They play important role in the detoxification of various industrial effluents like pulp and paper, brewery, textile, dye, tanneries, distilleries, oil refinery and pharmaceuticals industries.

Cyanobacteria such as *Aphanocapsa*, *Westiellopsis*, *Phormidium* and *Oscillatoria* are able to remove phosphate and nitrogen ions from wastewater (Blier et al. 1995). Removal of nitrogenous and phosphate ions by cyanobacteria is given in Table 3. Cyanobacteria can produce bioflocculants that protect from the toxic effects of heavy metals (Bender et al. 1994). The uptake of metal ions (Cd, Ni, Cr, Zn, Pb and Cu) in *Spirulina platensis* was related to an ion-exchange mechanism. *Spirulina platensis* was able to remove lead and mercury (Slotton et al. 1989). *Synechococcus* sp. can uptake cadmium, chromium, nickel, lead, and copper.

**Table 2.** Applications of cyanobacteria in bioremediation of various contaminants

| Cyanobacteria | Removal of contaminants | References |
|---|---|---|
| *Microcystis aeruginosa, Anacystics nidulans, Synechococcus elongatus* | organo-phosphorus and organo-chlorine insecticides | Vijayakumar 2012 |
| *Oscillatoria-Gammaproteo* bacteria consortium | dibenzothiophene, n-octadecane and Phenanthrene | Abed et al. 2005 |
| *Microcystis* sp., *Spirulina* sp., *Lyngya* sp., *Nostoc* sp., *Anabaena* sp. | glyphosate herbicide from agricultural sites | Forlani et al. 2008 |
| *Oscillatoria boryana* BDU 92181 | melanoidin from distillery effluents | Kalavathi et al. 2001 |
| *Microcoleus chthonoplastes* | decomposition of aliphatic heterocyclic organo-sulfur compounds and hydrocarbons | Sánchez et al. 2005 |
| *Halobacterium* US 101, *Aphanocapsa* sp. BDU 16, *Oscillatoria* sp. BDU 30501 | reduction of calcium and chloride content in water | Uma and Subramanian 1990 |
| consortium of *Chlorella kessleri* and *Anabaena oryzae* | crude oil | Hamouda et al. 2016 |

**Table 3.** Role of cyanobacteria in removal of phosphorous and nitrogen

| Cyanobacteria | References |
|---|---|
| *Aphanothece microscopica* | Queiroz et al. 2011 |
| *Spirulina platensis* | Lodi et al. 2003 |
| *Phormidium tenue* | Dumas et al. 1998 |
| *Synechococcus elongatus* | Aguilar-May et al. 2009 |
| *Phormidium bohneri* | Chevalier et al. 2000 |
| *Synechococcus* sp. PCC7942 | Hu et al. 2000 |
| *Planktothrix isothrix* | Silva-Benavides et al. 2012 |
| *Synechocystis salina* | Gonçalves et al. 2016 |

## Role of Cyanobacteria in Industrial Wastewater Treatment

Cyanobacteria help in bioremediation and detoxification of industrial effluents of chemical, bio-fertilizer, feed, fuel and food industries and wastewater treatment. Cyanobacteria *Cyanothece* sp., *Synechococcus* sp., *Anabaena* sp., *Nostoc* sp., *Nodularia* sp. and *Oscillatoria* sp., have high biodegradation, biosorption and are able to decolorize dye of industrial effluents (Dubey et al. 2011; Mona et al. 2011).

## Role of Cyanobacteria in the Removal of Dyes and Heavy Metals from Wastewaters

Cyanobacteria are able to remove heavy metals zinc (Zn), lead (Pb), copper (Cu), cadmium (Cd) and chromium (Cr) (Table 4). *Oscillatoria angustissima* has capacity to absorb copper (Ahuja et al. 1999). *Gloeothece magna* has manganese and cadmium adsorption capacity (Mohamed 2001). *Nostoc* sp., *Anabaena flos-aquae*, and *Cylindrica* accumulated nickel (Corder and Reeves 1994). The uptake and toxicity of copper, zinc and iron have been reported for *Oscillatoria perornata* by Mittal et al. (1992). Rangsayatorn et al. (2004) reported the biosorption of cadmium by silica gel and alginate immobilized cells of *Spirulina platensis* TISTR 8217 containing max biosorption capacities of 36.6 and 70.9 mg Cd/g biomass. Biosorption of remazol black B and chromium (VI) by *Phormidium* sp. has been reported by Aksu et al. (2009).

**Table 4.** Removal of heavy metals by cyanobacteria from domestic and industrial wastewater

| Heavy metals | Cyanobacteria | References |
|---|---|---|
| Zinc (Zn) | *Nostoc rivularis, Nostoc linckia, Synechocystis* sp. | El-Enany and Issa 2000 |
| Cobalt (Co) | *Anabaena subcylindrica, Nostoc muscorum, Oscillatoria quadripunctulata* | Ajayan et al. 2011, El-Sheekh and Ghareib 2012 |
| Lead (Pb) | *Gloeocapsa* sp., *Anabaena subcylindrica, Nostoc muscorum, Oscillatoria quadripunctulata* | Ajayan et al. 2011, El-Sheekh and Ghareib 2012, Raungsomboona et al. 2006 |
| Copper (Cu) | *Anabaena subcylindrica, Nostoc muscorum, Spirulina maxima* | Chan et al. 2013, Li et al. 2018, El-Sheekh and Ghareib 2012. |
| Cadmium (Cd) | *Tolypothrix tenuis, Nostoc linckia, Nostoc rivularis* | Travieso et al. 1999, El-Enany and Issa 2000. |
| Chromium (Cr) | *Chroococus sp., Nostoc calcicola, Spirulina* sp. | Rezaei 2016, Anjana et al. 2007 |
| Mangnese (Mn) | *Anabaena subcylindrica, Nostoc muscorum* | El-Sheekh and Ghareib 2012 |
| Mercury (Hg) | *Aphanothece flocculosa, Spirulina platensis* | Cain et al. 2008 |

*Synechococcus* sp. PCC7942, *Phormidium autumnale* UTEX1580 and *Anabaena flosaquae* UTCC64 were capable to degrade the textile dyes like Remazol Brilliant Blue R (RBBR), black sulfur and indigo (Dellamatrice et al. 2017). *Phormidium ceylanicum* and *Gloeocapsa pleurocapsoides* were able to decolorize FF Sky Blue and Acid Red 97 dyes. 55% decolorization of

Amido Black 10B was shown by *Chroococcus minutus* (Parikh and Madamwar 2005). *Spirulina* was able to completely remove nitrogen from wastewaters (Chevalier et al. 2000). 60-90% of wastewater was decolorized by *Phormidium tenue* (Nagasathya and Thajuddin 2008).

## Role of Cyanobacteria in Degradation of Pesticides

*Nostoc ellipsosporum* strain B1453–7, *Anabaena azotica* and *Anabaena* sp. PCC 7120 were able to degrade Lindane pesticide (Kuritz and Wolk 1995). *Nostoc* sp. MMI, *Nostoc* sp. MM2, *Nostoc* sp. MM3, *Nostoc muscorum* and *Anabaena* sp. can degrade Fenamiphos (an organophosphorus pesticide) by oxidation and hydrolysis (Ca'ceres et al. 2008). Three steps are involved in the degradation of pesticides. The first step is oxidation or reduction, hydrolysis in which the pesticides are converted into less toxic and water-soluble substances. In the second step, pesticides bind to the amino acids, glutathione or sugar and in this solubility is increased and toxicity is decreased. In step third, metabolites of step second get converted to form nontoxic substances (Hatzios 1991). *Microcystis* sp., *Spirulina* sp., *Nostoc* sp., *Lyngya* sp. and *Anabaena* sp. play role in the removal of the glyphosate herbicide effect from the agricultural fields (Zahra et al. 2020).

## Role of Cyanobacteria in Degradation of Crude Oil

Cyanobacteria play important role in the bioremediation of crude oil because they are able to grow under several environmental conditions and form blooms with other organisms. Crude oil was degraded by *Plectonema terenbans*, *Aphanocapsa* sp. and *Oscillatoria salina* (Raghukumar et al. 2001). Narro et al. 1992 reported that naphthalene, which is the main component of the water-soluble part of crude oil, was degraded by *Oscillatoria* sp. strain JCM. Crude oil was also degraded by *Microcoleus chthonoplastes* and *Phormidium corium* by consuming the individual n-alkanes (Al Hasan et al. 1994).

## Role of Cyanobacteria in Degradation of Plastics

Microbes grow on the plastic as a thin layer of biofilm which is called 'plastisphere' (Kirstein et al. 2019). The extracellular enzymes released by microbes in the plastisphere increase the physical and chemical degradation of plastics. *Rivularia* and *Phormidium* found in the marine environment had degraded plastics (Dussud et al. 2018).

## Mass Cultivation of Cyanobacteria

Large-scale cultivation of cyanobacteria and other microalgae is necessitated because of their useful properties in different sectors. The primary factor which determines the success rate of large-scale cultivation of commercially important products is economic sustainability. Crops are complemented by cyanobacterial biomass for the production of numerous essential products such as food, feeds, fuels, cosmetics, antibiotics, biofertilizers, drugs and chemicals and therefore, the large-scale production of cyanobacterial biomass is a pre-requisite for satisfying the above-mentioned purposes. Cultivation of cyanobacterial biomass implies better aerial productivity than plants and also they do not compete for resources with agricultural crops (Cañedo and Lizárraga 2016; Ooms et al. 2016).

Genetic manipulations of cyanobacteria are studied mainly in vitro for their usefulness in various fields, but for commercial application, the process must be standardized and applied outside the field. The concept of large-scale mass cultivation of cyanobacteria as well as microalgae was proposed around the 1950s (Cook 1951). The success rate of growth of cyanobacteria was determined by five crucial parameters and these are light, pH, temperature, water, $CO_2$ and nutrient supplements (C, N, P, S, K, Fe, etc.) (Flynn et al. 2010; Meena et al. 2017a; 2017b). Regulation of these factors involves numerous expertise and resources. Cyanobacterial sp. like *Arthrospira, Dunaliella*, *Chlorella,* and *Haematococcus,* are cultivated on a mass scale and are economically and commercially suited viable crops. Commercial application of photosynthetic microorganisms can be achieved mainly in three different ways:

(1) Cultivation requiring sunlight in an open system
(2) Cultivation requiring sunlight in a closed system
(3) Cultivation requiring artificial light

## Cultivation Requiring Sunlight in Open Systems

In general, the systems used for mass cultivation of cyanobacteria and microalgae are raceway or circular type shallow open ponds (Cañedo and Lizárraga 2016). Natural light is the source of energy usually utilized in large raceways or open ponds. The foremost advantageous attribute of this technique is that the source of light used is freed from cost. However, a list of major disadvantages compensates for this advantage and limit general biomass production. The net productivity is compromised on the price of unpreventable contaminations like algal contamination, grazers and other microorganisms in open systems. Avoiding such contamination problems for organisms requires unique growth conditions which can successfully prevent the expansion of other organisms. Moreover, this strategy limits the application of open systems as it can be used only for the cultivation of very few organisms (Costa et al. 2006; De Morais and Costa 2007). Example: *Spirulina* has been extensively cultivated using the open system under unique growth requirements in various countries like Mexico and Thailand (Lee 1997).

## Cultivation Requires Sunlight in a Closed System

Natural light is the source of light energy used in this system which is free of cost (Grima et al. 1999). The making of vessels requires transparent material such as plastic or glass which is placed outside in the field in natural light for illumination (Khatoon and Pal 2015).

Advantages:

- Prevention of contamination of algae, grazers, other microorganisms and competitors.
- Benefit of higher surface to volume ratio and cell densities obtained over open systems.

Disadvantages:

- Application of transparent materials significantly increases the cost of these systems (Carvalho et al. 2006).
- Critical observation of some factors like removal of photosynthetic oxygen and constancy of optimum temperature is required.

Several techniques are implied that have been designed and established to regulate these parameters; however, the cost advantage of using natural sunlight is counterbalanced by the associated costs (Khatoon and Pal 2015).

### Cultivation Using Artificial Light

Artificial light is the source of light energy used in this system (Ratchford and Fallowfield 1992; Lee and Palsson 1994). Cultivation of various organisms is obtained through photobioreactor. A photobioreactor is a closed vessel that utilizes a light source to cultivate phototrophic microorganisms. Culture parameters and real-time control are optimized using this software (Ratchford and Fallowfield 1992). Production of high-value products such as stable-isotope-labeled biochemicals requires efficient tools like photobioreactors (Apt and Behrens 1999). These systems are liable also for the idea of the cultivation of genetically engineered organisms. Making of vessel utilizes plastic or glass material which is the major drawback to this system as it increases the production cost and also, power consumption is costly. Production of high quality and quantity biomass using these systems can offset the cost. Four other techniques such as shallow metal troughs method cemented tank method, polythene lined pit method and field methods are implied for the mass cultivation of cyanobacteria. Small and marginal farmers mainly rely on the polythene lined method for the manufacturing of biofertilizers. In this process, small pits are prepared in the field and thick polythene sheets are used for lining, for example, for inoculum preparation *Anabaena, Cylindrospermum, Aulosira, Gloeotrichia, Nostoc, Plectonema,* etc. are being used (Singh et al. 2016).

## Role of Cyanobacteria in Production of Bioenergy

Cyanobacteria are the unique group of photosynthetic prokaryotes having a short generation time and faster growth rate due to their simple structure and bare minimum requirement of light, water and nutrients. These organisms are now emerging as an efficient bio-agent since various biochemical processes can be applied to them thereby enhancing their aptitude to produce a variety of bioenergy fuels such as bio-diesel, cellulosic ethanol, bio- or syngas, bio-hydrogen, and directly converted fuels etc. (Kumar and Singh 2016).

Cyanobacteria take up $CO_2$ through photosynthesis more efficiently as compared to land plants and convert them to carbon-rich lipids that can be further implied in the generation of biofuels and biodiesel. Additionally, biodiesel produced by cyanobacteria promotes environmental sustainability as it lowers sulfur emissions and increases oxygen content, also they have the aptitude to generate bio-hydrogen ($H_2$) which could be the better option and an apt substitute for fossil fuels. They can be useful in the production of different feedstock for energy generation like molecular $H_2$ which can be obtained through photosynthesis, carbohydrates for ethanol production, cellular lipids can be converted to FAMEs, which is the building blocks of biodiesel, etc. (Parmar et al. 2011; Rosgaard et al. 2012).

The raw materials consist of feedstocks that are rich in simple sugars and starch such as corn, rapeseed, potato, soybean, sugarcane, sunflower, wheat, switchgrass, rice and plant oil, peanuts, and sesame seeds are being utilized for first and second generations of biofuels. These raw materials are implied for the production of different energy sources such as ethanol, and vegetable oils (Quintana et al. 2011). The third generation of biofuel production uses microalgae for the production of different energy sources. This generation of biofuel production which uses microalgae as the feedstock has emerged as an alternative to avoid the competition with food and animal feed within limited available resources which is an advantage over the first and second generation of biofuels production, which compete with conventional food sources and other mineral resources. Furthermore, their aptitude to effectively capture solar energy and rapid conversion of biomass into fuels makes cyanobacteria a significant feedstock for the production of third-generation biofuels (Quintana et al. 2011; Rajneesh et al. 2017). The use of cyanobacteria for biofuels production is advantageous over land plants because of its features like rapid growth rate and ability to thrive in marginal areas with bare minimum requirements of light and nutrients cultivation (Singh 2014; Sarma et al. 2016). Particularly, cyanobacteria exhibit higher photosynthetic efficiency (~10%), in comparison to land plants (~3–4% maximum efficiency) (Melis 2009).

Molecular $H_2$ is taken into consideration as the potential future energy source which could be a significant solution to the problem of depleting fossil fuels resources. $H_2$ gas generation is possible through several conventional methods but also cyanobacteria are significant in producing molecular hydrogen photobiologically. Under stressed conditions, the cyanobacterial species instead of reducing $CO_2$, can consume biochemical energy and use it to produce molecular hydrogen. The cyanobacterial lignocellulosic biomass can be processed for molecular $H_2$ production. The first step involved in this process is syngas production which is followed by a Fischer–Tropsch process and gasification/liquefaction (Lawson et al. 2011). Using molecular $H_2$ as the clean fuel is advantageous as it is eco-friendly, renewable, efficient, and does not utilize $CO_2$ in its production. Moreover, at most only small amounts of $NO_x$ are generated during its production. The calculated heating value of $H_2$ based on weight is 141.65 MJ $Kg^{-1}$, which is the highest amongst known fuels (Ali and Basit 1993). Also, this has greater energy conversion efficiency than petroleum and can be stored as gas-metal hydride or as a liquid. Since water is the only by-product of $H_2$, it will not cause environmental pollution if used as a fuel. Certain cyanobacterial species such as *Anabaena cylindrica, Anabaena variabilis, Anabaena flos-aquae, Calothrix scopulorum, Oscillatoria brevis*, *Nostoc muscorum*, *Nostoc linckia*, *Synechococcus* sp., *Microcystis* sp., *Microcoleus* sp., are known for their aptitude to produce molecular $H_2$ under various culture conditions (Parmar et al. 2011; Nozzi et al. 2013) (Figure 4). Cyanobacteria generally use two sets of enzymes to generate mainly molecular $H_2$ and these are mentioned below (Pinzon-Gamez et al. 2005):

a. The first one is the nitrogenase enzyme complex which is present in the heterocysts of certain filamentous BGA. Molecular $H_2$ is produced as a by-product of nitrogen fixation into reduced form like ammonia.

   $N_2 + 10H^+ + 8\ e- + 16ATP + 16H_2O \rightarrow 2NH_4^+ + H_2 + 16ADP + 16\ Pi$

b. The second one is mediated by reversible action of hydrogen-metabolizing enzyme hydrogenases

   $H_2 \rightarrow 2H^+ + 2e-$

Though cyanobacterial $H_2$ production is a sustainable technology, this process is limited because of lesser $H_2$ production thereby making it unsuitable for being economically feasible (Tiwari and Pandey 2012). This process is associated with certain shortcomings which form an obstacle in the scale-up of cyanobacterial $H_2$ production. Concurrent production of $O_2$ poses a serious limitation to the cyanobacterial molecular $H_2$ production via hydrogen metabolizing enzyme hydrogenase because of the enzyme being strongly $O_2$-sensitive. The process implied in cyanobacterial $H_2$ production has its pros and cons both in terms of technology and efficacy. Investigations from previous research reported that this field is yet preliminary and theoretical and is without any potentially applied practical information. Evaluation and modification of these processes for productivity and cost of $H_2$ commercialization are still pending and require further study.

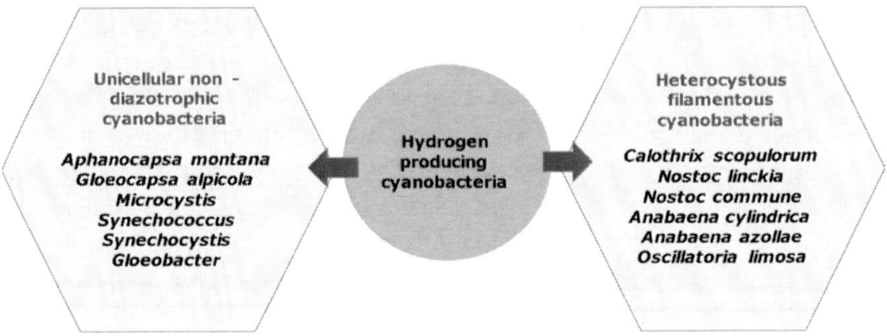


**Figure 4.** Hydrogen producing cyanobacteria.

Besides the production of bio-fuel and molecular $H_2$, cyanobacterial biomass can also be implied for biogas production through the process of anaerobic digestion or fermentation. Biogas is formed during anaerobic digestion and comprises of two main components: the larger constituent being the methane and the smaller one being carbon dioxide comprising around 55-70% and 30-40% by volume respectively. Cyanobacterial biomass comprises mainly proteins and also they lack a hard polysaccharidic wall. These specific features increase the digestibility of cyanobacterial species. For the production of biogas (mixture of $CH_4$ and $CO_2$), the organic biopolymers like carbohydrates, lipids, and proteins (constituents of cyanobacterial biomass) are first hydrolyzed and broken down into monomers followed by anaerobic digestion. $CO_2$ the second main component of biogas production can be removed foe producing bio-methane. Removing $CO_2$ during bio-methane production significantly increases the calorific value of biogas (Hankamer et al. 2007). $CH_4$ or bio-methane being more eco-friendly than gasoline/petrol and diesel are used as the CNGs in vehicles.

Biogas production and purification are completed in a two-step biological system, comprising of fed-batch pulse-feeding anaerobic digestion of mixed sludge, which is followed by the involvement of the cyanobacteria for $CH_4$ enrichment of biogas (Converti et al. 2009). Certain cyanobacterial genera mainly *Arthospira* and *Anabaena* are extensively studied for their significance as a feedstock for the anaerobic digestion process. The proportionate contribution of bio-methane and carbon dioxide ranges between 70.5–76.0% and 13.2–19.5%, respectively. The linear relationship between the rates of *A.*

*platensis* growth and $CO_2$ removal from biogas revealed the estimation of carbon utilization efficacy of cyanobacterial biomass which is almost to the extent of 95% (Converti et al. 2009).

Biogas has several applications such as heat production, power generation, electricity generation, which can be used as cooking gas in place of natural gas, production of chemicals and used as fuels for vehicles. The C/N proportion is an important factor that influences bio-methane production via anaerobic digesters (Zhong et al. 2012). The major factor involved in high ammonia released during anaerobic digestion is the high protein content of the cyanobacterial biomass in comparison to terrestrial plants. This protein content is chargeable for suppressing the growth of anaerobic micro-flora thereby inhibiting bio-methane production (Sialve et al. 2009). The protein-rich cyanobacterium *Spirulina maxima*, when anaerobically digested, can release around 7000 $mgL^{-1}$ of ammonia which is an extremely high concentration. The methanogenic microbes are therefore one amongst the most sensitive micro-flora to high ammonia content. The factor that genetic manipulation of cyanobacteria is much easier than any other algae makes it a feasible candidate in comparison to eukaryotic algae for the generation of targeted bio-fuels. Till date genomes of several cyanobacterial genera have been sequenced and are found to be relatively small (Rajneesh et al. 2017). The above-mentioned statements clarify the fact that cyanobacteria are an exceptional opportunist to be genetically manipulated and designed for improved biomass production which is a tough task to do with eukaryotic algae and other organisms (Rittmann 2008). Genetic manipulations have been performed mainly on *Synechocystis* sp. PCC 6803 and *S. elongatus* PCC 7942, for enhanced production of biofuel using molecular techniques which are well-established (Kaneko et al. 1996). Cyanobacteria are genetically manipulated and designed for the production of various fuels such as isobutanol, 2,3-butanediol, ethanol, fatty acids, isobutyraldehyde, 2-methyl-1-butanol, isoprene, etc. Significantly, genetically modified cyanobacteria are crucial for creating crude oil independent lifestyle and reducing $CO_2$ emissions, owing to the direct fixation of $CO_2$ to biofuel and other beneficial chemicals (Ducat et al. 2011; Oliver et al. 2016). The application of cyanobacteria in the field of biofuel production is limited in the sense that it cannot surpass the production rate of valuable chemicals in the sugar-utilizing systems like *S. cerevisiae* and *E. coli* which is much higher than the photoautotrophic cyanobacteria (Savakis and Hellingwerf 2014). In general, photoautotrophic cyanobacteria produces biochemicals of around ~100 mg/liter of cell culture, such a low concentration is insignificant for any commercial application (Gao et al. 2016). The production rate of targeted chemicals from cyanobacteria is evaluated on theoretical background under heterotrophic and autotrophic growth conditions to find out the limiting factors in the cyanobacterial metabolic network (Gudmundsson and Nogales 2015). Recent advancements imply on generation, designing, development, establishment and expression of the biosynthetic routes of technologies for genetic and metabolic manipulations in cyanobacteria which will ultimately result in economic profitability by extending the production of existing and targeted chemicals and biofuels (Wang et al. 2012; Oliver and Atsumi, 2014; Gudmundsson and Nogales 2015; Markley et al. 2015).

## Conclusion

Cyanobacteria play important role in the conservation of natural resources and maintenance of the diversity and complexity of ecosystems in a sustainable manner. It supports and assists ample food production for the increasing world population, assure safer living for both humans and other livestock. It enhances the organic nitrogen and carbon of soil and increases the bioavailability of phosphorus to the plants. They also help in the reclamation of saline alkali soils. They are bio-agents for safe and eco-friendly agriculture and environmental sustainability, along with various other uses. Cyanobacteria are capable of degradation of several environmental contaminants such as pesticides, dyes, heavy metals, plastics and crude oil. They can be used for zero waste resources for various applications such as biofuels, biofertilizers, pigments, secondary metabolites and bioremediation. Therefore, cyanobacteria can be explored for beneficial applications in agriculture and the environment for the benefit of future generations.

## Acknowledgements

Varsha K. Singh (09/0013(12862)/2021-EMR-I) and Neha Kumari (09/013(0819)/2018-EMR-I) are thankful to the Council of Scientific and Industrial Research, New Delhi, India, for the financial support in the form of fellowship. Jyoti Jaiswal (926/CSIR-UGC-JRF DEC, 2018) is grateful to the University Grants Commission (UGC), New Delhi, India, for research grants.

## References

Abed RMM, Köster J (2005) The direct role of aerobic heterotrophic bacteria associated with cyanobacteria in the degradation of oil compounds. *Int Biodeterior Biodegrad* 55: 29-37.

Aguilar-May M, del Pilar Sánchez-Saavedra M (2009) Growth and removal of nitrogen and phosphorus by free-living and chitosan-immobilized cells of the marine cyanobacterium *Synechococcus elongates*. *J Appl Phycol* 21: 353-360.

Ahuja P, Gupta R, Saxena RK (1999) Sorption and desorption of cobalt by *Oscillatoria anguistissim*. *Curr Microbiol* 39: 49-52.

Ajayan KV, Selvaraju M, Thirugnanamoorthy K (2011) Growth and heavy metals accumulation potential of microalgae grown in sewage wastewater and petrochemical effluent. *Pak J Biol Sci* 14: 805-811.

Aksu Z, Ertugrul S, Donmez G (2009) Single and binary chromium (VI) and Remazol black B biosorption properties of *Phormidium* sp. *J Hazard Mater* 168: 310-318.

Al Hasan R, Sorkhoh N, Al Bader D, Radwan S (1994) Utilization of hydrocarbons by cyanobacteria from microbial mats on oily coasts of the Gulf. *Appl Microbiol Biotechnol* 41: 615-619.

Ali I, Basit MA (1993) Significance of hydrogen content in fuel combustion. *Inter J Hydrogen Energ* 18: 1009-1011.

Anjana K, Kaushik A, Kiran B, Nisha R (2007) Biosorption of Cr (VI) by immobilized biomass of two indigenous strains of cyanobacteria isolated from metal contaminated soil. *J Hazard Mater* 148: 383-386.

Apt KE, Behrens PW (1999) Commercial developments in microalgal biotechnology. *J Phycol* 3: 215-226.

Bender J, Rodriguez-Eaton S, Ekanemesang UM, Phillips P (1994) Characterization of metalbinding bioflocculants produced by the cyanobacterial component of mixed microbial mats. *Appl Environ Microbiol* 60: 2311-2315.

Benson D, Kerry K, Malin G (2014) Algal biofuels: impact significance and implications for EU multi-level governance. *J Clean Prod* 72: 4-13.

Biondi N, Piccardi R, Margheri MC, Rodolfi L, Smith GD, Tredici MR (2004) Evaluation of *Nostoc* strain ATCC 53789 as a potential source of natural pesticides. *Appl Environ Microbiol* 70: 3313-3320.

Blier R, Laliberte G, dela Noue J (1995) Tertiary treatment of cheese factory anaerobic effluent with *Phormidium bohneri* and *Micractinum pusillum*. *Bioresour Technol* 52: 152-155.

Bose P, Nagpal U S, Venkataraman GS, Goyal SK (1971) Solubilization of tricalcium phosphate by blue-green algae. *Curr Sci* 40: 165-166.

Bowker MA, Reed SC, Maestre FT, Eldridge DJ (2018) Biocrusts: The living skin of the earth. *Plant Soil* 429: 1-7.

Ca'ceres T, Megharaj M, Naidu R (2008) Biodegradation of the pesticide fenamiphos by ten different species of green algae and cyanobacteria. *Curr Microbiol* 57: 643-646.

Cain A, Vannela R, Woo LK (2008) Cyanobacteria as a biosorbent for mercuric ion. *Bioresour Technol* 99: 6578-6586.

Cameron HJ, Julian GR (1988) Utilisation of hydroxyapatite by cyanobacteria as their sole source of phosphate and calcium. *Plant Soil* 109: 123-124.

Cañedo JCG, Lizárraga GLL (2016) Considerations for photobioreactor design and operation for mass cultivation of microalgae. In: Thajuddin N, Dhanasekaran D, editors. *Algae-Organisms for Imminent Biotechnology:* InTech.

Cardellina JH, Moore BS (2010) Editorial: Richard E. Moore (1933–2007). *J Nat Prod* 73: 301-302.

Carvalho AP, Meireles LA, Malcata FX (2006) Microalgal reactors: a review of enclosed system designs and performances. *Biotechnol Prog* 22: 1490-1506.

Chamizo S, Adessi A, Certini G, De Philippis R (2019a) Cyanobacteria inoculation as a potential tool for stabilization of burned soils. *Restor Ecol.*

Chamizo S, Adessi A, Mugnai G, Simiani A, De Philippis R (2019b) Soil Type and Cyanobacteria Species Influence the Macromolecular and Chemical Characteristics of the Polysaccharidic Matrix in Induced Biocrusts. *Microb Ecol* 78: 482-493.

Chamizo S, Mugnai G, Rossi F, Certini G, De Philippis R (2018) Cyanobacteria inoculation improves soil stability and fertility on different textured soils: gaining insights for applicability in soil restoration. *Front Env Sci* 6: 49.

Chan A, Salsali H, McBean E (2013) Heavy Metal Removal (Copper and Zinc) in Secondary Effluent from Wastewater Treatment Plants by Microalgae. *ACS Sustain Chem Eng* 2: 130 137.

Chevalier P, Proulx D, Lessard P, Vincent WF, De La Noüe J (2000) Nitrogen and phosphorus removal by high latitude mat-forming cyanobacteria for potential use in tertiary wastewater treatment. *J Appl Phycol* 12: 105-112.

Chevalier P, Proulx D, Lessard P, Vincent WF, De La Noüe J (2000) Nitrogen and phosphorus removal by high latitude mat-forming cyanobacteria for potential use in tertiary wastewater treatment. *J Appl Phycol* 12: 105-112.

Chittapun S, Limbipichai S, Amnuaysin N, Boonkerd R, Charoensook M (2018) Effects of using cyanobacteria and fertilizer on growth and yield of rice, *Pathum Thani*: a pot experiment. *J Appl Phycol* 30: 79-85.

Chittora D, Meena M, Barupal T, Swapnil P, Sharma K (2020) Cyanobacteria as a source of biofertilizers for sustainable agriculture. *Biochem Biophys Rep* 22: 100737.

Chua M, Erickson TE, Merritt DJ, Chilton AM, Ooi MKJ, Muñoz-Rojas M (2019) Bio-priming seeds with cyanobacteria: Effects on native plant growth and soil properties. *Restor Ecol.*

Converti A, Oliveira RP, Torres BR, Lodi A, Zilli M (2009) Biogas production and valorization by means of a two-step biological process. *Bioresour Technol* 100: 5771-5776.

Cook PM (1951) Chemical engineering problems in large scale culture of algae. *Ind Eng Chem* 43: 2385-2389.

Corder SL, ReevesM (1994) Biosorption of Ni in complex aqueous streams by cyanobacteria. *Appl Biochem Biotech* 45: 847-859.

Costa JAV, De Morais MG, Dalcanton F, Reichert CC, Durante AJ (2006) Simultaneous cultivation of *Spirulina platensis* and the toxigenic cyanobacteria *Microcystis aeruginosa*. *Z. Naturforsch* 61: 105-110.

Dahms HU, Xu Y, Pfeiffer C (2006) Antifouling potential of cyanobacteria: a mini-review. *Biofouling* 22: 317-327.

De Caire GZ, De Cano MS, De Mule MCZ, De Halperin DR (1990) Antimycotic products from the cyanobacterium *Nostoc muscorum* against *Rhizoctonia solani*. *Phyton* 51: 1-4.

De Morais MG, Costa JAV (2007) Carbon dioxide fixation by *Chlorella kessleri*, *C. vulgaris*, *Scenedesmus obliquus* and *Spirulina* sp. cultivated in flasks and vertical tubular photobioreactors. *Biotechnol Lett* 29:1349-1352.

Dellamatrice PM, Silva-Stenico ME, de Moraes LAB, Fiore MF, Monteiro RTR (2017) Degradation of textile dyes by cyanobacteria. *Braz J Microbiol* 48: 25-31.

Deviram G, Pant G, Prasuna R (2013) Biochemical changes induced by fungicides in nitrogen fixing *Nostoc* sp. *J Environ Sci Eng* 55: 81.

Dubey SK, Dubey J, Mehra S, Tiwari P, Bishwas AJ (2011) Potential use of cyanobacterial species in bioremediation of industrial effluents. *Afr J Biotechnol* 10: 1125-1132.

Ducat DC, Way JC, Silver PA (2011) Engineering cyanobacteria to generate high-value products. *Trends biotechnol* 29: 95-103.

Dumas A, Laliberte G, Lessard P, De La Noüe J (1998) Biotretament of fish farm effluents using the cyanobacterium *Phormidium bohneri*. *Aquacult Eng* 17: 57-68.

Dussud C, Hudec C, George M, Fabre P, Higgs P, Bruzaud S, Delort AM, Eyheraguibel B, Meistertzheim AL, Jacquin J, Cheng J, Callac N, Odobel C, Rabouille S, Ghiglione JF (2018) Colonization of non-biodegradable and biodegradable plastics by marine microorganisms. *Front Microbiol* 9: 1571.

El-Sheekh MM, Ghareib MM, EL-Souod GWUA (2012) Biodegradation of phenolic and polycyclic aromatic compounds by some algae and cyanobacteria. *J Bioremed Biodegrad* 3: 133.

El-Enany AE, Issa AA (2000) Cyanobacteria as a biosorbent of heavy metals in sewage water. *Envir Toxicol Pharmacol* 8: 95-101.

El-Eslamboly AASA, Abd El-Wanis MM, Amin AW (2019) Algal application as a biological control method of root-knot nematode *Meloidogyne incognita* on cucumber under protected culture conditions and its impact on yield and fruit quality. *Egypt J Biol Pest Control 29*: 18.

El-Sohaimy SA (2012) Functional foods and nutraceuticals – modern approach to food science, World Appl. *Sci J* 20 (5): 691-708.

Flynn KJ, Greenwell HC, Lovitt RW, Shields RJ (2010) Selection for fitness at the individual or population levels: modelling effects of genetic modifications in microalgae on productivity and environmental safety. *J Theor Biol* 263: 269-280.

Forlani G, Pavan M, Gramek M, Kafarski P, Lipok J (2008) Biochemical bases for a widespread tolerance of cyanobacteria to the phosphonate herbicide glyphosate. *Plant Cell Physiol* 49: 443-456.

Foster RA, Zehr JP (2019) Diversity, Genomics, and Distribution of Phytoplankton-Cyanobacterium Single-Cell Symbiotic Associations. *Annu Rev Microbiol* 73: 435-456.

Gao X, Sun T, Pei G, Chen L, Zhang W (2016) Cyanobacterial chassis engineering for enhancing production of biofuels and chemicals. *Appl Microbiol Biotechnol* 100: 3401-3413.

Gaysina LA, Saraf A, Singh P (2019) Cyanobacteria in Diverse Habitats. In: Mishra AK, Tiwari DN, Rai AN, editors. *Cyanobacteria from Basic Science to Applications*: Academic Press: Cambridge, MA, USA, pp. 1-28.

Gonçalves AL, Pires JCM, Simões M (2016) Biotechnological potential of *Synechocystis salina* co-cultures with selected microalgae and cyanobacteria: Nutrients removal, biomass and lipid production. *Bioresour Technol* 200: 279-286.

Grima EM, Fernández FA, Camacho FG, Chisti Y (1999) Photobioreactors: light regime, mass transfer, and scale up. *J Biotechnol* 70: 231-247.

Gudmundsson S, Nogales J (2015) Cyanobacteria as photosynthetic biocatalysts: a systems biology perspective. *Mol Biosyst* 11: 60-70.

Hagmann L, Juttner F (1996) Fischerellin A, a novel photosystem-II-inhibiting allelochemical of the cyanobacterium *Fischerella muscicola* with antifungal and herbicidal activity. *Tetrahedron Lett* 37: 6539-6542.

Hamouda RAEF, Sorour NM, Yeheia DS (2016) Biodegradation of crude oil by *Anabaena oryzae, Chlorella kessleri* and its consortium under mixotrophic conditions. *Int Biodeterior Biodegrad* 112: 128-134.

Hankamer B, Lehr F, Rupprecht J, Mssgnug JH, Posten C, Kruse O (2007) Photosynthetic biomass and $H_2$ production by green algae: from bioengineering to bioreactor scale-up. *Physiol Plant* 131: 10-21.

Hatzios KK (1991) Biotransformation of herbicides in higher plants. In: Cessna AJ, Grover R editors. *Environmental chemistry of herbicides.* CRC Press, Boca Raton, pp. 141-185.

He M, Hu R, Jia R (2019) Biological soil crusts enhances the recovery of nutrient levels of surface dune soil in arid desert regions. *Ecol Indic* 106.

Hu Q, Westerhoff P, Vermaas W (2000) Removal of nitrate from groundwater by cyanobacteria: quantitative assessment of factors influencing nitrate uptake. *Appl Environ Microbiol* 66: 133-139.

Ibraheem IB (2007) Cyanobacteria as alternative biological conditioners for bioremediation of barren soil, Egypt. *J Phycol* 8 (100): 99-116.

Issa AA, Abd-Alla MH, Ohyama T (2014) Nitrogen fixing cyanobacteria: future prospect. In: Ohyama T, editors. *Advances in Biology and Ecology of Nitrogen Fixation*: Rijeka: In Tech.

Kalavathi DF, Uma L, Subramanian G (2001) Degradation and metabolization of the pigment—Melanoidin in distillery effluent by the marine cyanobacterium *Oscillatoria boryana* BDU 92181. *Enzyme Microb Technol* 29: 246-251.

Kaneko T, Sato S, Kotani H, Tanaka A, Asamizu E, Nakamura Y (1996) Sequence analysis of the genome of the unicellular cyanobacterium *Synechocystis* sp. 6803. II. Sequence determination of the entire genome and assignment of potential protein-coding regions. DNA *Res* 3(3): 109-136.

Kaushik BD (1997) Blue green algal biofertilizers. Indian Agricultural Institute, New Delhi.

Khatoon N, Pal R (2015) Microalgae in biotechnological application: a commercial Approach. In: Bahadur B, Rajam MV, Sahijram L, Krishnamurthy KV, editors. *Plant Biology and Biotechnology*, Springer, pp. 27-47.

Kirstein IV, Wichels A, Gullans E, Krohne G, Gerdts G (2019) The Plastisphere - uncovering tightly attached plastic "specific" microorganisms. *PLoS One* 14: e0215859.

Kulik MM (1995) The potential for using cyanobacteria (blue-green algae) and algae in the cyanobacterial control of plant pathogenic bacteria and fungi. *Eur J Plant Path* 101: 585-599.

Kumar A, Singh JS (2016) Microalgae and cyanobacteria biofuels: a sustainable alternate to crop-based fuels. In: Singh JS, Singh DP, editors. *Microbes and Environmental Management*, New Delhi: Studium Press Pvt. Ltd. 1-20.

Kuritz T, Wolk CP (1995) Use of filamentous cyanobacteria for biodegradation of organic pollutants. *Appl Environ Microbiol* 61: 234-238.

Lawson PA, Allen TD, Caldwell ME, Tanner RS (2011) Anaerobes: a piece in the puzzle for alternative bio-fuels. *Anaerobe* 17: 206-210.

Lee CG, Palsson BO (1994) High-density algal photobioreactors using light-emitting diodes. *Biotechnol Bioeng* 44: 1161-1167.

Lee RE (1989) Phycology. 2nd Ed. Cambridge University Press. New York. 645pp.

Lee YK (1997) Commercial production of microalgae in the Asia Pacific rim. *J Appl Phycol* 9: 403-411.

Li Y, Yang X, Geng B (2018) Preparation of Immobilized Sulfate-Reducing Bacteria-Microalgae Beads for Effective Bioremediation of Copper-Containing Wastewater. *Water Air Soil Pollut* 229: 54.

Lodi A, Binaghi L, Solisio C, Converti A, Del Borghi M (2003) Nitrate and phosphate removal by *Spirulina platensis*. *J Ind Microbiol Biotechnol* 30: 656-660.

Mahanty T, Bhattacharjee S, Goswami M, Bhattacharyya P, Das B, Ghosh A, Tribedi (2017) Biofertilizers: a potential approach for sustainable agriculture development. *Environ Sci Pollut Res* 24: 3315–3335.

Malam Issa O, Défarge C, Le Bissonnais Y, Marin B, Duval O, Bruand A, D'Acqui LP, Nordenberg S, Annerman M (2007) Effects of the inoculation of cyanobacteria on the microstructure and the structural stability of a tropical soil. *Plant Soil* 290(1): 209-219.

Malliga P, Uma L, Subramanian G (1996) Lignolytic activity of the cyanobacterium *Anabaena azollae* ML2 and the value of coir waste as a carrier for BGA bio-fertilizer. *Microbios* 86: 175-183.

Mandal B, Das SC, Mandal LN (1992) Effect of growth and subsequent decomposition of blue-green algae in the transformation of phosphorus in submerged soils. *Plant Soil* 143: 289-297.

Mandal B, Vlek PLG, Mandal LN (1999) Beneficial effects of blue-green algae and Azolla, excluding supplying nitrogen, on wetland rice fields: a review. *Biol Fertil Soils* 28: 329-342.

Markley AL, Begemann MB, Clarke RE, Gordon GC, Pfleger BF (2015) Synthetic biology toolbox for controlling gene expression in the cyanobacterium *Synechococcus* sp. strain PCC 7002. *ACS Synth Biol* 4(5): 595-603.

Mason J (2003) Sustainable Agriculture, 2nd Edn. Collingwood, VIC: Landlinks Press.

Meeks JC, Elhai J (2002) Regulation of cellular differentiation in filamentous cyanobacteria in free-living and plant-associated symbiotic growth states. *Microbiol Rev* 66: 94–121.

Meena M, Prasad V, Zehra A, Gupta VK, Upadhyay RS (2015) Mannitol metabolism during pathogenic fungal–host interactions under stressed conditions. *Front Microbiol* 6: 1019-1026.

Meena M, Swapnil P, Zehra A, Dubey MK, Upadhyay RS (2017a) Antagonistic assessment of *Trichoderma* spp. by producing volatile and non-volatile compounds against different fungal pathogens, Arch. *Phytopathol Plant Protect* 50(13-14): 629-648.

Meena M, Zehra A, Swapnil P, Dubey MK, Patel CB, Upadhyay RS (2017b) Effect on lycopene, β-carotene, ascorbic acid and phenolic content in tomato fruits infected by *Alternaria alternata* and its toxins (TeA, AOH and AME). *Arch Phytopathol Plant Protect* 50: 317-329.

Meena M, Aamir M, Vikas K, Swapnil P, Upadhyay RS (2018a) Evaluation of morphophysiological growth parameters of tomato in response to Cd induced toxicity and characterization of metal sensitive NRAMP3 transporter protein. *Environ Exp Bot* 148: 144-167.

Meena M, Swapnil P, Zehra A, Dubey MK, Aamir M, Patel CB, Upadhyay RS (2018b) Virulence factors and their associated genes in microbes. In: Singh HB, Gupta VK, Jogaiah S, editors. *New and Future Developments in Microbial Biotechnology and Bioengineering:* Elsevier, pp. 181-208.

Melis A (2009) Solar energy conversion efficiencies in photosynthesis: minimizing the chlorophyll antennae to maximize efficiency. *Plant Sci* 177: 272-280.

Michalak I, Chojnacka K, Dmytryk A, Wilk R, Gramza M, Rój E (2016). Evaluation of Supercritical Extracts of Algae as Biostimulants of Plant Growth in Field Trials. *Front Plant Sci* 7: 1591.

Mishra U, Pabbi S (2004) Cyanobacteria: a potential biofertilizer for rice. *Resonance* 9: 6-10.

Mittal S, Sengar RMS, Kaushik BD (1992) Uptake and toxicity of heavy metals to algae. *Ind J Microbiol* 31: 1-26.

Mohamed ZA (2001) Removal of cadmium and manganese by a non-toxic strain of the freshwater cyanobacterium. *Water Res* 35: 4405-4409.

Mohan A, Kumar B (2017) Growth performance and yield potential of cereal crops (wheat, maize and barley) in association with cyanobacteria. *Int J Curr Microbiol App Sci* 6: 744-758.

Mona S, Kaushik A, Kaushik CP (2011) Hydrogen production and metal-dye bioremoval by a *Nostoc linckia* strain isolated from textile mill oxidation pond. *Bioresour Technol* 102: 3200-3205.

Muñoz-Rojas M, Román JR, Roncero-Ramos B, Erickson TE, Merritt DJ, Aguila-Carricondo P, Cantón Y (2018) Cyanobacteria inoculation enhances carbon sequestration in soil substrates used in dryland restoration. *Sci Total Environ* 636: 1149-1154.

Nagasathya A, Thajuddin N (2008) Decolourization of paper mill effluent using hypersaline cyanobacterium. *Res J Environ Sci* 2: 408-414.

Narro M, Cerniglia C, Van Baalen C, Gibson D (1992) Evidence for an NIH shift in oxidation of naphthalene by the marine cyanobacterium *Oscillatoria* sp. strain JCM. *Appl Environ Microbiol* 58: 1360-1363.

Nisha R, Kaushik A, Kaushik CP (2007) Effect of indigenous cyanobacterial application on structural stability and productivity of an organically poor semi-arid soil. *Geoderma* 138: 49-56.

Nozzi NE, Oliver JWK, Atsumi S (2013) Cyanobacteria as a platform for biofuel production. *Front Bioeng Biotechnol* 1: 7.

Oliver JW, Atsumi S (2014) Metabolic design for cyanobacterial chemical synthesis. *Photosynth Res* 120: 249-261.

Oliver NJ, Rabinovitch-Deere CA, Carroll AL, Nozzi NE, Case AE, Atsumi S (2016) Cyanobacterial metabolic engineering for biofuel and chemical production. *Curr Opin Chem Biol 35: 43-50.*

Ooms MD, Dinh CT, Sargent EH, Sinton D (2016) Photon management for augmented photosynthesis. *Nat Comm* 7: 12699.

Pandey KD, Shukla PN, Giri DD, Kashyap AK (2005) Cyanobacteria in alkaline soil and the effect of cyanobacteria inoculation with pyrite amendments on their reclamation. *Biol Fert Soil* 41(6): 451-457.

Panjiar N, Mishra S, Yadav AN, Verma P (2018) Functional foods from cyanobacteria: an emerging source for functional food products of pharmaceutical importance. In: Gupta VK, Treichel H, Shapaval VO, Antonio de Oliveira L, Tuohy MG, editors. *Microbial Functional Foods and Nutraceuticals*: first ed., John Wiley & Sons Ltd.

Parikh A, Madamwar D (2005) Textile dye decolorization using cyanobacteria. *Biotechnol Lett* 27: 323-326.

Park C-H, Li X-R, Jia R-L, Hur J-S (2017a) Combined application of cyanobacteria with soil fixing chemicals for rapid induction of biological soil crust formation. Arid Land Res Manag 31(1): 81-93.

Park CH, Li XR, Zhao Y, Jia RL, Hur JS (2017b) Rapid development of cyanobacterial crust in the field for combating desertification. *PLoS One* 12: e0179903.

Parmar A, Singh NK, Pandey A, GnansounouE, Madamwar D (2011) Cyanobacteria and microalgae: a positive prospect for biofuels. *Bioresour Technol* 102: 10163-10172.

Pathak J, Rajneesh, Pandey A, Singh SP, Sinha RP (2017) World agriculture and impact of biotechnology. In: Dubey SK, Pandey A, Sangwan RS, editors. *Current Developments in Biotechnology and Bioengineering: Crop Modification, Nutrition, and Food Production:* Radarweg, Amsterdam, Elsevier, 1-22.

Perera I, Subashchandrabose SR, Venkateswarlu K, Naidu R, Megharaj M (2018) Consortia of cyanobacteria/microalgae and bacteria in desert soils: An underexplored microbiota. *Appl Microbiol Biotechnol* 102: 7351-7363.

Pignantello JJ, Porwoll J, Carson RE, Xavier A, Gleason FK, Wood JM (1983) Structure of the antibiotic cyanobacterin, a chlorine containing $\gamma$-lactone from the freshwater cyanobacterium Scytonema hofmanni. *J Org Chem* 48: 4035-4038.

Pinzon-Gamez NM, Sundaram S, Ju LK (2005) Heterocyst differentiation and $H_2$ production in $N_2$-fixing cyanobacteria. In: *Proceedings of the AIChE Annual Meeting*: Cincinnati, OH.

Prasanna R, Adak A, Verma S, Bidyarani N, Babu S, Pal M, Shivay YS, Nain L (2015a) Cyanobacterial inoculation in rice grown under flooded and SRI modes of cultivation elicits differential effects on plant growth and nutrient dynamics. *Ecol Eng* 84: 532-541.

Prasanna R, Hossain F, Babu S, Bidyarani N, Adak A, Verma S, Shivay YS, Nain L (2015b) Prospecting cyanobacterial formulations as plant-growth-promoting agents for maize hybrids. *S Afr J Plant Soil* 32: 199-207.

Queiroz MI, Hornes MO, da Silva-Manetti AG, Jacob-Lopes E (2011) Single-cell oil production by cyanobacterium *Aphanothece microscopica* Nägeli cultivated heterotrophically in fish processing wastewater. *Appl Energ* 88: 3438-3443.

Quintana N, Van der Kooy F, Van de Rhee MD, Voshol GP, Verpoorte R (2011) Renewable energy from cyanobacteria: energy production optimization by metabolic pathway engineering. *Appl Microb Biotechnol* 91: 471-490.

Raghukumar C, Vipparty V, David J, Chandramohan D (2001) Degradation of crude oil by marine cyanobacteria. *Appl Microbiol Biotechnol* 57: 433-436.

Rajneesh, Singh SP, Pathak J, Sinha R P (2017) Cyanobacterial factories for the production of green energy and value-added products: an integrated approach for economic viability. *Renew Sustain Energy Rev* 69: 578-595.

Rangsayatorn N, Pokethitiyook P, Upatham ES, Lanza GR (2004) Cadmium biosorption by cells of *Spirulina platensis* TISTR 8217 immobilized in alginate and silica gel. *Environ Int* 30: 57-63.

Ratchford IA, Fallowfield HJ (1992) Performance of a flatplate air-lift reactor for the growth of high biomass algal culture. *J Appl Phycol* 4: 1-9.

Raungsomboona S, Chidthaisonga A, Bunnagb B, Inthornc D, Harveya NW (2006) Production, composition and $Pb^{2+}$ adsorption characteristics of capsular polysaccharides extracted from a cyanobacterium *Gloeocapsa gelatinosa*. *Water Res* 40: 3759-3766.

Reddy RS, Triveni S (2016) Biofertilizers for sustainable production in oil seed crops. *Sch J Agric Vet Sci* 3: 435-441.

Renuka N, Prasanna R, Sood A, Bansal R, Bidyarani N, Singh R, Shivay YS, Nain L, Ahluwalia AS (2017) Wastewater grown microalgal biomass as inoculants for improving micronutrient availability in wheat. *Rhizosphere* 3: 150-159.

Rezaei H (2016) Biosorption of chromium by using *Spirulina* sp. *Arab J Chem* 9: 846-853.

Rittmann BE (2008) Opportunities for renewable bioenergy using microorganisms. *Biotechnol Bioeng* 100: 203-212.

Rodríguez AA, Stella AM, Storni MM, Zulpa G, Zaccaro M (2006) Effects of cyanobacterial extracellular products and gibberellic acid on salinity tolerance in *Oryza sativa L. Saline Syst* 2(1): 7.

Roger PA, Reynaud PA (1982) Free—living blue—green algae in tropical soils. *Microbiology of Tropical Soils and Plant Productivity*: Springer, Dordrecht, pp. 147-168.

Román JR, Roncero-Ramos B, Chamizo S, Rodríguez-Caballero E, Cantón Y (2018) Restoring soil functions by means of cyanobacteria inoculation: importance of soil conditions and species selection. *Land Degrad Dev* 29(9): 3184-3193.

Roncero-Ramos B, Román JR, Gómez-Serrano C, Cantón Y, Acién FG (2019) Production of a biocrust-cyanobacteria strain (*Nostoc commune*) for large-scale restoration of dryland soils. *J Appl Phycol* 31: 2217-2230.

Rosgaard L, de Porcellinis AJ, Jacobsenc JH, Frigaardc NU, Sakuragia Y (2012) Bioengineering of carbon fixation, biofuels, and biochemicals in cyanobacteria and plants. *J Biotechnol* 162: 134-147.

Rossi F, Li H, Liu Y, De Philippis R (2017) Cyanobacterial inoculation (cyanobacterisation): perspectives for the development of a standardized multifunctional technology for soil fertilization and desertification reversal. *Earth-Sci Rev* 171: 28-43.

Roychoudhury P, Kaushik BD (1989) Solubilization of Mussorie rock phosphates by cyanobacteria. *Curr Sci* 58: 569-570.

Saadatnia H, Riahi H (2009) Cyanobacteria from paddy fields in Iran as a biofertilizer in rice plants. *Plant Soil Environ* 55(5): 207-212.

Saha KC, Mandal LN (1979) Effect of algal growth on the availability of P, Fe, and Mn in rice soils. *Plant Soil* 52: 139-149.

Sánchez O, Diestra E, Esteve I, Mas J (2005) Molecular characterization of an oil-degrading cyanobacterial consortium. *Microb Ecol* 50: 580-588.

Sarma MK, Kaushik S, Goswami P (2016) Cyanobacteria: a metabolic power house for harvesting solar energy to produce bio-electricity and biofuels. *Biomass Bioenerg* 90: 187-201.

Savakis P, Hellingwerf KJ (2014) Engineering cyanobacteria for direct biofuel production from $CO_2$. Curr Opin Biotechnol 33C: 8-14.

Scherer S, Almon H, Boger P (1988) Interaction of photosynthesis, respiration and nitrogen fixation in cyanobacteria. *Photosynth Res* 15: 95-114.

Shetty K, Paliyath G, Pometto A, Robert EL (2006) Food Biotechnology, second ed., CRC Press, Boca Raton.

Sialve B, Bernet N, Bernard O (2009) Anaerobic digestion of microalgae as a necessary step to make microalgal biodiesel sustainable. *Biotech Adv* 4: 409-416.

Silva-Benavides AM, Torzill G (2012) Nitrogen and phosphorus removal through laboratory batch cultures of microalga *Chlorella vulgaris* and cyanobacterium *Planktothrix isothrix* grown as monoalgal and as co-cultures. *J Appl Phycol* 24: 267-276.

Singh AK, Singh PP, Tripathi V, Verma H, Singh SK, Srivastava AK, Kumar A (2018) Distribution of cyanobacteria and their interactions with pesticides in paddy field: A comprehensive review. *J Environ Manag* 224: 361-375.

Singh JS (2011) Methanotrophs: the potential biological sink to mitigate the global methane load. *Curr Sci* 100: 29-30.

Singh JS (2014) Cyanobacteria: a vital bio-agent in eco-restoration of degraded lands and sustainable agriculture. *Climate Change Environ Sustain* 2: 133-137.

Singh JS, Kumar A, Rai AN, Singh DP (2016) Cyanobacteria: a precious bio-resource in agriculture, ecosystem, and environmental sustainability. *Front Microbiol* 7: 529.

Singh JS, Pandey VC, Singh DP (2011) Efficient soil microorganisms: a new dimension for sustainable agriculture and environmental development. *Agric Ecosyst Environ* 140: 339-353.

Singh JS, Strong PJ (2016) Biologically derived fertilizer: a multifaceted bio-tool in methane mitigation. *Ecotoxicol Environ Saf* 124: 267-276.

Singh R, Parihar P, Singh M, Bajguz A, Kumar J, Singh S, Singh VP, Prasad SM (2017a) Uncovering potential applications of cyanobacteria and algal metabolites in biology, agriculture and medicine: current status and future prospects. *Front Microbiol* 8: 515.

Singh S, Kant C, Yadav RK, Reddy YP, Abraham G (2019) Cyanobacterial Exopolysaccha-rides: Composition, Biosynthesis, and Biotechnological Applications. In: Mishra AK, Tiwari DN, Rai AN, editors. *Cyanobacteria from Basic Science to Applications*: Academic Press: Cambridge, MA, USA, pp. 347-358.

Singh SP, Pathak J, Sinha RP (2017b) Cyanobacterial factories for the production of green energy and value-added products: an integrated approach for economic viability, Renew. *Sustain Energy Rev* 69: 578-595.

Slotton DG, Goldman CR, Frank A (1989) Commercially grown Spirulina found to contain low levels of mercury and lead. *Nutr Rep Int* 40: 1165-1172.

Soletto D, Binaghi L, Lodi A, Carvalho JCM, Converti A (2005) Batch and fed-batch cultivations of *Spirulina platensis* using ammonium sulphate and urea as nitrogen sources. *Aquaculture* 243: 217-224.

Song T, Mårtensson L, Eriksson T, Zheng W, Rasmussen U (2005) Biodiversity and seasonal variation of the cyanobacterial assemblage in a rice paddy field in Fujian, China. *FEMS Microbiol Ecol* 54(1): 131-140.

Sood A, Renuka N, Prasanna R, Ahluwalia AS (2015) Cyanobacteria as potential options for wastewater treatment. In: Ansari AA, Gill SS, Gill R, Lanza GR, Newman, editors. *Phytoremediation: Management of Environmental Contaminants, Volume 2:* Springer International Publishing, Cham, pp 83-93.

Stewart WDP, Rowell P (1986) Biochemistry and physiology of nitrogen fixation with particular emphasis on nitrogen fixing phototrophs. *Plant Soil* 90: 167-191.

Subramanian G, Sundaram SS (1986) Induced ammonia release by the nitrogen fixing cyanobacterium *Anabaena*. *FEMS Microbiol Lett* 37: 151–154.

Tassara C, Zaccaro MC, Storni MM, Palma M, Zulpa G (2008) Biological control of lettuce white mold with cyanobacteria. *Int J Agric Biol* 10: 487-492.

Thajuddin N, Subramanian G (2005) Cyanobacterial biodiversity and potential applications in biotechnology. *Curr Sci* 89: 47-57.

Tharifkhan SA, Garlapati D, Arulraj D, Lakshmanan U, Prabaharan D (2018) Growth and nitrogen (N) metabolizing enzymes of mesophilic and psychrophilic heterocystous cyanobacteria—in response to temperature regimes. Res Rev: *J Life Sci* 8(3): 31-43.

Tiwari A, Pandey A (2012) Cyanobacterial hydrogen- a step towards clean environment. *Inter J Hydrogen Energ* 37: 139-150.

Travieso L, Canizares RO, Borja R, Benitez F, Dominguez AR, Valiente V (1999) Heavy metal removal by microalgae. *Bull Environ Contam Toxicol* 62: 144-151.

Uma L, Subramanian G (1990) Effective use of cyanobacteria in effuent treatment. In: *Proceedings of the National Symposium on Cyanobacterial N2 Fixation;* Indian Agricultural Research Institute. New Delhi, India, pp. 437-444.

Vijayakumar S, Manoharan C (2012) Treatment of Dye Industry Effluent Using Free and Immobilized Cyanobacteria. *J Bioremed Biodeg* 3: 165.

Wang B, Wang J, Zhang W, Meldrum DR (2012) Application of synthetic biology in cyanobacteria and algae. *Front Microbiol* 3:44.

Whitton BA (2002) Soils and rice-fields. In: Whitton BA, Potts M, editors. *The Ecology of Cyanobacteria: Their Diversity in Time and Space:* Springer Netherlands, Dordrecht, pp 233-255.

Whitton BA, Potts M (2000) Introduction of cyanobacteria. In: Whitton BA, Potts M, editors. *The Ecology of Cyanobacteria: Their Diversity in Time and Space*: Kluwer Academic, Dordrecht, pp. 1-10.

Williams W, Chilton A, Schneemilch M, Williams S, Neilan B, Driscoll C (2019) Microbial biobanking—Cyanobacteria-rich topsoil facilitates mine rehabilitation. *Biogeosciences* 16: 2189-2204.

Wilson LT (2006) Cyanobacteria: a potential nitrogen source in rice fields, *Texas Rice* 6: 9-10.

Wolf AM, Baker DE, Pionke HB, Kunichi HM (1985) Soil test for estimating labile, soluble and algal available phosphorus in agricultural soils. *J Environ Qual* 14: 341-348.

Zahra Z, Choo DH, Lee H, Parveen A (2020) Cyanobacteria: Review of current potentials and applications. *Environments* 7: 13.

Zhong W, Zhang Z, Luo Y, Qiao W, Xiao M, Zhang M (2012) Biogas productivity by co-digesting Taihu blue algae with corn straw as an external carbon source. *Bioresour Technol* 114: 281-286.

Zuncan S, Trevisan R, Paoletti MO (2006) Soil algae composition under different agro-ecosystems in North-Eastern Italy. *Agri Environ* 112: 1-12.

## Chapter 5

# Cyanobacteria as a Promising Bioresource for Applications in Sustainable Agriculture, Medicine, and Green Nanotechnology

**Pritam Kaur[1], Preeti Agarwal[1], Darakshan Zabin[2], Farheen Islam[3], Aadrita Das[4] and Garvita Singh[1,*]**

[1]Department of Botany, Gargi College, University of Delhi, New Delhi, India
[2]School of Life Sciences, Arizona State University, USA
[3]Department of Botany, University of Delhi, Delhi, India
[4]Department of Environmental Science, Tezpur University, Assam, India

## Abstract

Keeping in mind the ever-increasing demand for sustainable development, biotechnology is now providing a reliable approach to effectively deal with a multitude of problems in the agricultural, pharmaceutical, and industrial sectors. The long evolutionary history and rich bioresource of various bioactive compounds make cyanobacteria a role model and the most successful and sustained prokaryotic organisms studied so far. The remarkable growth rate, and oxygenic photosynthesis enables their potential use in a wide range of applications to contribute as bio-fertilizers for safe agriculture enhancing soil fertility, promising source of sustainable energy production, as food supplements, other valuable co-products, including antioxidants, biodegradable polymers, and sunscreens. With ever increasing global demand of food supply followed by the depletion of natural resources have forced humans to look for the alternate source of food and energy. Cyanobacteria has emerged as a promising green technology for ecofriendly, sustainable

[*] Corresponding Author's Email: Garvita.singh@gargi.du.ac.in.

In: Cyanobacteria
Editor: Rajeshwar P. Sinha
ISBN: 979-8-88697-169-9

production of biomass of very high value which can be used to support both environment and human needs of food, energy and medicine. Their acceptance in the field of nanotechnology and medical sciences has opened the door to their numerous application potentials on a wide range of issues that can result in greater sustainability. Apart from their pre-eminent applications, cyanobacteria produce cyanotoxins which can have adverse effects on human health. Recent developments have established their potential in cancer therapeutics. We believe that cyanobacteria are yet to be tapped for their repository of applications. Moreover, the metabolites produced by the cyanobacteria show some very unique and exciting biological properties including anticancer, antimicrobial, anti-HIV, anticoagulant, anti-inflammatory, immunosuppressant, antiprotozoal, antimalarial, antitumor, and antiviral activities. The prevalence of antibiotic-resistant genes in microorganisms is a substantial threat to the crop, in lieu of the pesticides, scientists are now exploiting the idea of using cyanobacteria either combined with nanoparticles or alone in the crop field as nano biosensors to detect the presence of heavy metals, nanofertilizers to supplement growth. This chapter incorporates comprehensive knowledge about the role and applications of cyanobacteria and how these microorganisms act as a potential bioresource in the field of sustainable agriculture, green nanotechnology, and medicines.

**Keywords:** bioactive compounds, cyanobacteria, green nanotechnology, sustainability

## Introduction

Cyanobacteria, also known as Blue-Green Algae (BGA), are prokaryotic, Gram-negative bacteria that inhabited the Earth 3.5 billion years ago and were the first organisms to carry out oxygenic photosynthesis (Mazard et al. 2016; Carpine et al. 2020). Cyanobacteria have the ability to fix atmospheric nitrogen and therefore can be used in agriculture as potential biofertilizers. Cyanobacteria help in reducing environmental risk and production cost of crops by substituting chemical nitrogen-based fertilizers. Their use as biofertilizers has been reported for the cultivation of some economically important staple crops like rice, consumed by the majority of the world population (Singh et al. 2016). Some other unique features of cyanobacteria include oxygenic photosynthesis, the ability to grow on non-agricultural lands and make the land cultivable, generation of bio-fuels and other useful derivatives along with decreasing greenhouse gas emissions making them a reliable source of biofertilizers for sustainable agriculture (Singh et al. 2016). Additionally, numerous bioactive compounds in cyanobacteria pose their potential role in the pharmaceutical industry (Singh et al. 2017; Chittora et al. 2020).

By 2050, according to the UN, the human population is proposed to increase by 9 billion with majorly population expansion will be in developing countries of Asia and Africa (Wood 2001; Singh et al. 2016). With the increasing population, the demand for food security will further increase and create pressure on agriculture systems. Increasing demands also pose the use of chemical fertilizers at excessive levels however, these are effective for only the short term and have long-term lethal effects on both humans and other organisms sustaining in ecosystems. Many microbes in our diverse ecosystems such as cyanobacteria can improve and enhance crop yields in the agriculture system (Chittora et al. 2020). The wide uses of biofertilizers, green manure and biocontrol agents are being studied and reported. The number of microbes that lives in close association with eukaryotic plants as symbionts either endophytic or exophytic, are used in crop fields and help plants procure elements like nitrogen, sulphur, phosphorous, carbon, etc. Besides using microbes as biofertilizers, cyanobacteria have a huge role as plant growth promoters; enhance nutrients and metabolites availability in soil, in the reclamation of salt-affected soil, and bioremediation thereby improving soil fertility and overall growth of plants and crop yields (Jangir et al. 2015). Cyanobacteria are also a very well documented and studied source of bioenergy. However, to improve the production of crops by making them stress-resistant and improving photosynthetic rates, and for bioremediation purposes, integration of prokaryotic desired genes in eukaryotic plants is a recently growing and promising technology and has been undergoing extensive research (Jangir et al. 2015).

All these benefits in agriculture can lead to food security and better food quality, healthy agroecosystems and help in various kinds of stress management. Besides, in the environment, they can help in producing biofuels and food supplements. Cyanobacteria have potential beneficial and nutritional values, thus making it possible for them to be widely used as food supplements, in the forms of tablets, juices, etc. Additionally, they have convenient methods of propagation, owing to their fewer requirements of water and land, possibilities of large-scale productions, and wide pH and temperature ranges tolerance in which the products/metabolites/bioactive compounds remain stable. Therefore, their use can lead to overall sustainable agriculture and can provide environmental benefits (Singh et al. 2016).

As the lifestyle habits of people are changing day by day, the world mortality and morbidity rate is increasing rapidly because of several diseases caused by such habits. Some of these diseases are metabolic syndromes, cardiovascular diseases, chronic diabetes, cancers, stroke, obesity, and chronic respiratory diseases. Over the last few decades, the use of natural compounds in drug development and research related to the application of such compounds in maintaining health and curing the number of human diseases has been vastly increasing. Cyanobacteria are an important target due to their cosmopolitan distribution and being a rich source of a large number of bioactive metabolites, with varying bioactivities (Singh et al. 2017). They act as an important and rich source of some novel bioactive metabolites which have interesting antimicrobial, anticarcinogenic, and antitumor activities (Demay et al. 2019; Khalifa et al. 2021). Moreover, the metabolites produced by the cyanobacteria show some very unique and interesting biological properties including anti-HIV, anticoagulant, anti-inflammatory, immunosuppressant, antiprotozoal, antimalarial, and antiviral activities and are involved in the management of many other diseases (Demay et al. 2019; Gondim et al. 2019, Khalifa et al. 2021). Some available information also suggests that these microorganisms have been used since 1500 BC for the treatment of fistula, gout, and various other forms of cancer (Tiwari and Tiwari 2020; Carpine and Sieber 2021; Qamar et al. 2021) The bioactive compounds from cyanobacteria that are applied to the research in the medical sector have demonstrated an interesting future with high potential of their use in developing medicinal drugs. Most of these cyanobacterial compounds exhibit strong pharmacological activities, therefore, these compounds can also be used as promising candidates for the treatment of COVID-19, a global pandemic (Khalifa et al. 2021). Therefore, cyanobacteria act as promising natural resource offering a range of chemical substances with medicinal properties for the discovery of new drugs.

Cyanobacteria have been proposed among the most instrumental natural candidates that have application in human health and other biological activities (Bajpai et al. 2018). The various applications of cyanobacteria pertain to the chemically diverse and biologically active compounds produced by them, such as alkaloids, saccharides, fatty acids, terpenes, sterols, and peptides. Also, these bioactive compounds and other pigments in cyanobacteria act as reducing and stabilizing agents. Hence, the potential of cyanobacteria to be used as therapeutics and to resolve varying drug-related problems is immense.

More than half of the total cyanobacterial population could be used to extract bioactive compounds, but only 10% of it is currently utilized due to limitations on solubility which needs further understanding (Panjiar et al. 2017). To address this issue, nanotechnology can be used to resolve the poor solubility of cyanobacterial bioactive compounds and natural marine drugs obtained from various cyanobacterial species.

Nanotechnology is a rapidly growing discipline that entails the synthesis, characterization, and development of a wide range of nanomaterials (Dubchak et al. 2010) that play essential roles in everyday life by delivering valuable goods that improve industrial output, agriculture, environment, and medicine. Nanomaterials themselves have many applications; combined with cyanobacteria; they can reshape the dynamics of agricultural, pharmaceutical, and industrial sectors. Nanomaterials incorporated with cyanobacterial metabolites can be used as biopesticides, biofertilizers, nanocarriers of drugs, and other therapeutic purposes bringing sustainability in all spheres of life.

Cyanobacteria are biofactories used to synthesize nanoparticles (NPs) because of their ease of culture, the ability to take up heavy metals, development of sustainable products, and high growth rate. The use of cyanobacteria to synthesize NPs has become a popular research topic in recent years. They are differentiated by their ability to repair atmospheric nitrogen by employing nitrogen reductase enzymes to convert nitrogen gas to ammonia. This is a crucial advantage in their involvement in the biotransformation of metals to NPs since they can remove heavy metal ions from their surroundings (Khanna et al. 2019). Besides these characteristics, cyanobacteria have a fast growth rate that allows for large biomass output. As a result, they are essential nanotechnology-mediated microorganisms that can act as nano-biofactories. Although various extensive reviews of biological synthesis utilizing microorganisms have been published, little research has concentrated on cyanobacterial nanoparticle synthesis (Hamida et al. 2020d). Nanoparticles can be synthesized in different ways by following different methodologies. Chemical and physical ways are the most widely used and accepted methods to synthesize nanoparticles, but these methods are highly toxic, expensive, and cumbersome. Green synthesis of nanomaterials is an unrivalled alternative for producing inexpensive and environmentally friendly nanomaterials. New strategies to prevent tumor progression and metastasis have been developed with advances in nanoscience. Magnetic iron oxide NPs, for example, have improved cellular imaging, hence improving the quality and speed of the diagnostic procedure. Injecting Au-NPs intravenously can improve radiotherapy (X-rays). Many studies have been published on the biological synthesis of NPs utilizing cyanobacteria and their effects on various cancer cell lines (Bajpai et al. 2018). This chapter aims to highlight cyanobacteria as a role model organism that can be explored further for the benefits of sustainable agriculture, promising pharmaceutical drugs in treatment lines of various chronic diseases and exploring its potential in the emerging field of nanoscience. These primitive microorganisms can be mini-factories for various sectors on which human growth and development are dependent.

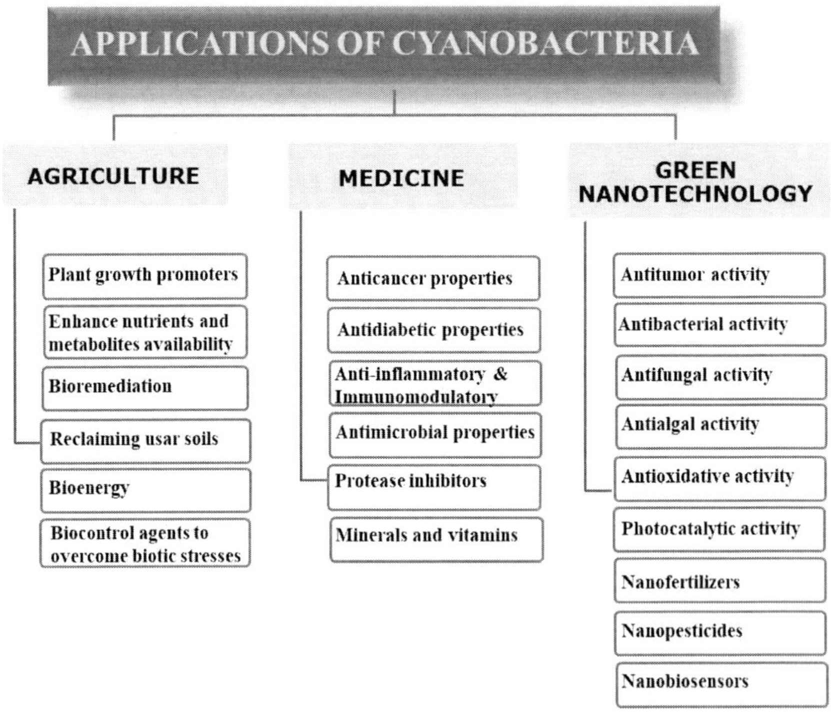


**Figure 1.** Applications of cyanobacteria in agriculture, medicine and nanotechnology.

## Role of Cyanobacteria in Agriculture

### Sustainable Agriculture

Sustainable agriculture is in urgent need of alternative ways to chemical nitrogenous fertilizers. Biological nitrogen fixation (BNF) is one of the promising methods that fix atmospheric elemental nitrogen into a plant useable form through well-orchestrated biological processes. The deficiency of nitrogen in the soil and so in crops can be fulfilled by biofertilizers such as cyanobacteria through BNF. Nitrogen-fixing symbiotic associations with plants are natural, economically and ecologically beneficial means that can help in reducing dependency on external high-cost inorganic nitrogen fertilizer inputs (Chittora et al. 2020). Cyanobacteria can easily survive in wet soils, garden soil, rock, and desert to freshwater bodies, marine water bodies, rivers, salt marshes, lakes, and swamps. The majority of cyanobacteria is either alkalophilic along with or without halophytic characteristics, and can survive under high salt concentration and pH (Lopez-Archilla et al. 2004). Cyanobacteria can also tolerate extreme environments of desiccation, temperature, light intensity, and nutrients (Whitton and Potts 2007). Cyanobacteria are a promising source of biofertilizers, as they are non-pathogenic, the most primitive photosynthetic prokaryote with minimum nutritional requirements for growth, and therefore easily cultivable and capable for outdoor/field survival (Ruffing 2011). These requirements provide a wide scope for the commercial application of cyanobacterial species. Soil improvement and crop management practice like soil fertility restoration, biocontrol of plant diseases, and crop rotation if used adequately with biofertilizers such as cyanobacteria can improve the rate of crop production. The well-known cyanobacterial nitrogen fixers are *Anabaena, Nostoc, Trichodesmium, Calothrix, Aulosira, Plectonema* and *Tolypothrix* etc. BNF is a well-coordinated mechanism where filamentous cyanobacteria aid in reducing the elemental nitrogen into an inorganic useable form and is facilitated by an enzyme, Nitrogenase in-thick-walled specialized cells called heterocysts. The nitrogen fixed into ammonia is available to the crop plants either by the exudates from the organism, autolysis or after the death of blue green algal species from their decaying body. The term "algalization" was coined for the mass production of cyanobacteria to be used as biofertilizers. Focusing on paddy cultivation in Indian agriculture, blue-green algae on the nitrogen economy led to a 25% increase in grain yield. Other plants apart from rice benefitted are- tomato, barley, oats, sugarcane, wheat, maize and cotton (Jangir et al. 2015). As demonstrated in various studies, the use of inoculum of various species of cyanobacteria (*Anabaena* sp., *Nostoc* sp., *Oscillatoria* sp.) has improved the grain, straw yields, increased the availability of nitrogen in the soil, photosynthetic pigments and overall improved the rice crops by increasing plant height, leaf length, improved the seeds germination rate and grain size when compared to chemical fertilizers. Considering the fact that improvement in crop plants is due to nitrogen availability in soil, cyanobacteria could be the potential candidates to replace chemical nitrogen fertilizers (Gonçalves 2021).

## Plant Growth Promoters

Cyanobacteria play a significant role in increasing agricultural productivity as well as an important source of the biologically-active wide diversity of compounds. Cyanobacteria produce plant growth-promoting substances like auxin, gibberellins, cytokinin, and abscisic acid. Cyanobacteria are excellent sources of vitamins particularly vitamin B, antibiotics, toxins, minerals and essential amino acids (Singh et al. 2016). As pure extracts or as crude algal compost, these metabolites improve crops' production and stimulate plant growth by improving soil quality and by enhancing protection against biotic and abiotic stress conditions. Phytohormones are signal molecules that regulate several cellular processes in plants. Typically, the wide diversity of these molecules is identified in higher plants and cyanobacteria have a similar function. For example, in cyanobacteria, abscisic acid controls cells' response to various stress conditions namely salt, oxidative, change in osmoticum, nutrients' stresses, and drought, and induces carotenoids and other antioxidative compounds which act as the protecting agents against oxidative stress (Gonçalves 2021). Cytokinins have a crucial role in plant growth and development and mediate several processes in higher plants, namely cell division, growth, regulation of seed dormancy, biogenesis, chloroplast differentiation and plant tolerance to adverse environmental conditions. Gibberellins from cyanobacteria mainly play role in the induction of pigments, growth stimulation and proteins accumulation. The cyanobacteria in paddy are capable of synthesizing and liberating gibberellic acid for the growth of plants, also able to overcome salt stress (Rodríguez et al. 2006). Indole-3-acetic acid (IAA) production by 34 different free-living and symbiotic cyanobacterial strains was also analyzed playing a major role in the growth and development of these essential microbes (Jangir et al. 2015). As hormones from plants and cyanobacteria have the same function hence for additional benefits, these organisms (or their extracts) can be used in agricultural fields for improving plant growth and protection against biotic and abiotic stress factors (Gonçalves 2021).

## Enhance Nutrients and Metabolites Availability in Soils

Cyanobacteria in soils can be the source of essential nutrients for growth, redox reactions, metabolism and development of plants and can help in overcoming the use of chemical fertilizers (Ronga et al. 2019). Cyanobacteria are able to uptake nitrogen and phosphorus and other essential microelements from different environments (even from nutrients limiting sites) and store them in their biomass. Therefore, cyanobacteria can improve the growth of crops by improving the soil nutrient concentrations. Cyanobacteria fix the atmospheric $CO_2$, converting it into organic biomass and serving as a source of organic matter for higher plants. Moreover, the organic carbon content in agricultural fields may increase as cyanobacteria excrete extracellular polymeric substances, and also by degradation of cyanobacterial biomass present (Gonçalves 2021). Soil improvement also resulted in an increased grain yield with improved nutritional composition.

Additionally, cyanobacteria produce metabolites like phenolic compounds, terpenoids, free fatty acids (FFAs), polysaccharides, carotenoids and phytohormones promoting plant growth that can be used as biofertilizers and biopesticides. Phenolic aromatic hydrocarbon compounds can be broadly classified into: phenols (can be phenolic acids, flavonoids and phenylpropanoids) and polyphenols (Singh et al. 2017; Pan et al. 2019). Phenolic compounds with antioxidant properties have an important role in growth, reproduction and protection against various stresses such as biotic and abiotic stress conditions (Esquivel-Hernández et al. 2017). Another metabolite, terpenoids are organic compounds that possess antioxidant, antibacterial, and anticancer properties and in agriculture used as biopesticides for crops' protection against bacteria, insects and other pathogenic organisms and to enhance plant growth (Pavela and Benelli 2016; Rodriguez-Garcia et al. 2017; Awasthi et al. 2018; Betterle and Melis 2019).

Cyanobacterial free fatty acids (FFAs) are non-esterified unsaturated and saturated short, medium, long and very-long-chain fatty acids. FFAs possess antioxidant, antifungal, antimicrobial, antiviral, antibiotic, and anticarcinogenic activities, and so are potential candidates to provide crop protection in agriculture fields (Desbois and Smith 2010; Singh et al. 2017). Carotenoids are light-harvesting pigments protein assembly in photosystems in cyanobacteria and provide protection against photo-induced free radicals exposure. Carotenoids increase the availability of provitamin A, lead to the development of biofortified crops (Guedes et al. 2011; Sakamoto et al. 2017). Polysaccharides secreted by cyanobacteria–are used as carbon and energy sources under both normal and stress conditions. Polysaccharides act as soil conditioners, nutrient carriers and can improve the soil quality (Guilherme et al. 2015; Chanda et al. 2019; Gonçalves 2021). Different genera of cyanobacteria including *Nostoc*, *Aphanothece*, *Scytonema*, and *Arthrospira* have been reported to produce a variety of extracellular polysaccharides (arabinoxylans, cellulose, chitin and pectins) and have structural functions in the cell walls of different organisms and exhibit antibacterial, anticancer, anticoagulant, anti-inflammatory and antioxidant properties (Gonçalves 2021).

## Reclaiming Usar (Unproductive) Soils

Exposure to extreme environmental conditions such as very cold and hot temperatures, drought and high salinity conditions can affect crop production and development. Extreme-salt conditions and drought create osmotic stress in plants and affect the nutritional composition of plants, so affects plants' metabolism and growth (Górka et al. 2018; Bulgari et al. 2019). The algal-based products as biostimulants are applied in food crops under temperature, salt and drought stress conditions that promote the accumulation of antioxidant compounds, thus increasing the tolerance to stress conditions; and stimulating the overall growth of plants (Du Jardin 2015; Bulgari et al. 2019; Gonçalves 2021).

Cyanobacteria have a great role to play in the reclamation of damaged and affected soils by modulating pH, controlling salinity and removing heavy metals contaminants thereby improving quality and fertility (Prasanna et al. 2014; Priya et al. 2014; Kumar et al. 2015; Chatterjee et al. 2017). Cyanobacteria can also do physical and chemical amendments in the soil to improve soil quality and fertility. Moreover, salt-affected soils are rigid soils impermeable to water due to the presence of excessive salts in the upper layers, high pH, high exchangeable Na, measurable amounts of carbonates, and it undergoes extensive clay dispersion. Therefore, reduced soil aeration and poor hydraulic conductivity make the soil less productive and infertile. Cyanobacteria aid in conserving organic C, N, and P as well as moisture, and convert the $Na^+$ clay to $Ca^{2+}$ clay by forming a thick stratum on the soil surface and therefore help in binding the soil particles (Singh et al. 2016).

Cyanobacteria can aid in maintaining soil pH by reducing the soils' salinity and increasing the water retention capacity of soils or by producing gibberellic acid to increase tolerance of plants to high salinity levels (Gonçalves 2021). The exopolysaccharides (EPS) produced by cyanobacteria may aid in the aggregation of soil particles, organic matter accumulation, porosity, permeability and increase the water-holding capacity of the upper layer of soil that further helps in maintaining soils' temperature, pH and salinity and thus avoids soil erosion (Nisha et al. 2007; Chatzissavvidis and Therios 2014; Baweja et al. 2019). By restoring the soil properties, cyanobacteria have been widely used in improving soils' fertility, boosting soil microbial activity, stimulating root growth and significantly enhancing crops' productivity in cost-effective manner maintaining soil sustainability.

## Bioremediation

Cyanobacteria are self-sufficient for growth and are able to survive even in diverse polluted environments. Cyanobacteria pose a great potential either through their accumulation or degradation to treat various types of environmental contaminants namely crude oil, naphthalene, pesticides, phenol and catechol, heavy metals, phenanthrene, and xenobiotics. Also, due to high metal sorption capacity and high multiplication rate cyanobacteria can aid in the detoxification of various industrial effluents namely sugar mills, dye and pharmaceuticals industries, oil refineries, brewery distilleries and paper mill industries (Singh et al. 2016). Cyanobacteria may be used in agro-industrial effluents and urban treatment, and thereby can help in mitigating metal toxicity problems and eutrophication in aquatic ecosystems. Some cyanobacterial species offer additional advantages like higher interior pH that prevents transfer of pollutants in water and thus helps in the removal of heavy metals from the wastewaters. Due to the low cost, cyanobacteria have been used as bioremediating agents for the treatment of dairy wastewaters rich in nitrogen and phosphorous and convert them into biomass cost (Singh et al. 2016).

Cyanobacteria also can accumulate pesticides in very high concentrations. *Synechococcus elongatus*, *Anacystis nidulans*, and *Microcystis aeruginosa* may degrade many organo-phosphorus and organochlorine insecticides from the polluted aquatic systems (Vijayakumar 2012). Cyanobacteria such as *Anabaena* sp., *Microcystis* sp., and *Nostoc* sp. degrade the broad range organo-phosphorous herbicide glyphosate, and the mineralized glyphosate (Forlani et al. 2008). However, *Spirulina* sp. could degrade the glyphosate herbicide

(Lipok et al. 2007) and *Synechocystis* sp. can mineralize the aniliofos herbicides. Cyanobacteria in wastewater lagoons thus may have great potential to degrade the pollutants and pesticides that could help in supporting the growth of other microbial populations.

According to the previous studies, the naturally occurring cyanobacterial–bacterial associations can be used to remediate oil-contaminated soils successfully. *Plectonema* sp., *Oscillatoria* sp., and *Synechococcus* sp., have been successfully used in the bioremediation of oil spills in different parts of the world. *Microcoleus chthonoplastes* and *Phormidium corium* were isolated and reported to degrade *n*-alkanes from Arabian Gulf oil-rich sediments, *Oscillatoria* sp. and *Agmenellum* sp. oxidize naphthalene to 1-naphthol, *Agmenellum* sp. metabolizes phenanthrene and *Oscillatoria* sp. oxidize biphenyl to 4-hydroxybiphenyl (Singh et al. 2016). Thus above studies indicate, that cyanobacteria play a major role in detoxifying various pollutant sites *via* simple means leading towards their selection in vast treatment sites.

## Biocontrol Agents to Combat Biotic Stresses

Though plants have their defense and resistance mechanisms against various stresses, external bioactive compounds from cyanobacteria can be used for crop protection. Pathogenic factors that can infect plant and impact crop productivity include insects, nematodes, bacteria and fungi (Chanda et al. 2019). These cyanobacterial bioactive compounds possess antifungal, antibacterial, antiviral, and antialgal properties. These bioactive compounds include fatty acids, amides, polyketides, lipopeptides, alkaloids, and indoles. The cyanobacteria can reduce infection of *Botrytis cinerea* on strawberries and *Erysiphe polygoni* causing powdery mildew on turnips and damping off disease in tomato seedlings, along with can reduce growth of plant pathogens such as *Rhizoctonia solani* and *Sclerotinia sclerotiorum.* Other several researchers reported antifungal properties of cyanobacterial compounds against several plant pathogenic fungi such as brown rust, powdery mildew, rice blast, but not much effective against brown rot and stem break (Singh et al. 2016).

Among cyanobacteria, *Nostoc* sp. is the source of natural pesticides (produce cryptophycin) against fungi, insects, and nematodes (Biondi et al. 2004). *Nostoc muscorum* has shown antifungal properties against soil fungi such as *S. sclerotiorum* (cottony rot of vegetables and flowers), and *Rhizoctonia solani* (root and stem rots). *N. muscorum* also inhibited the growth of certain dark-colour fungi such as *Aureobasidium, Cladosporium, Alternaria* (Zulpa et al. 2003). Polysaccharides recognize signaling molecules present in the cell wall of pathogenic organisms and induce several defense responses (Pardee et al. 2004; Singh et al. 2016; Chanda et al. 2019). Their promising role as biocontrol agents has attracted several researchers to explore their potential compounds effective against several plant diseases causing loss.

## Cyanobacterial Genes in Transgenics

However, crop improvement using cyanobacterial transgenics has also been used in wheat and rice for improving their photosynthesis and water-use efficiency (WUE) as compared with C3 and C4 plants, the autotrophic cyanobacteria are more efficient in performing photosynthesis. Therefore, it was observed through the genetic approach by introducing two cyanobacterial bicarbonate transporters genes- *bicA* and *sbtA*, into C3 plant chloroplasts improved photosynthesis by 28% (Price et al. 2013). Additionally, cyanobacterial fructose-1,6-biphosphatase and sedoheptulose-1,7-bisphosphatase showed enhanced photosynthetic efficiency and growth characteristics (Tamoi et al. 2006). Another report suggested that the specific genes synthesizing the osmoregulators can be integrated to higher plants such as Δ12-acyl-lipid desaturase gene from the cyanobacterium *Synechocystis* sp. PCC 6803 was inserted into potato leaf cells. This integration of the desaturase gene stabilized the physical properties and composition of biomembranes by promoting polyunsaturation of fatty acids and these changes in the membrane confer cold resistance to potato plants (Demin et al. 2013). Therefore, research studies are in progress, and many trials are going on to develop genetically engineered plants utilizing cyanobacterial genes that are healthy, robust and could tolerate extreme environments (Jangir et al. 2015).

## Bioenergy

Cyanobacteria with minimum nutrient requirements have the ability to produce bio-energy such as bio- or syngas, bio-hydrogen and bio-diesel (Singh et al. 2016). During photosynthesis, carbon dioxide ($CO_2$) taken up by cyanobacteria is converted to carbon-rich lipids used for biofuels production. Additionally, cyanobacteria produce molecular hydrogen ($H_2$) is an ideal substitute for fossil fuels. The molecular $H_2$ is one of the most abundant elements in the universe, with a maximum energy per unit weight (122 KJ $g^{-1}$), is also a clean fuel that will not cause environmental pollution because its only by-product is water. Also, the molecular $H_2$ has greater energy conversion efficiency than petroleum and can be stored as gas-metal hydride or as a liquid. Several cyanobacterial genera have the ability to produce $H_2$ under various culture conditions namely *Gloeobacter, Anabaena, Nostoc, Calothrix, Oscillatoria, Microcystis,* and *Microcoleus* (Parmar et al. 2011; Nozzi et al. 2013). Through anaerobic digestion or fermentation, the cyanobacterial biomass can also be used to produce biogas. In cyanobacterial biomass, the organic biopolymers are hydrolyzed into monomers, and anaerobic digestion produces biogas i.e., a mixture of $CH_4$ and $CO_2$, where the second main component $CO_2$ (approximately 25–50%) is removed to obtain bio-methane and the calorific value of biogas can be significantly enhanced (Hankamer et al. 2007). Bio-methane can be compressed as natural gas in vehicles and can be more environmentally friendly than gasoline/petrol and diesel. It has been reported that the estimation of carbon utilization efficiency of cyanobacterial biomass is to an extent of almost 95% (Converti et al. 2009; Singh et al. 2016).

At large the beneficial effects of cyanobacteria and their extracts (interest metabolites) of agricultural use have been demonstrated, however, the autotrophic growth of cyanobacteria can be expensive and energy-demanding, though the biomass productivities achieved are low (Renuka et al. 2018). The costs can be higher than the actual price of chemical-based compounds. Another concern is the sustainability of the most commonly used extraction procedures of cyanobacterial extracts which disposes large amounts of organic solvents. Therefore, very little information about the interactions established between cyanobacteria (and their extracts), plants, the environment and the wide diversity of metabolites in them pose limitations in the broad applications of cyanobacterial products in agriculture. Therefore, extensive research and technology are required to conduct studies on cyanobacterial extracts, metabolites and to make use of them in agriculture is highly required. Also, cyanobacterial biodegradation potential is a cost-effective, economical and maintenance-free remediation technology for contaminated eco-systems and can be enhanced through genetic engineering. Though, regulatory constraints and ecological concerns are the major obstacles in releasing the transgenic cyanobacteria for bioremediation purposes under field conditions.

**Table 1.** Important bioactive compounds from different cyanobacterial species with their use in agriculture

| Sl. No | Properties | Compound Name | Cyanobacterial Species/Sources | Types of Activity | References |
|---|---|---|---|---|---|
| 1. | Sustainable Agriculture | Nitrogenase | *Trichodesmium, Calothrix, Aulosira, Anabaena sphaerica, Nostoc hatei, Cylindrospermum majus, Westiellopsis prolifica, Oscillatoria angustissima Plectonema, Tolypothrix,* | Reduce elemental nitrogen into an inorganic useable form - nitrogen fixation | Singh et al. 2016; Chittora et al. 2020; Gonclaves 2021 |
| 2. | Plant growth promoters | Indole-3-acetic acid, Gibberellins, Cytokinins, Abscisic acid | *Anabaena, Anabaenopsis, Calothrix, Chlorogloeopsis, Cylindrospermum, Nostoc, Plectonema, Synechocystis* | Growth and development, cell division and growth, response to various stress conditions | Rodríguez et al. 2006; Jangir et al. 2015; Singh et al. 2016; Goncalves 2021 |
| 3. | Enhance nutrients and metabolites availability in soils | FFAs | *Anabaena, Spirulina* | Antioxidant, antifungal, antimicrobial, antiviral, antibiotic, anticarcinogenic activities | Desbois et al. 2010; Singh et al. 2017; Goncalves 2021 |
| | | Carotenoids | *Spirulina* sp. | Provide protection against photo-induced free radicals exposure | Guedes et al. 2011; Sakamoto et al. 2017; Goncalves 2021 |
| | | Polysaccharides | *Nostoc, Aphanothece, Scytonema, Arthrospira* | Nutrient carriers and improve the soil quality | Guilherme et al. 2015; Chanda et al. 2019, Goncalves 2021 |
| 4. | Reclaiming Usar soils | Exopolysaccharides | *Nostoc, Aphanothece, Scytonema, Arthrospira* | Aggregation of soil particles, organic matter accumulation, porosity, permeability and increase the water-holding capacity of the upper layer of soil, maintain soils' temperature, pH and salinity, prevents soil erosion | Nisha et al. 2007; Chatzissavvidis and Therios 2014; Baweja et al. 2019 |

**Table 1.** (Continued)

| Sl. No | Properties | Compound Name | Cyanobacterial Species/Sources | Types of Activity | References |
|---|---|---|---|---|---|
| 5. | Bioremediation | | *Synechococcus elongatus, Anacystis nidulans, Microcystis aeruginosa* | Degrade many organo-phosphorus and organochlorine insecticides from the polluted aquatic systems | Vijayakumar 2012 |
| | | | *Anabaena* sp., *Microcystis* sp., *Nostoc* sp. | Degrade the broad range organo-phosphorous herbicide glyphosate, and the mineralized glyphosate | Forlani et al. 2008 |
| | | Phosphorus-containing metabolites | *Spirulina* sp. | Degrade the glyphosate herbicide | Lipok et al. 2007 |
| | | | *Synechocystis* sp., *Plectonema* sp., *Oscillatoria* sp., *Synechococcus* sp., *Microcoleus chthonoplastes, Phormidium corium, Agmenellum* sp. | Mineralize the aniliofos herbicides, Bioremediation of oil spills, Degrades n-alkanes, Oxidize naphthalene to 1-naphthol, Metabolizes phenanthrene, Oxidize biphenyl to 4-hydroxybiphenyl | Singh et al. 2016 |
| 6. | Biocontrol agents to combat biotic stresses | Extracellular products | *Nostoc* sp. | Against soil fungi such as *S. sclerotio-rum* (Cottony rot of vegetables and flowers), *Rhizoctonia solani* (root and stem rots). | Zulpa et al. 2003 |
| | | Phenolic substances | *N. muscorum* | *Aureobasidium, Cladosporium, Alternaria* | Zulpa et al. 2003 |
| 7. | Bioenergy | | *Gloeobacter, Anabaena, Nostoc, Calothrix, Oscillatoria* | Produce $H_2$ under various culture conditions | Parmar et al. 2011; Nozzi et al. 2013 |

However, most of the research on engineered cyanobacteria for their potential role in bioremediation is a growing need for the regulatory, security or economical-beneficial systems.

## Role of Cyanobacteria in Medicine

### Bioactive Compounds

Advancement in microbiology has led to understanding and exploring that cyanobacteria are an extremely rich source of the secondary metabolites (Cardellina and Moore 2010). These secondary metabolites consist of several bioactive compounds including antitumor, antimicrobial, cytotoxic, and other compounds including herbicides, antimycotics, antimalarials, multidrug resistance reversers, algaecides, insecticides, and immunosuppressive agents (Burja et al. 2001). The natural molecules not only functions directly as drugs, but can also be used as prototypes for the synthesis of new drugs. Therefore, the compounds produced by the cyanobacteria continue to be explored for application in various biological areas and can also be an exceptional source of molecules for drug discovery and possible cure to many trivial diseases (Singh et al. 2005, Bajpai et al. 2010; Nunnery et al. 2010; Singh et al. 2011; Tiwari and Tiwari 2020).

### Anticancer

Screening of cyanobacterial extracts for new anticancer compounds was initiated by Moore and Gerwick in the 1990s. Several bioactive compounds of blue-green algae possess the potential of killing the cancerous cells. Many cyanobacterial compounds like Curacin A and cryptophycins have cytotoxic activity *via* an interaction with the microtubule-binding proteins, that arrest the cell division and hence induce apoptosis (Tiwari and Tiwari 2020; Qamar et al. 2021). Curacin A isolated from *Lyngbya majuscula*, acts as an inhibitor of cell growth and mitosis activity by interacting with tubulin (Tiwari and Tiwari 2020; Qamar et al. 2021). Cryptophycin was first isolated from the *Nostoc sp.* and is an effective fungicide. Cryptophycin from *Nostoc* sp. GSV 224 showed anti-cancer activity against the KB human nasopharyngeal cells and LoVo human colorectal cancer cells (Yasin et al. 2019). Cryptophycin molecule acts as a tubulin- destabilizing agent and therefore arrests the tumor cells into G-M phase of the cell cycle, that results in apoptotic signaling and hyper-phosphorylation of Bcl-2 (Drew et al. 2002; Dias et al. 2015; Tiwari and Tiwari 2020). This compound was found to be extremely toxic and is also disregarded as a natural product, and exhibits strong cytotoxicity against tumor cell lines in humans. By attacking the microfilaments found in the eukaryotic cell, this molecule prevents the cell division and reproduction. It has a recognizing property which makes it recognize the cancerous tumor cells and even those of the solid tumors such as colon, brain, pancreas, prostate, breast, and lung cancers. It was found to be effective in demolishing the multidrug resistant tumor cells. Its mechanism of cytotoxicity is tubulin-interaction, with disordered tubulin-dynamics, leading to the apoptosis of tumor cells (Panda et al. 1998). A cyanobacterium, *Symploca* sp. possesses a metabolite named Dolastatin-10, which is a strong anti-proliferative agent and also induces apoptosis (Khalifa et al. 2021). The boron-containing metabolite, Borophycin, produced by the marine *Nostoc spongiaeforme* var. *tenue* shows a strong cytotoxic effect in human carcinoma and human colorectal adenocarcinoma cell lines and also has been found to exhibit the antimicrobial activity (Gupta et al. 2012; Burja et al. 2001). An analogue of apratoxin i.e., apratoxin D (apratoxin family of cytotoxins isolated from *Lyngbya* sp.) derived from *Lyngbya majuscula* is cytotoxic to the human lung cancer cells and arrests apoptosis by inducing the G1-phase cell cycle (Gutierrez et al. 2008; Thornburg et al. 2013; Masuda et al. 2014). In particular, apratoxin A has also been found to be a potent cytotoxin for cancer cell. However, there is no definite understanding of the mechanism of how this cytotoxin is effective against tumor. Another molecule, symplocamide A from *Symploca* sp. exhibits strong cytotoxic activity against the neuroblastoma and lung cancer cells (Linington et al. 2008; Tan 2010). The indolophenanthridine from the *Calothrix* sp. i.e., calothrixin A also induces the cell cycle arrest in the G2/M phase in the human cancer cell lines (Chen 2003; Tiwari and Tiwari 2020; Qamar et al. 2021). An understanding and consistent study of these anti cancerous compounds from various cyanobacterial species could be a boon to cancer treatment lines. Therefore, role of cyanobacteria in cancer treatment is an advancement in phycology and medical sector which still needs more attention and further research.

## Antidiabetic

After cancer and other cardiovascular diseases, diabetes, is a major threat to the world population that affects human health in various ways (Senthil kumar and John 2008). In this metabolic disorder due to insulin deficiency, the prolonged rise in the blood sugar level causes retinopathy, nephropathy, hypertension, and cardiomyopathy in patients (Senthilkumar and John 2008; Pandurangan and Kim 2016). Several active cyanobacterial metabolites have been studied in terms of their application in treating diabetes in different ways (Layam and Reddy 2006; Pandurangan and Kim 2016; Priatni et al. 2016). The various phytochemicals such as phytopigments, flavonoids and sterols produced by cyanobacteria like *Arthrospira pltatensis,* showed strong hypoglycaemic effects that can reduce the blood sugar, Serum Glutamic Pyruvic Transaminase (SGPT), Serum Glutamic Oxaloacetate Transaminase (SGOT), and Alkaline phosphate (ALP) levels (Senthilkumar and John 2008). The renormalization of the glucose level, insulin, C-peptide, lipid peroxidation and glycolytic enzymes were attained after the cyanobacterium *Spirulina platensis* administration in diabetic rats was tested in trials. The administration of cyanobacterium leads to regaining body weight in diabetic rats and also reversed the hepatic damage by renormalizing of serum hepatic marker enzymes (Pandurangan and Kim 2016). The cyanobacterial extracts from *Lyngbya*, *Microcoleus,* and *Synechocystis* sp. containing C-phycoery-thrin (C-PE), C-phycocyanin (C-PC), lycopene, and myxoxanthophyll were investigated to inhibit carbohydrate hydrolysis through enzymatic activity thus helping in the successful management of post-prandial hyperglycemia. Moreover, *in vitro* studies also showed the antihyperglycemic effects of cyanobacterial pigments such as C-PE, C-PC, myxoxanthophyll and lycopene due to the inhibition of α glucosidase by these cyanobacterial pigments (Ghosh et al. 2016). This study elaborated the role and antioxidative properties of various pigments and compounds abundantly present in cyanobacteria which could be trusted in treatment of diabetes.

## Anti-Inflammatory and Immunomodulatory Effects

Cyanobacteria consist of enormous carotenoids (lycopene, β-carotene, and lutein) that have antioxidant properties. The carotenoid also possesses anti-inflammatory properties *via* the quenching action on reactive oxygen species.

The anti-inflammatory activity is also due to a photo-harvesting pigment, phycocyanin that inhibits the formation of leukotriene (an inflammatory metabolite of arachidonic acid) (Romay et al. 1999). Similarly, C-phycocyanin, which is a free radicle scavenger is reported to possess a significant hepatoprotective activity in past research studies (Vadiraja et al. 1998; Bhat and Madyastha 2000). Several different studies suggested the anti-inflammatory and immunomodulatory activities associated with *Spirulina* and other genera of cyanobacteria (Abdel-Daim et al. 2013; Ou et al. 2013; Finamore et al. 2017). The report has shown the immunomodulatory and anti-inflammatory effects of *Spirulina* in acetic acid-induced ulcerative colitis rats (Abdel-Daim et al. 2013). Further, another study indicated the active components of *Spirulina* like β-carotene and phycocyanin and their regulatory effects on JNK, ERK1/2, IkB, and p38 signaling pathways resulted in anti-inflammatory and immunomodulatory responses. Several other products prepared from *Spirulina* shows the property of influencing the immune systems in different ways like stimulating antibody, increasing the phagocytic activity of macrophages and activating the T- and B- cells. The supplementation of *Arthospira platensis* powder to mice reduces tumor necrosis factor-alpha (TNF- α) in serum and therefore alters the expression of some other oxidative stress markers was reported (Abdel-Daim et al. 2013). Also, oral doses of *Aphanizomenon flos-aquae* in healthy humans can result in a slight reduction of phagocytic activity of polymorph nucleated cells (Jensen et al. 2000). In earlier studies, it was reported that when the immunized mice were fed with the algae supplemented diet, they showed a surge in the numbers of splenic IgM antibody-producing cells. During the secondary immune response, the antibody production was hardly affected therefore this discovery is only true for the primary immune response (Hayashi et al. 1994). It has also been reported that when the cyanobacterial extract was injected intraperitoneally an hour prior to the allergic challenge, results in decreased the mortality induced by the anaphylactic compound (Yang et al. 1997; Kim et al. 1998). Along with all the mentioned immunomodulatory effects, a neurotoxin, kalkitoxin from marine cyanobacterium *Lyngbya majuscule*, blocks the sodium channels hence preventing the nerves from firing off their electrical signals. It can also treat disorders like neurodegenerative by selectively activating and blocking the sodium channels. Hence it is a valuable tool and a useful pharmaceutical molecule to understand the working of the sodium channels and the effect of diseases on them (Wu et al. 2000). The above reports strongly demonstrated the role of cyanobacteria as anti-inflammatory and immunomodulatory sources and highlighted the functions in treating various related diseases.

## Antimicrobial Properties

### Antibacterial

Recently, multidrug-resistant bacteria that cause nosocomial infections like vancomycin-resistant *Enterococci*, Ampicillin C β-lactamase producing Enterobacteriaceae, and methicillin-resistant *Staphylococcus aureus* gave rise to the therapeutic challenges which are of high concern worldwide (Reinert et al. 2007). Scientists have screened and identified extracts of cyanobacteria for their antibacterial activity in order to develop and counter these drugs naturally resisting bacterial strains (Kreitlow et al. 1999; Skulberg 2000; Biondi et al. 2008). A compound from *Nostoc*, Nostocarboline inhibits the growth and development of green algae and other cyanobacteria (Blom et al. 2006). Another compound, hapalindole (alkaloids) was isolated from *Fischerella* sp. and *Nostoc* CCC537 having antimicrobial activity against *Salmonella typhi* MTCC3216, *Mycobacterium tuberculosis* H37Rv, *E. coli* ATCC25992, *Pseudomonas aeruginosa* ATCC27853, *Enterobacter aerogenes* MTCC2822, and *Staphylococcus aureus* ATCC25923 has been reported (Asthana et al. 2009). Further, intense research is required to understand and investigate compounds and metabolites from cyanobacteria and also to explore their response mechanism as antibacterial therapeutics. As the resistance of bacteria against antibiotics is increasing day by day, there is an immense need to search for new bioactive compounds possessing antibacterial activity and for that purpose cyanobacteria will be potentially a promising candidate.

### Antiviral

In many past studies, antiviral substances isolated from cyanobacteria generally possess bioactivity *via* blocking the viral penetration or absorption and by penetrating into the cells they inhibit the replication stages of progeny viruses (Luescher-Mattli 2003). Several past and recent studies have also demonstrated anti-HIV activity in many bioactive compounds as antivirals from *Phormidium tenue* and *Lyngbya lagerheimii* and the extracts of *Spirulina* possess the potential of inhibiting the replication of viruses as - measles, HIV-1 and *Herpes simplex* virus type 2 (HSV-2) (Schaeffer and Krylov 2000; Jha and Zi-Rong 2004). Cyanovirin-N is 101 amino-acid-long polypeptides, discovered and isolated from *Nostoc ellipsosporum.* It shows efficient *in vitro* and *in vivo* activities against HIV and various other lentiviruses and also inhibits the herpes simplex virus-6 *in vitro*. Cyanovirin-N is virucidal and recombinant *Escherichia coli* can also produce CV-N and purification results in the monomeric protein (Burja et al. 2001; Colleluori 2005; Gondim et al. 2019; Carpine and Sieber 2021). Many species of cyanobacteria release acidic polysaccharides, like calcium spirulan (Ca-SP) from *Arthrospira platensis,* act as a strong inhibitor of *Herpes simplex* virus (Hayashi et al. 1996; Kanekiyo et al. 2005). The polysaccharides, Ca-SP shows the potent and broad-spectrum effect against many viruses and inhibits from replicating and exhibits the antiviral activity against human immunodeficiency virus type 1 (HIV-1), HIV-2, influenza virus, mumps virus, measles virus, HSV-1, and human cytomegalovirus (HVMV) and a series of other enveloped viruses (Carpine and Sieber 2021). These sulphated polysaccharides prevent the reverse transcriptase activity against HIV-1 (like azidothymidine) and also inhibit virus-cell attachment and fusion with host cells. In a mechanism that greatly enhances the viral infection, these compounds also inhibit the fusion between HIV-uninfected CD4+ lymphocytes and HIV- infected lymphocytes. Moreover, four other cyanobacterial species such as *Chroococcus turgidus, Anabaena sphaerica, Arthrospira platensis, and Oscillatoria limnetica* showed antiviral activity against adenovirus type 40 (Abdo et al. 2012). Some lectins, AML, MEL and SfL were also studied from algae including cyanobacteria *Oscillatoria agardhii* (OAA) which inhibited HIV infection through binding to high mannose (HM) oligosaccharides of enveloped glycoprotein group and thereby influenced the anti-HIV activity (Gondim et al. 2019). The research on antiviral cyanobacterial compounds requires extensive efforts to commercialize them as medical supplies.

## Anti-SARS-CoV-2

Recently, in December 2019, a highly contagious disease 'COVID-19' caused by the virus SARS-CoV-2, became pandemic within a very short period of time. For like a year or so there was a lack of measures, drugs or vaccines availability that can control this rapidly increasing, deadly disease, which mainly attacks the immune system. To manage this, various researches were initiated related to drug discovery with certain anti-viral properties. Several antiviral drugs from different microorganisms were also being used for the treatment of this deadly disease. Ca-spirulan (polysaccharides isolated from the cyanobacterium *Spirulina*), exhibits long-lasting antiviral activity; however, their role against COVID-19 remains limited. The other antiviral polysaccharide nostoflan, from *Nostoc flagelliforme* possesses antiherpes simplex virus type 1 (HSV-1) activity, which could be a potential compound for the disease as it possesses 50% cytotoxic concentration/50% inhibitory concentration against viral replication (Kanekiyo et al. 2005; Khalifa et al. 2021). Further, by utilizing *in vitro* enzymatic assay and *in silico* molecular docking screenings, it has been found that the compound phycocyanobilin (PCB) and the cyanobacterial metabolites like cyclindrospermopsin are strong phytochemical inhibitors of main protease ($M^{pro}$) and papain-like proteases ($PL^{pro}$) (proteases of Covid-19, a key CoV enzyme and an attractive drug target) of SARS-CoV-2 (Khalifa et al. 2021). In another research study, 23 metabolites including antillatoxin, curacin A and cryptophycin 52 were investigated from different cyanobacterial strains that were found effective against SARS-CoV-2 $M^{pro}$ and $PL^{pro}$ proteases (Naidoo et al. 2020). So, these metabolites and chemical compounds from cyanobacteria can be investigated further in terms of their antiviral activities and can be studied for their potential role and application in the treatment of COVID-19.

## Antiprotozoal

More than a billion of people throughout the world are suffering from tropical diseases caused by *Trypanosoma, Plasmodium, Schistosoma,* and *Leishmania* and others (Simmons et al. 2008). Due to the resistance developed by these protozoa, the treatment of diseases such as leishmaniasis and malaria seen failure (Lanzer and Rohrbach 2007; Priotto et al. 2007). The Panamanian International Co-operative Biodiversity group is screening the extracts of marine and terrestrial sources, in order to encourage the development of affordable and effective treatment for these diseases. With this respect, an alkaloid has been isolated from *Nostoc* sp., protease inhibitor nostocarboline which was found to be active against *Trypanosoma cruzi, Trypanosoma brucei, Plasmodium falciparum, and Leishmania donovani* (Barbaraus et al.

2008). Hence, the application of secondary metabolites obtained from cyanobacteria could be an effective solution for several protozoal diseases.

## Protease Inhibitors

From the toxic genera of cyanobacteria, five classes of protease have been identified, and these are micropeptins, microginins, microverdins, aerugenosins, and abaenopeptins. Few cyanopeptolins are inhibitors of serine proteases including elastases, extremely important in a number of diseases such as lung emphysema and are of great pharmaceutical value. High blood pressure can easily be treated by microginins.

## Minerals and Vitamins

Cyanobacteria are also known for being a rich source of vitamins and minerals, can act as strong antioxidants and fight disease. Also, they can be taken as food supplements surpassing other foods. Levels of calcium, potassium, iron, magnesium, and trace elements are high in *Spirulina* as compared to the other available foods (Shetty et al. 2006). *Spirulina* has been used as a potential food supplement since ancient times, was first commercially available in 1970s and in Africa, was used as a protein supplement since 1940 (Abdulqader et al. 2000). These have the strong potential of being applied in the medical fields as it is good for teeth, bones, and blood. Several other minerals are also found in *Spirulina* are copper, zinc, phosphorus, selenium, and manganese. It contains various other vitamins also, that includes B1, B2, B3, B6, A, E, H, folacin, pantothenic acid, and inositol. Being a rich source of vitamin B12, β-carotene and pantothenic acid, cyanobacteria can surpass other foods and can meet the required daily intake amount (Shetty et al. 2006; Panjiar et al. 2017). Other cyanobacteria are also being commercialized and consumed as food supplements like *Nostoc commune* in Asia, *N. flagelliforme* in China, *N. punctiforme* in China and South America, Mongolia and South America, *Aphanotheca sacrum* in Japan as food supplements (Fujishiro et al. 2004; Singh et al. 2011; Panjiar et al. 2017).

**Table 2.** Important bioactive compounds from different cyanobacterial species with their use in medicine

| Sl. No | Properties | Compound Name | Cyanobacterial Species/ Sources | Types of Activity | References |
|---|---|---|---|---|---|
| 1. | Anticancer | Curacin A | *Lyngbya majuscula* | Inhibits cell growth and mitosis activity by interacting with tubulin | Tiwari and Tiwari 2020, Qamar et al. 2021 |
| | | Cryptophycin | *Nostoc* sp. | Tubulin- destabilizing agent, arrests the tumor cells into G-M phase of the cell cycle, cause apoptotic signaling and hyper-phosphorylation of Bcl-2 | Yasin et al. 2019, Drew et al. 2002; Dias et al. 2015, Tiwari and Tiwari 2020 |
| | | Dolastatin 10 | *Symploca* sp. | anti-proliferative agent, induces apoptosis | Khalifa et al. 2021 |
| | | Borophycin | *Nostoc spongiaeforme var. tenue* | Cytotoxic effect in human carcinoma and human colorectal adenocarcinoma cell lines | Gupta et al. 2012; Burja et al. 2001 |
| | | Apratoxin-D | *Lyngbya majuscula* | Arrest apoptosis by inducing the G1- phase cell cycle | Gutierrez et al. 2008; Masuda et al. 2014; Thornburg et al. 2013 |
| | | Symplocamide A | *Symploca* sp. | Cytotoxic activity against the neuroblastoma and lung cancer cells | Tan 2010; Linington et al. 2008 |
| | | Calothrixin A | *Calothrix* sp. | Induces the cell cycle arrest in the G2/M phase in the human cancer cell lines | Chen et al. 2003, Tiwari and Tiwari 2020, Qamar et al. 2021 |
| 2. | Antidiabetic | Phytopigments, flavonoids and sterols | *Arthrospira pltatensis* | Reduce the blood sugar, Serum Glutamic Pyruvic Transaminase (SGPT), Serum Glutamic Oxaloacetate Transaminase (SGOT), and Alkaline phosphate (ALP) levels | Senthilkumar and John 2008 |
| | | | *Spirulina platensis* | Renormalize the glucose level, insulin, C-peptide, lipid peroxidation and glycolytic enzymes | Pandurangan and Kim 2016 |
| | | C-phycoerythrin (C-PE), C-phycocyanin (C-PC), lycopene, and myxoxanthophyll | *Lyngbya, Microcoleus, Synechocystis* sp. | Role in post- prandial hyperglycemia | Ghosh et al. 2016 |

**Table 2.** (Continued)

| Sl. No | Properties | Compound Name | Cyanobacterial Species/ Sources | Types of Activity | References |
|---|---|---|---|---|---|
| 3. | Anti-inflammatory and Immunomodulatory effects | C-phycocyanins, b carotene, vitamins, minerals, proteins, lipids and carbohydrates | *Spirulina platensis* | Alters the expression of some other oxidative stress markers | Abdel-Daim et al. 2013 |
| | | | *Aphanizomenon flos-aquae* | Reduce phagocytic activity of polymorph nucleated cells | Jensen et al. 2000 |
| | | kalkitoxin | *Lyngbya majuscule* | blocks the sodium channels hence preventing the nerves from firing off their electrical signals | Wu et al. 2000 |
| 4. | Antibacterial | Nostocarboline | *Nostoc* | Inhibits the growth and development of green algae and other cyanobacteria | Blom et al. 2006 |
| | | Hapalindole | *Fischerella* sp., *Nostoc* CCC537 | Antimicrobial activity against *Salmonella typhi* MTCC3216, *Mycobacterium tuberculosis* H37Rv, *E. coli* ATCC25992, *Pseudomonas aeruginosa* ATCC27853, *Enterobacter aerogenes* MTCC2822, and *Staphylococcus aureus* ATCC25923 | Asthana et al. 2009 |
| 5. | Antiviral | Cyanovirin-N | *Nostoc ellipsosporum* | *In vitro* and *in vivo* activities against HIV and various other lentiviruses, inhibits the herpes simplex virus-6 | Burja et al. 2001; Colleluori 2005; Gondim et al. 2019; Carpine and Sieber 2021 |
| | | Calcium spirulan (Ca-SP) | *Arthrospira platensis* | Inhibits replication and exhibits the antiviral activity against human immunodeficiency virus type 1 (HIV-1), HIV-2, influenza virus, mumps virus, measles virus, HSV-1, and human cytomegalovirus (HVMV) | Carpine and Sieber 2021 |
| | | Lectins | *Oscillatoria agardhii* | Inhibited HIV infection through binding to high mannose (HM) oligosaccharides of enveloped glycoprotein gp, pose the anti-HIV activity | Gondim et al. 2019 |

| Sl. No | Properties | Compound Name | Cyanobacterial Species/ Sources | Types of Activity | References |
|---|---|---|---|---|---|
| 6. | Anti-SARS-CoV-2 | Ca-spirulan | *Spirulina* | Long-lasting antiviral activity | Khalifa et al. 2021 |
| | | Nostoflan | *Nostoc flagelliforme* | Affect viral replication | Khalifa et al. 2021 |
| | | Phycocyanobilin | | Phytochemical inhibitor of main protease ($M^{pro}$) and papain-like proteases ($PL^{pro}$) | Khalifa et al. 2021 |
| | | Antillatoxin, curacin A and cryptophycin 52 | | Effective against SARS-CoV-2 $M^{pro}$ and $PL^{pro}$ proteases | Naidoo et al. 2020 |
| 7. | Antiprotozoal | Nostocarboline | *Nostoc* sp. | Active against *Trypanosoma cruzi, Trypanosoma brucei, Plasmodium falciparum, and Leishmania donovani* | Barbaraus et al. 2008 |

## Ever Beneficial Bio-Economic Aspects of Cyanobacterial Products in Medicine

The continuous increase in the global population will have a great impact on health and food resources. Therefore, to meet the increasing demands of growing world's population in terms of nutrition and healthcare, we need to discover some cost-effective alternative resources. In that case, blue-green algae can also be useful to fulfil the need of the increasing population (Mulkidjanian et al. 2006; Pisciotta et al. 2010). Though the use of allopathic medicines provides quick relief against many health-related problems, it comes out as a cause of severe side effects on other cells and ultimately on the whole human body, affecting the immune system. So, the use of natural therapy or ayurvedic treatment proved to be a useful and cost-effective way of handling and fighting health issues. It has no side effects, takes a relatively longer time but has a great potential of curing several diseases and is of low cost than the allopathic medicines. Among several medicinal species, cyanobacteria being a storehouse of bioactive and natural compounds can also be used in natural therapy and in the ayurvedic medicine system. As cyanobacteria do not require much nutrition for their growth and therefore can easily be grown with no/little residual nutrient with the high production of proteins, lipids, ~~and~~ carbohydrates etc., possess fast-growing potential with no or very less requirement of agricultural land. These features of blue-green algae make it bio-economically a good potential candidate for additional exploration towards therapeutic purposes.

## Cyanobacteria: A Potential Candidate for Green Nanotechnology

Nanobiotechnology is emerging as one of the important tools in the field of agriculture, industry as well as medicine (Dubchak et al. 2010). At present, several nanoproducts in various forms are commercially available in world markets as paints varnishes, cosmetics, medicines, textile etc. (Roco 2010). The term "nano" refers to particles having dimension in range of 1–100 nm. Nanoparticles (NPs) differ considerably from bulk materials with regard to their physical, chemical and biological properties. These variations are primarily accounted to their high surface area to volume ratio, which leads to substantial differences in their mechanical properties, optical absorption, conductivity, etc. (Nanotechnology 2008; Thakkar et al. 2010; Schmid 2011). The interest has now shifted toward the biological or green synthesis of NPs, as there are various disadvantages associated with the physical and chemical synthesis of NPs (Khanna et al. 2019). Cyanobacteria are the most preferred candidates for NPs synthesis as they possess various advantages like ability to accumulate heavy metals, and contain a number of bioactive compounds like pigments and enzymes that help in stabilizing and act as reducing agents. The cyanobacteria-based NPs also display a large number of chemical, physical and biological properties that further increases the range of their applicability. These NPs have been found to exhibit various properties like antibacterial, antifungal, antialgal, anticancer and photocatalytic activities (MubarakAli et al. 2012). Thus, the use of cyanobacteria to synthesize NPs holds a remarkable value as it is a safe and eco-friendly synthesis, involves low cost, time and energy saving approach and produces novel NPs with various shapes and sizes. Hence developing and enhancing the biosynthetic processes using cyanobacteria may results in discovery of novel biosynthetic NPs with exclusive properties that would be widely applicable and would be useful in large number of sectors (Hamida et al. 2020d). The current chapter mainly focuses on application of cyanobacteria in the field of nanobiotechnology to boost agriculture, industrial and pharmaceutical production.

## Cyanobacteria as "Biofactories" for Biosynthesis of Nanoparticles

Cyanobacteria occur in a wide range of habitats including extreme harsh conditions (Gaysina et al. 2019). Cyanobacteria are easy to culture (at ambient temperature and atmospheric pressure), display high rate of growth, possess several biocompounds inside their cells and have the ability to absorb heavy metal ions. Thus, cyanobacteria are recommended as biofactories to biosynthesize NPs either by extracellular or intracellular methods (Sharma et al. 2016). Different cyanobacterial strains are found to form distinct NPs with different properties and this includes metallic NPs (Ag-NPs and Au-NPs), metal oxide NPs (CuO NPs and ZnO NPs), and others like Ag-Au nanoalloys and semiconductor NPs (CdS NPs) (Lengke et al. 2006; MubarakAli et al. 2013; Saran et al. 2017; Ebadi et al. 2019).

## Cyanobacteria-Based Biosynthesis of Nanoparticles

The mode of biosynthesis of NPs differs from organism to organism. However, the mode of synthesis using microorganism has been categorized into two types: intracellular and extracellular synthesis (Hamida et al. 2020d).

## Intracellular Synthesis of Nanoparticles

Intracellular synthesis of NPs refers to reduction of bulk material into NPs inside the host cells and is controlled by several biological factors. It is reported that in cyanobacteria, the ions are reduced intracellularly by electrons that are produced by the electron transport system that includes photosynthetic reaction that generates energy and respiratory electron transport system due to presence of enzymes like NADH-dependent reductases and redox reactions that takes place in the cytoplasm, cell membranes and thylakoid membranes (Jeffryes et al. 2015; Sharma et al. 2016). Several studies on cyanobacteria have established that the organism's metabolic status and the growth phase are accountable for their ability to synthesize the NPs (Lengke et al. 2006).

## Extracellular Synthesis of Nanoparticles

The extracellular synthesis refers to synthesis that occurs outside the host cell. It has been found that exuded molecules like ions, enzymes, proteins, pigments and antioxidants play a major role in the biosynthesis of NPs. Various methods have been reported for the extracellular synthesis of NPs, like use of biomolecules, cell biomass extracts, and cell-free culture media (Vijayan et al. 2014; Keskin et al. 2016; Hamida et al. 2020a). In cell-free media method, the NPs are made from the culture media of microorganisms after removing the biomass *via* centrifugation. This media is then mixed with solution of precursor metal under particular conditions. Keskin et al. (2016) demonstrated that under light conditions, cell-free media of *Synechococcus* sp. was able to biosynthesize Ag-NPs. Singh et al. (2018) also showed that Ag-NPs can be synthesized extracellularly using the supernatant. Biocompounds like enzymes, antioxidants, phenolic, etc. found in the cell-free culture are reported to play a crucial role in the bioreduction process of NPs (Ashmore et al. 2018). Several experiments have established that the cell biomass filtrate can be used for synthesis of NPs and active biomoieties like enzymes and proteins present in the cells filtrate also play a significant role in biofabrication and stabilization of NPs (Hamida et al. 2020a).

There are several studies on the synthesis NPs by using biomolecules. Cyanobacteria are highly rich in different types of pigments that have many pharmaceutical uses, including the synthesis NPs. The role of C-phycocyanin for NP synthesis and its fabrication is well known. Patel et al. (2015) showed that the C-phycocyanin from cyanobacteria species (*Limnothrix* and *Spirulina* species) can be used to make Ag-NPs of various shapes and sizes. Phycoerythrin extracted from *Nostoc carneum* can be used as a reducing agent to make Ag-NPs (El-Naggar et al. 2017). MubarakAli et al. (2012) demonstrated the synthesis of cadmium sulfide (CdS) NPs using C-phycoerythrin from *Phormidium tenue* NTDM05.

Previous studies have suggested that amino acids and proteins act as reducing and stabilizing agent in the biofabrication of MNPs (metal nanoparticles) (Khalifa et al. 2016). The presence of amino, carboxyl, sulfate groups in the cyanobacterial proteins facilitates the bioreduction process that involves synthesis of very small NPs with regular particle size allocation. Besides AgNPs and Au-NPs, the extracellular synthesis of silver-gold core-shell NPs has been reported by using *Spirulina plantensis* single cell protein (Govindaraju et al. 2008, Hamida et al. 2020d).

Various studies have also indicated the role of enzymes like nitrate reductases and NADH-dependent enzymes in the bioreduction process for NPs synthesis. These enzymes act as a reducing agent in the biofabrication process of MNPs. It was found that in *Chlorella pyrenoidosa,* nitrate reductase facilitates the NADH-dependent extracellular synthesis of Au-NPs (Oza et al. 2012). Brayner et al. (2007) documented the synthesis of Ag-NPs, Au-NPs, Pd-NPs and Pt-NPs using nitrogenases in the presence of cyanobacteria.

## Applications of NPs Synthesized Using Cyanobacteria

Various studies in the past have confirmed the bioactivities of NPs synthesized using strains of cyanobacteria. These include antibacterial, antialgal, antifungal, anticancer, antioxidant, antihemolytic activities, etc. Besides this, it has other applications like nanofertilizer, nanopesticide that can help in improving agricultural production and can also be used for biolabeling and decolorizing agents.

## Antitumor Activity

Cancer is one of the major causes of death worldwide and accounts for around 10 million deaths in 2020. However, the advancement in field of nanoscience has led to development of various new methods that would help in preventing the progression of tumor and metastasis. It has been reported that magnetic iron oxide NPs have helped in improving cellular imaging, as well as the quality and speed of the diagnostic method (Park et al. 2008; Hamida et al. 2020a). Marine cyanobacteria (more than 50%) are known to be used for the extraction of bioactive compounds that are successful in killing the cancer cells *via* instigating apoptotic death. Thus nano-formulations using the diversity of NPs along with their polymerized forms could be employed as an effective strategy to develop new drugs to treat cancer (Bajpai et al. 2018). Antitumor properties were reported for Ag-NPs *Anabaena* sp. against Dalton's lymphoma and colo205 cancer cells. Ag-NPs from *Nostoc* sp. HKAR-2 showed strong antitumor activity against cell lines of MCF-7 (Singh et al. 2014; Sonker et al. 2017). In another study by El-Naggar et al. (2018), Ag-NPs synthesized from *Nostoc* sp. was found to repress the MCF-7 cancer cells proliferation. *Nostoc* Silver-NPs were found to decrease the viability of Caco-2 cell by 50%, with an IC50 (150 μg/mL) (Bin-Meferij et al. 2019). Thus, these sets of observations clearly suggest the potential applications of cyanobacterial produced NPs in the treatment of cancerous cells.

## Antibacterial Activity

There are also various reports on antibacterial activity of biosynthesized NPs. For example, Keskin et al. (2016) found that Ag-NPs from *Synechococcus* sp. possess antibacterial activity to various strains, like *Bacillus subtilis*, *Escherichia coli* and *Staphylococcus aureus*. It was reported that Ag NPs from *Anabaena* sp. displayed high antibacterial activity to multidrug-resistant bacteria (*K. pneumoniae*, *E. coli*, and *S. aureus*) (Singh et al. 2014). Patel et al. (2015) observed that NPs synthesized by cyanobacterial strain (*Anabaena* sp. 66-2, *Cylindrospermopsis* sp. USC-CRB3, *Lyngbya* sp. 15-2, *Synechococcus* sp. 145-6 and *Synechocystis* sp. 48-3) and by algal strains (*Botryococcus* sp. and *Coelastrum* sp. 143-1) showed antibacterial activity.

However, Ag-NPs by *Anabaena* sp. 66-2, *Cylindrospermopsis* sp. USC-CRB3, *Synechocystis* sp. 48-3 and *B. braunii* were found to be efficient to all the bacterial strains used in experiment. Ag-NPs produced *via* phycocyanin from *N. linckia* considerably inhibited the pathogenic bacterial strains (*E. coli, K. pneumonia, P. aeruginosa,* and *S. aureus*) (El-Naggar et al. 2017). Silver NPs from *Nostoc* sp. was found to be effective against *Ralstonia solanacearum* and *Xanthomonas campestris* (Sonker et al. 2017). D-SNPs-Silver-NPs from *Desertifilum* sp. displayed considerable antibacterial activity against clinical isolates of *Bacillus cereus, B. subtilis, Pseudomonas aeruginosa, Salmonella enterica* and *Shigella flexneri* (Hamida et al. 2020a). D-SNPs was also found to be successful against clinical isolates of *E. coli, Salmonella typhimurium, Klebsiella pneumoniae* (drug-resistant), *Streptococcus mutans*, and methicillin-resistant *S. aureus* (MRSA). MRSA however was found to respond best when treated with D-SNPs (Hamida et al. 2020b). Silver-NPs produced from *Nostoc* sp. (N-SNPs) was examined against *K. pneumoniae* (drug-resistant). N-SNPs lead to decline of the genes (hns, hcp-1, VgrG-1, and VgrG-3) that induces the secretion system (Hamida et al. 2020c). Thus, these sets of results clearly indicate the antibacterial properties of NPs produced from cyanobacteria.

## Antialgal Activity

El-Sheekh et al. (2014) examined the suppressive activity of Ag-NPs produced from *Spirulina platensis*, *Chlorella vulgaris*, and *Scenedesmus obliquus* against *Microcystis aeruginosa*. *M. aeruginosa* is known to secrete toxins that cause liver cancer and the results obtained suggested the role of Ag-NPs as an antialgal agent against *M. aeruginosa* and other toxin producing algal species. In another study, El-Sheekh et al. (2020) synthesized Ag-NPs from cyanobacterium *Nostoc careneum* and macroalgae (*Caulerpa racemsoa*, *Jania rubens*, *Sargassum cinereum*, and *Turbinaria decurrens*) and further analyzed the efficiency of synthesized Ag-NPs against the cyanobacterium *M. aeruginosa.* It was observed that there was remarkable decline in growth rates and photosynthetic pigments of *M. aeruginosa.* The inhibition of the growth of *M. aeruginosa* was found to increase with increasing concentration of Ag-NPs and the incubation time.

### Antifungal Activity

Ag-NPs produced by *Nostoc* sp. HKAR-2 were found to inhibit the *Aspergillus niger* and *Trichoderma harzianum* (Sonker et al. 2017). Morsy et al. (2014) also reported the antifungal property of Ag-NPs produced by EPSs of *N. commune* against various strains of fungi that infect seeds of maize, sesame, sorghum and soybean. It was observed that there was total suppression of *Fusarium nygamai* on seeds of sorghum and soyabean when sterilized with Ag-NP when compared with sterilization of total *Fusarium* (14.63% and 17.24%) and total fungi (7.5% and 8.7%) with NaOCl.

### Antioxidative Activity

Previous studies have also reported the antioxidant properties of cyanobacterial NPs. For example, Singh et al. (2014) described simple method for making of ZnONPs-(zinc oxide nanoparticles) using *Anabaena* strain L31 cell extract and also conjugated the synthesized ZnONPs with shinorine (potential natural sunscreen compound from cyanobacterium). It was found that ZnO NP–shinorine conjugates exhibited antioxidative activity and it could be used in making effective sunscreen filters. Au-NPs from *Phormidium* sp. displayed antioxidant properties (MubarakAli et al. 2013).

## Application and Significance of Biosynthesized Nanoparticles in Agriculture

The agricultural production is limited by various factors including both biotic and abiotic stresses as well as environmental problems associated with the chemical inputs like fertilizers and pesticides. Therefore, there is need to adopt and replace the traditional agricultural system with new technologies. The application of nanotechnology, particularly green technology holds great potential in enhancing the productivity as well as overcoming the environmental degradation associated with chemical fertilizers and pesticides (Bahrulolum et al. 2021). Advancement in nanotechnology has tremendously revolutionized the agricultural sector that includes its application in form of pesticides, fertilizers, biosensors, remediation, food safety, food packaging, and many more (Gupta et al. 2022).

**Table 3.** Important bioactive compounds from different cyanobacterial species with their use in nanotechnology

| Sl. No | Properties | Compound Name | Cyanobacterial Species/ Sources | Types of Activity | References |
|---|---|---|---|---|---|
| 1. | Antitumor Activity | Ag-NPs | *Anabaena* sp. | Against Dalton's lymphoma and colo205 cancer cells | Singh et al. 2014 |
| | | | *Nostoc* sp. HKAR-2 | Repress the MCF-7 cancer cells proliferation | Sonker et al. 2017 |
| | | | *Nostoc* sp. | Decreases Caco-2 cell viability | Bin-Meferij et al. 2019 |
| 2. | Antibacterial Activity | Ag-NPs | *Synechococcus* sp. | Against various strains, like *Bacillus subtilis, Escherichia coli* and *Staphylococcus aureus* | Keskin et al. 2016 |
| | | | *Anabaena* sp. | Against *K. pneumoniae*, *E. coli*, and *S. aureus* | Singh et al. 2014 |
| | | | *Anabaena* sp. 66-2, *Cylindrospermopsis* sp. USC-CRB3, *Synechocystis* sp. 48-3 | Effective against all bacterial strains used in experiment | Patel et al. 2015 |
| | | | *N. linckia* | Inhibited *E. coli*, *K. pneumonia, P. aeruginosa,* and *S. aureus* | El-Naggar et al. 2017 |
| | | | *Nostoc* sp. | *Ralstonia solanacearum* and *Xanthomonas campestris* | Sonker et al. 2017 |
| | | | *Nostoc* sp. | Against *K. pneumoniae* It leads to decline of the genes (hns, hcp-1, VgrG-1, and VgrG-3) | Hamida et al. 2020c |
| 3. | Antifungal Activity | Ag-NPs | *Nostoc* sp. HKAR-2 | Inhibited the *Aspergillus niger* and *Trichoderma harzianum* | Sonker et al. 2017 |
| | | | *N. commune* | Effective against various strains of fungi that infect seeds of maize, sesame, sorghum and soybean. Complete suppression of Fusarium nygamai on seeds of sorghum and soyabean | Morsy et al. 2014 |
| 4. | Antialgal Activity | Ag-NPs | *Nostoc carneum* | Decreases in growth rates and photosynthetic pigments of *Microcystis aeruginosa* | El-Sheekh et al. 2020 |
| | | | *Spirulina platensis* | Against *Microcystis aeruginosa* | El-Sheekh et al. 2014 |

**Table 3.** (Continued)

| Sl. No | Properties | Compound Name | Cyanobacterial Species/ Sources | Types of Activity | References |
|---|---|---|---|---|---|
| 5. | Antioxidative Activity | ZnONPs | *Anabaena* strain L31 | Antioxidative activity and it could be used in making effective sunscreen filters | Singh et al. 2014 |
| 6. | Photocatalytic Activity | Ag-NPs | *Synechococcus* sp. | Photodegradation of methylene blue | Keskin et al. 2016 |
| | | | *Microchaete* sp. NCCU-342 | Dye decolorization ability | Husain et al. 2019 |
| 7. | Anti-myocardial infarction activities | Au-NPs | *L. majuscule* | Inhibit the isoproterenol-induced changes in serum cardiac injury markers, ECG, arterial pressure indices, and antioxidant capabilities of the heart | Bakir et al. 2018 |
| | Antihemolytic activity | Ag-NPs | *Nostoc* sp. | Ag- NPs inhibit hemolytic activity by 96.4% | El-Naggar et al. 2018 |
| 8. | Biolabeling | C-phycoerythrin CdS nanoparticles | *Phormidium tenue* NTDM05 | Ultrasensitive detection of DNA | Mubarak Ali et al. 2012 |
| | | C-phycoerythrin-CdS NPs and Au-NP | *Phormidium willei* and *Coelastrella* | the biogenic nanoparticles were actively bound to DNA in increased concentration | Mubarak Ali et al. 2013 |
| 9. | Agriculture | SeNPs | *Gloeocapsa gelatinosa* NCCU -430 | Improves the stress induced by PQ | Fatma et al. 2022 |
| | | Ag-NPs | *Leptolyngbya* sp. WUC 59 | Enhance seed germination and early development of wheat (*Triticum aestivum* L.) seedling at low concentration | Singh et al. 2020 |
| | | Nanosilicon particles (silicon and titanium nanoparticles) | *Oscillatoria agardhii* | Can be applied as biofertilizers and biocontrol agents. It enhanced the resistance of wheat plant to biotic stress like diseases, environmental stress, soil and water irrigation | Haggag et al. 2018 |

MtNPs (metal nanoparticles) *via* green-synthesis are of immense importance in agriculture to augment agricultural productivity. These MtNPs are commonly used for synthesizing various products related to agriculture like nanofertilizers, nanofungicides, nanopesticides and nanobiosensors that can enhance the quality as well as the yield, minimize the pollution and damages associated with it on the agriculture~~al~~ and environment (Usman et al. 2020). The invent of nanobiosensors, a modification of biosensor is also gaining a lot of importance as they provide fast, precise and specific detection of plant pathogens and toxins. The nanobiosensors include enzymatic biosensors, immunosensors, aptasensors and genosensors (Mocan et al. 2017). Nanopesticides are emerging as one of nanobiotechnological advancements that are used to encapsulate pesticides to control its release, enhance its selectivity and stability (Usman et al. 2020). This provides numerous advantages like improved efficiency, longevity and decreased amount of active ingredient. It has been found that these nano-formulations of pesticides with MtNPs display considerable effect against pathogens, insects and other pests that attack the plants and affect their growth and yield (Shang et al. 2019). The application of nanofungicides has been found to be efficient to fungal pathogens. MtNPs are very commonly used in nanofungicides formulation and it provides precise delivery and high bioavailability because of high permeability and solubility, low dose, and their restricted release (Haq et al. 2019). The nutrient uptake is enhanced by ~~be~~ encapsulating nanofertilizer that reduces the loss of nutrients, promotes growth as well as improves the quality of crops. It has several advantages as nano-formulations help in gradual and restricted release of nutrients to the target sites. This prevents the interaction of nutrients with soil, air, water, and microorganisms and thereby reducing the environmental degradation (Bahrulolum et al. 2021)

The use of microalgae for synthesis of the MtNPs, is an emerging approach and is referred to as "phyco-nanotechnology" (Chaudhary et al. 2020). Though the synthesis of MtNPs by cyanobacteria takes place in various forms however it is known that these NPs can be easily synthesized from cyanobacterial biomass *via* intracellular and extracellular methods. Many species of cyanobacteria have been employed for making MtNPs by using their extracted biomolecules (Govindaraju et al. 2008). The synthesis of MtNPs using cyanobacteria offers several benefits like it is cost effective and can be synthesized at low temperatures with advanced energy efficiency, low toxicity and lowers the risk of environmental degradation (Agarwal et al. 2019). To produce Au-NPs, the algal biomass is initially lyophilized and is followed by RP-HPLC (reverse-phase-high performance liquid chromatography) that is performed to purify GSP (gold-shaped protein) that influences the shape of the NPs. This protein is then treated with solution of $HAuCl_4$ to produce NPs of varying shapes. For synthesis of AgNPs, low and high weight proteins (PLW and PHW) in algal biomass accounts for reducing of silver ions to its metal. AgNPs can be synthesized by using cell-free supernatants of cyanobacterium cell lysates (Govindaraju et al. 2008). Filamentous cyanobacteria can synthesize AuNPs structures of varying shapes (spherical, cubic, and octagonal) using $Au^{+}$-$S_2O^{-2}{}_3$ and $Au^{3+}$-NaCl complexes (Lengke et al. 2006). It was demonstrated by Haggag et al. (2018) that nanosilicon particles (silicon and titanium nanoparticles) derived from ~~of~~ *Oscillatoria agardhii* enhanced the resistance of wheat plant to biotic stress like diseases, environmental stress, soil and water irrigation. It was observed to improve the quality, reduce the losses and increase the yield in various semi-arid regions and under normal conditions. It was concluded that *O. agardhii* based nanosilicon particles can be used as biofertilizers and biocontrol agents that would reduce the plant pathogens and thus would enhance plant resistance to environmental stress, and diseases as well as enhance the growth and productivity of wheat plant. Another cyanobacterium-based nanoparticle, Ag-NPs synthesized from cell extract of *Leptolyngbya* WUC 59 was found to display antibacterial activity against *B. subtilis* and *E. coli*. It was also found to enhance seed germination and early development of wheat (*Triticum aestivum* L.) seedling when given at low concentration of 25 mg L−1 (Singh et al. 2020). Hence it was concluded that Ag-NPs from *Leptolyngbya* WUC 59 cell-free extract can be of immense importance as biocontrol agents in agricultural as well as in pharmaceutical industries. Fatma et al. (2022) synthesized selenium nanoparticles, SeNPs from *Gloeocaspa gelatinosa* NCCU -430 and analyzed its application as a mitigating agent for PQ (paraquat, a pesticide) toxicity in *Anabaena variabilis* NCCU-442, cyanobacterial based biofertilizer. When a comparative analysis of control, PQ, SeNPs and SeNPs+PQ was performed to study the role SeNPs in mitigating the toxicity induced by PQ treatment, it was observed that culture treated with PQ when supplemented with SeNPs enhanced the activity of antioxidant enzymes like SOD (superoxide dismutase) by 7.55%, CAT (catalase) by 57.94%, APX (ascorbate peroxidase) by 17.45% and GR (glutathione reductase) by 14.72% in comparison to only PQ treated culture. Thus, it was concluded that SeNPs derived from *Gloeocaspa gelatinosa* NCCU -430 can improve the stress induced by PQ and can be applied in sustainable rice cultivation (Fatma et al. 2022).

Green nanotechnology thus offers a wide range of new agrochemicals and delivery systems that can help in improving the crop productivity and decreases the dependence on chemical agricultural inputs (Gupta et al. 2022). Green synthesis of NPs provides an efficient, non-toxic, clean, simple, and eco-friendly approach for the synthesis of MtNPs. Various studies have suggested that it can be used as a vital tool to address the various problems associated with conventional agriculture like targeted delivery of nutrient and fertilizer and identification of disease at early stages to protect the crops (Bahrulolum et al. 2021). Thus, it is observed that green nanotechnology offers a promising future that can help in transforming the agricultural sector and thus can serve as alternatives to attain sustainable agricultural growth and development.

## Miscellaneous Applications

In addition to above mentioned properties, NPs produced by cyanobacteria also exhibits photocatalytic activity. For example, Ag-NPs from *Synechococcus* sp. display photocatalytic property (Keskin et al. 2016). These NPs leads to photodegradation of methylene blue. Similarly, Ag-NPs by *Microchaete* sp. NCCU-342 was observed to degrade methyl red dye more efficiently in comparison to extract from cyanobacteria (Husain et al. 2019). Au-NPs synthesized by *L. majuscule* have anti-myocardial infarction activity (Bakir et al. 2018). Ag-NPs produced from phycobiliproteins from *Nostoc* sp. displayed antihemolytic activity (El-Naggar et al. 2018). It was found that C-phycoerythrin-CdS NPs from *Phormidium tenue* NTDM05 can be used in biolabeling as fluorescence probes for ultrasensitive detection of DNA (MubarakAli et al. 2012). MubarakAli et al. (2013) demonstrated that C-phycoerythrin-CdS NPs and Au-NP synthesized by *Phormidium willei* and *Coelastrella* sp. can be used in biolabeling.

Green nanotechnology is recommended as a promising tool in the field of modern science that has revolutionized all the sectors, particularly medicine and agriculture. This technology facilitates the synthesis of NPs by eco-friendly and inexpensive processes. There are several natural resources however cyanobacteria hold special importance for their ease of growth and active metabolites found only in these microbes (Bahrulolum et al. 2021; Gupta et al. 2022). Cyanobacteria are highly rich in various types of biomolecules like polysaccharides, lectins, pigments, essential lipids, proteins, vitamins, antioxidants, steroids, MAAs, and halogenated compounds. These compounds exhibit multiple usages and thus are of immense importance in pharmaceuticals, biomedical and nutraceuticals industries (Bajpai et al. 2018; Hamida et al. 2020d). However, some aspects essentially have safety issues and its toxicity requires thorough study and assessment. There is still a need to unravel the biocatalytic activities and decipher the mechanism of NPs synthesis using microorganisms. Also, there is a need to explore the cause why different strains of the same species synthesize different types of NPs that vary morphologically (Hamida et al. 2020d). Advancements in the biosynthesis methods based on cyanobacteria may lead to discovery of novel NPs with distinct properties with multiple applications (Gupta et al. 2022).

## Conclusion

The continual increasing human population is posing a huge demand on food security and the medicine sector hence the emerging need for alternative, natural, environmentally safe and bioeconomic agricultural practices and pharmaceutical supplies is need of the time. Cyanobacteria now have emerged as a potential source for sustainable agriculture, better reliable non-toxic medical supplies, food supplements, and nano-bioconjugates without putting a negative impact on the environment. Cyanobacteria are populating diverse habitats worldwide; they are photosynthetic and can fix the atmospheric $CO_2$, converting it into organic biomass and serving as a source of organic matter to higher plants and bioenergy. The cyanobacteria as a bio-fertilizer enhance the availability of nutrients to plants such as through nitrogen fixation, it can enhance soil fertility and thereby crop productivity. These microbes can be the source of essential nutrients and various metabolites for the growth of plants in soil and can thereby enhance protection against environmental stresses. Also, cyanobacteria can enhance soil quality and fertility by improving physio chemical features like water holding capacity and mineral nutrient conditions of the damaged lands by modulating pH, controlling salinity and absorbing heavy metals contaminants. Along with agricultural uses, during the last two decades, the potential of cyanobacteria as a very good source of novel therapeutic lead compounds has been realized due to the presence of various secondary metabolites and bioactive compounds including anticancer, antidiabetic, anti-inflammatory, immunomodulatory, antimicrobial (antibacterial, antiprotozoal), and antiviral (including against HIV and SARS COV-2) properties. Also, cyanobacteria are a rich source of vitamins and minerals that can act as strong antioxidants and fight disease and therefore provide an eventual mixture of nutrition in the right quantities of a single food. Another benefit of cyanobacteria for their use in drug development lies in the lower value of their products as compared to some other microorganisms. The cultivation of cyanobacteria is simple, economical and does not compete with the crops for land and water. The latest exhaustive research on the use of cyanobacterial bioproducts is required to meet the increased consumer demands. Although cyanobacteria are a promising source that offers diverse functional foods, as a natural resource these are yet to be explored for benefits at ground level and commercial scale. The lack of knowledge of biochemical pathways and purification and characterization techniques that led to the production of cost-effective and diverse metabolites has caused the hindrance in the commercial exploitation of cyanobacteria. But surely, there is a need for further research in the field of cyanobacteria for its potential usage in agriculture, medicines and the discovery of new drugs. Apart from these, numerous studies available on the successful use of various NPs synthesized using cyanobacteria and their metabolites are being utilized in the agriculture sector in the form of nanobiosensors, nanopesticides and nanofertilizers. It offers a wide range of new agrochemicals and delivery methods that can improve productivity and can also serve as a significant tool to solve the various issues related to traditional agriculture practice. Subsequently, exhaustive studies are required to examine and find out their long-term effects, and their safety. If found safe cyanobacteria could prove to be an alternative source to boost agricultural productivity without causing much damage to the environment. In the medical field, NPs can have a significant role as possessing crucial antimicrobial properties. Thus, green nanotechnology holds great potential that can enormously transform the various sectors including the agricultural and pharmaceutical sectors. The present study aims to highlight potential of different species of cyanobacteria for advancement in the medical and agricultural fields in light of advanced studies as nanotechnology utilizing cyanobacteria and their active metabolites. Thus, cyanobacteria though primitive and prokaryotic can be a promising tool in resolving many important hurdles in all spheres of life including medicine and agriculture.

## Acknowledgements

The authors would like to acknowledge and extend thanks to Gargi College, University of Delhi, for all the support during manuscript preparation. We are thankful to Prof. R. P. Sinha, Department of Botany, BHU, for his positive suggestions in improving the manuscript and providing us opportunity to write this manuscript.

## References

Abdel-Daim MM, Abuzead SMM, Halawa SM (2013) Protective role of *Spirulina platensis* against acute deltamethrin-induced toxicity in rats. *PloS One* 8: e72991.

Abdo S, Hetta M, El-Senousy WM, Salah El-Din R, Ali G (2012) Antiviral activity of freshwater algae. *J Appl Pharm Sci* 2: 21-25.

Abdulqader G, Barsanti L, Tredici MR (2000) Harvest of *Arthrospira platensis* from Lake Kossorom (Chad) and its household usage among the Kanembu. *J Appl Phycol* 12: 493-498.

Agarwal P, Gupta R, Agarwal N (2019) Advances in synthesis and applications of microalgal nanoparticles for wastewater treatment. *J Nanotechnol* 2019: 1-9.

Ali DM, Sasikala M, Gunasekaran M, Thajuddin N (2011) Biosynthesis and characterization of silver nanoparticles using marine cyanobacterium, *Oscillatoria willei* NTDM01. *Dig J Nanomater Biostruct* 6(2): 385-390.

Ashmore DA, Chaudhari A, Barlow B, Barlow B, Harper T, Vig K, Pillai S (2018) Evaluation of *E. coli* inhibition by plain and polymer-coated silver nanoparticles. *Rev Inst Med Trop São Paulo* 60: 1-11.

Asthana RK, Tripathi MK, Srivastava A, Singh AP, Singh SP, Nath G, Srivastava BS (2009) Isolation and identification of a new antibacterial entity from the Antarctic cyanobacterium *Nostoc* CCC 537. *J Appl Phycol* 21(1): 81-88.

Awasthi M, Upadhyay AK, Singh S, Pandey VP, Dwivedi UN (2018) Terpenoids as promising therapeutic molecules against Alzheimer's disease: amyloid beta-and acetylcholinesterase-directed pharmacokinetic and molecular docking analyses. *Mol Simul* 44(1): 1-11.

Bahrulolum H, Nooraei S, Javanshir N, Tarrahimofrad H, Mirbagheri VS, Easton AJ, Ahmadian G (2021) Green synthesis of metal nanoparticles using microorganisms and their application in the agrifood sector. *J Nanobiotechnol* 19(1): 1-26.

Bajpai R, Sharma NK, Rai AK (2010) Anticancerous/antiviral compounds from cyanobacteria. *Algas* 43: 4-6.

Bajpai VK, Shukla S, Kang SM, Hwang SK, Song X, Huh YS, Han YK (2018) Developments of cyanobacteria for nano-marine drugs: Relevance of nanoformulations in cancer therapies. *Mar Drugs* 16(6): 179.

Bakir EM, Younis NS, Mohamed ME, El Semary NA (2018) Cyanobacteria as nanogold factories: chemical and anti-myocardial infarction properties of gold nanoparticles synthesized by *Lyngbya majuscula*. *Mar Drugs* 16(6): 217.

Barbaraus D, Kaiser M, Brun R, Gademann K (2008) Potent and selective antiplasmodial activity of the cyanobacterial alkaloid nostocarboline and its dimers. *Bioorg Med Chem Lett*: 18: 4413-4415.

Baweja P, Kumar S, Kumar G (2019) Organic fertilizer from algae: a novel approach towards sustainable agriculture. *Biofertilizers for sustainable agriculture and environment*: Springer, Cham, pp. 353-370.

Betterle N, Melis A (2019) Photosynthetic generation of heterologous terpenoids in cyanobacteria. *Biotechnol Bioeng* 116(8): 2041-2051.

Bhat VB, Madyastha KM (2000) C-phycocyanin: a potent peroxyl radical scavenger in vivo and in vitro. *Biochem Biophys Res Commun* 275: 20-25.

Bin-Meferij MM, Hamida RS (2019) Biofabrication and antitumor activity of silver nanoparticles utilizing novel *nostoc* sp. Bahar M. *Int J Nanomedicine* 14:9019-9029.

Biondi N, Piccardi R, Margheri MC, Rodolfi L, Smith GD, Tredici MR (2004) Evaluation of *Nostoc* strain ATCC 53789 as a potential source of natural pesticides. *Appl Environ Microbiol* 70(6): 3313-3320.

Biondi N, Tredici MR, Taton A, Wilmotte A, Hodgson DA, Losi D, Marinelli F (2008) Cyanobacteria from benthic mats of Antarctic lakes as a source of new bioactivities. *J Appl Microbiol* 105(1): 105-115.

Blom JF, Brütsch T, Barbaras D, Bethuel Y, Locher HH, Hubschwerlen C, Gademann K (2006) Potent algicides based on the cyanobacterial alkaloid nostocarboline. *Organic Lett* 8(4): 737-740.

Brayner R, Barberousse H, Hemadi M, Djedjat C, Yéprémian C, Coradin T, Couté A (2007) Cyanobacteria as bioreactors for the synthesis of Au, Ag, Pd, and Pt nanoparticles via an enzyme-mediated route. *J Nanosci Nanotechnol* 7(8): 2696-2708.

Bulgari R, Franzoni G, Ferrante A (2019) Biostimulants application in horticultural crops under abiotic stress conditions. *Agronomy* 9(6): 306.

Burja AM, Banaigs B, Abou-Mansour E, Burgess JG, Wright PC (2001) Marine cyanobacteria prolific source of natural products. *Tetrahedron* 57: 9347-9377.

Cardellina JH, Moore BS (2010) Editorial: Richard E Moore (1933–2007). *J Nat Products* 73: 301-302.

Carpine R, Olivieri G, Hellingwerf KJ, Pollio A, Marzocchella A (2020) Industrial production of poly-β-hydroxybutyrate from $CO_2$: can cyanobacteria meet this challenge? *Processes* 8(3): 323.

Carpine R, Sieber S (2021) Antibacterial and antiviral metabolites from cyanobacteria: their application and their impact on human health. *Curr Res Biotechnol* 3:65-81.

Chanda MJ, Merghoub N, Arroussi HE (2019) Microalgae polysaccharides: the new sustainable bioactive products for the development of plant bio-stimulants? *World J Microbiol Biotechnol* 35(11): 1-10.

Chatterjee A, Singh S, Agrawal C, Yadav S, Rai R, Rai LC (2017) Role of algae as a biofertilizer. *Algal Green Chem*: Elsevier pp. 189-200.

Chatzissavvidis C, Therios I (2014) *Role of algae in agriculture. Seaweeds* (Ed. Pomin VH): Nova Science Publishers Inc pp. 1-37.

Chaudhary R, Nawaz K, Khan AK, Hano C, Abbasi BH, Anjum S (2020) An overview of the algae-mediated biosynthesis of nanoparticles and their biomedical applications. *Biomolecules* 10(11): 1498.

Chen XXS, Smith GD, Waring P (2003) Human cancer cell (Jurkat) killing by the cyanobacterial metabolite calothrixin A. *J Appl Phycol* 15: 269-277.

Chittora D, Meena M, Barupal T, Swapnil P, Sharma K (2020) Cyanobacteria as a source of biofertilizers for sustainable agriculture. *Biochem Biophys Rep* 22: 100737.

Colleluori DM, Tien D, Kang F, Pagliei T, Ryan Kuss R, McCormick T, Watson K, McFadden K, Chaiken I, Buckheit Jr RW, Romano JW (2005) Expression, purification, and characterization of recombinant cyanovirin-N for vaginal anti-HIV microbicide development. *Protein Expr Purif* 39: 229-236.

Converti A, Oliveira RPS, Torres BR, Lodi A, Zilli M (2009) Biogas production and valorization by means of a two-step biological process. *Bioresour Technol* 100(23): 5771-5776.

Demay J, Bernard C, Reinhardt A, Marie B (2019) Natural Products from Cyanobacteria: Focus on Beneficial Activities. *Mar Drugs* 17(6): 320.

Demin IN, Naraikina NV, Tsydendambaev VD, Moshkov IE, Trunova TI (2013) The effect of potato plant transformation with the gene encoding Δ12-acyl-lipid desaturase on the $CO_2$ exchange and activities of antioxidant enzymes under hypothermia. *Russ J Plant Physiol* 60(3): 367-374.

Desbois AP, Smith VJ (2010) Antibacterial free fatty acids: activities, mechanisms of action and biotechnological potential. *Appl Microbiol Biotechnol* 85(6): 1629-1642.

Dias E, Paulino S, Pereira P (2015) Cyanotoxins: From poisoning to healing–a possible pathway? *Limnetica* 34: 159-172.

Drew L, Fine RL, Do TN, Douglas GP, Petrylak DP (2002) The novel antimicrotubule agent cryptophycin 52 (LY355703) induces apoptosis via multiple pathways in human prostate cancer cells. *Clin Cancer Res* 8: 3922-3932.

Du Jardin P (2015) Plant biostimulants: definition, concept, main categories and regulation. *Sci Hortic* 196: 3-14.

Dubchak S, Ogar A, Mietelski J, Turnau K (2010) Influence of silver and titanium nanoparticles on arbuscular mycorrhiza colonization and accumulation of radiocaesium in *Helianthus annuus*. *Spanish J Agri Res* 8(1):103-108.

Ebadi M, Zolfaghari MR, Aghaei SS, Zargar M, Shafiei M, Zahiri HS, Noghabi KA (2019) A bio-inspired strategy for the synthesis of zinc oxide nanoparticles (ZnO NPs) using the cell extract of cyanobacterium *Nostoc* sp. EA03: from biological function to toxicity evaluation. *RSC Adv* 9(41): 23508-23525.

El-Naggar NE, Hussein MH, El-Sawah AA (2017) Bio-fabrication of silver nanoparticles by phycocyanin, characterization, in vitro anticancer activity against breast cancer cell line and *in vivo* cytotoxicity. *Sci Rep* 7(1): 10844.

El-Naggar NE, Hussein MH, El-Sawah AA (2018) Phycobiliprotein mediated synthesis of biogenic silver nanoparticles, characterization, *in vitro* and *in vivo* assessment of anticancer activities *Sci Rep* 8(1): 8925.

El-Sheekh MM, El-Kassas HY (2014) Application of biosynthesized silver nanoparticles against a cancer promoter cyanobacterium, *Microcystis aeruginosa*. *Asian Pac J Cancer Prev* 15(16): 6773-6779.

El-Sheekh MM, Alwaleed EA, Kassem WM, Saber H (2020) Antialgal and antiproliferative activities of the algal silver nanoparticles against the toxic cyanobacterium. *Microcystis aeruginosa* and human tumor colon cell line. *Environ Nanotechnol Monit Manag* 14: 100352.

Esquivel-Hernández DA, Ibarra-Garza IP, Rodríguez-Rodríguez J, Cuéllar-Bermúdez SP, Rostro-Alanis MDJ, Alemán-Nava GS, Parra-Saldívar R (2017) Green extraction technologies for high-value metabolites from algae: a review. *Biofuel Bioprod Biorefin* 11(1): 215-231.

Fatma T, Afzal B, Naaz H, Sami N, Yasin D, Khan NJ (2022) Mitigative effect of biosynthesized SeNPs on cyanobacteria under paraquat toxicity. *Chemosphere* 293: 133562.

Finamore A, Palmery M, Bensehaila, S, Peluso I (2017) Antioxidant, immunomodulating, and microbial-modulating activities of the sustainable and ecofriendly *Spirulina*. *Oxid Med Cell Longev* 2017: 1-14.

Forlani G, Pavan M, Gramek M, Kafarski P, Lipok J (2008) Biochemical bases for a widespread tolerance of cyanobacteria to the phosphonate herbicide glyphosate. *Plant Cell Physiol* 49(3): 443-456.

Fujishiro T, Ogawa T, Matsuoka M, Nagahama K, Takeshima Y, Hagiwara H (2004) Establishment of a pure culture of the hitherto unicellular cyanobacterium *Aphanotheca sacrum*, and phylogenetic position of the organism. *Appl Environ Microbiol* 70: 3338-3345.

Gaysina LA, Saraf A, Singh P (2019) Cyanobacteria in diverse habitats. *Cyanobacteria*: Academic Press, pp. 1-28.

Ghosh T, Bhayani K, Paliwal C, Maurya R, Chokshi K, Pancha I, Mishra S (2016) Cyanobacterial pigments as natural anti-hyperglycemic agents: An in vitro study. *Front Mar Sci* 3: 146.

Gonçalves AL (2021) The use of microalgae and cyanobacteria in the improvement of agricultural practices: a review on their biofertilising, biostimulating and biopesticide roles. *Appl Sci* 11(2): 871.

Gondim AC, da Silva SR, Mathys L, Noppen S, Liekens S, Sampaio AH, Balzarini J (2019) Potent antiviral activity of carbohydrate-specific algal and leguminous lectins from the Brazilian biodiversity. *Med Chem Comm* 10(3): 390-398.

Górka B, Korzeniowska K, Lipok J, Wieczorek PP (2018) The Biomass of algae and algal extracts in agricultural production. *Algae Biomass: Characteristics and Applications*: Springer, Cham, pp. 103-114.

Govindaraju K, Basha SK, Kumar VG, Singaravelu G (2008) Silver, gold and bimetallic nanoparticles production using single-cell protein (*Spirulina platensis*) Geitler. *J Mater Sci* 43(15): 5115-22.

Guedes AC, Amaro HM, Malcata FX (2011) Microalgae as sources of carotenoids. *Mar Drugs* 9(4): 625-644.

Guilherme MR, Aouada FA, Fajardo AR, Martins AF, Paulino AT, Davi MF, Muniz EC (2015) Superabsorbent hydrogels based on polysaccharides for application in agriculture as soil conditioner and nutrient carrier: A review. *Eur Polym J* 72: 365-385.

Gupta C, Prakash D, Garg AP, Gupta S (2012) Whey Proteins: A novel source of Bioceuticals. *Middle East J Sci Res* 12: 365-375.

Gupta V, Verma NK, Choudhary KK (2022) Cyanobacterial nanoparticles: Application in agriculture and allied sectors. In: *Cyanobacterial Lifestyle and its Applications in Biotechnology*: Academic Press, pp. 23-40.

Gutierrez M, Suyama TL, Engene N, Wingerd JS, Matainaho T, Gerwick WH (2008) Apratoxin D, a potent cytotoxic cyclodepsipeptide from Papua New Guinea collections of the marine cyanobacteria *Lyngbya majuscula* and *Lyngbya sordida*. *J Nat Prod* 71: 1099-1103.

Haggag WM, Hoballah MM, Ali RR (2018) Applications of nano biotechnological microalgae product for improve wheat productivity in Semai Aird areas. *Int J Agri Technol* 14(5): 675-92.

Hamida RS, Abdelmeguid NE, Ali MA, Bin-Meferij MM, Khalil MI (2020a) Synthesis of silver nanoparticles using a novel cyanobacteria *desertifilum* sp. extract: their antibacterial and cytotoxicity effects. *Int J Nanomedicine* 15: 49-63.

Hamida RS, Ali MA, Goda DA, Khalil MI, Al-Zaban MI (2020b) Novel biogenic silver nanoparticle-induced reactive oxygen species inhibit the biofilm formation and virulence activities of methicillin-resistant *Staphylococcus aureus* (MRSA) strain. *Front Bioeng Biotechnol* 8: 433.

Hamida RS, Ali MA, Goda DA, Khalil MI, Redhwan A (2020c) Cytotoxic effect of green silver nanoparticles against ampicillin-resistant *Klebsiella pneumoniae*. *RSC Adv* 10(36): 21136-21146.

Hamida RS, Ali MA, Redhwan A, Bin-Meferij MM (2020d) Cyanobacteria – A Promising Platform in Green Nanotechnology: A Review on Nanoparticles Fabrication and Their Prospective Applications. *Int J Nanomedicine* 15: 6033-6066.

Hankamer B, Lehr F, Rupprecht J, Mussgnug JH, Posten C, Kruse O (2007) Photosynthetic biomass and $H_2$ production by green algae: from bioengineering to bioreactor scale-up. *Physiol Plant* 131(1): 10-21.

Haq IU, Ijaz S (2019) Use of metallic nanoparticles and nanoformulations as nanofungicides for sustainable disease management in plants. *Nanobiotechnology in Bioformulations*: Springer pp. 289-316.

Hayashi K, Hayashi T, Kojima I (1996) A natural sulfated polysaccharide, calcium spirulan, isolated from *Spirulina platensis*: In vitro and ex vivo evaluation of anti-herpes simplex virus and anti-human immunodeficiency virus activities. *AIDS Res Hum Retrovir* 12: 1463-1471.

Hayashi O, Katoh T, Okuwaki Y (1994) Enhancement of antibody production in mice by dietary *Spirulina platensis*. *J Nutr Sci Vitaminol* (Tokyo) 40: 431-441.

Husain S, Afreen S, Yasin D, Afzal B, Fatma T (2019) Cyanobacteria as a bioreactor for synthesis of silver nanoparticles-an effect of different reaction conditions on the size of nanoparticles and their dye decolorization ability. *J Microbiol Methods* 162: 77-82.

Jangir MM, Kashyap JS, Vani B, Chowdhury S (2015) The role of cyanobacteria in agriculture. *J Agroecol Nat Resour Manag* 2(5): 345-350.

Jeffryes C, Agathos SN, Rorrer G (2015) Biogenic nanomaterials from photosynthetic microorganisms. *Curr Opin Biotechnol* 33: 23-31.

Jensen GS, Ginsberg DI, Huerta P, Citton M, Drapeau C (2000) Consumption of *Aphanizomenon flos aquae* has rapid effects on the circulation and function of immune cells in humans. A novel approach to nutritional mobilization of the immune system. *JANA* 2: 50-58.

Jha RK, Zi Rong X (2004) Biomedical compounds from marine organisms. *Mar Drugs* 2: 123-146.

Kanekiyo K, Lee JB, Hayashi K, Takenaka H, Hayakawa Y, Endo S, Hayashi T (2005) Isolation of an antiviral polysaccharide, nostoflan, from a terrestrial cyanobacterium, *Nostoc flagelliforme*. *J Nat Prod* 68: 1037-1041.

Keskin S, Oya N, Koçberber Kılıç N, Dönmez G, Tekinay T (2016) Green synthesis of silver nanoparticles using cyanobacteria and evaluation of their photocatalytic and antimicrobial activity. *J Nano Res* 40: 120-127.

Khalifa K, Hamouda R, Hanafy D, Hamza A (2016) In vitro antitumor activity of silver nanoparticles biosynthesized by marine algae. *Digest J Nanomater Biostruct* 11(1): 213-221.

Khalifa SA, Shedid ES, Saied EM, Jassbi AR, Jamebozorgi FH, Rateb M, El-Seedi HR (2021) Cyanobacteria—From the Oceans to the Potential Biotechnological and Biomedical Applications. *Mar Drugs* 19(5): 241.

Khan Z, Bhadouria P, Bisen P (2005) Nutritional and therapeutic potential of *Spirulina*. *Curr Pharm Biotechnol* 6: 373- 379.

Khanna P, Kaur A, Goyal D (2019) Algae-based metallic nanoparticles: synthesis, characterization and applications. *J Microbiol Methods* 163: 105656.

Kim HM, Lee EH, Cho HH, Moon YH (1998) Inhibitory effect of mast cell-mediated immediate-type allergic reactions in rats by *Spirulina*. *Biochem Pharmacol* 55: 1071-1076.

Kreitlow S, Mundt S, Lindequist U (1999) Cyanobacteria – a potential source of new biologically active substance. *J Biotechnol* 70: 61-73.

Kumar KS, Dahms HU, Won EJ, Lee JS, Shin KH (2015) Microalgae–a promising tool for heavy metal remediation. *Ecotoxicol Environ Saf* 113: 329-352.

Lanzer MR, Rohrbach P (2007) Subcellular pH and $Ca^{2+}$ in *Plasmodium falciparum:* implications for understanding drug resistance mechanisms. *Curr Sci* 92: 1561-1570.

Layam A, Reddy CLK (2006) Antidiabetic property of *Spirulina*. *Diabetol Croat* 35: 29-33.

Lengke MF, Ravel B, Fleet ME, Wanger G, Gordon RA, Southam G (2006) Mechanisms of gold bioaccumulation by filamentous cyanobacteria from gold (III)-chloride complex. *Environ Sci Technol* 40(20): 6304-6309.

Linington RG, Edwards DJ, Shuman CF, McPhail KL, Matainaho T, Gerwick WH (2008) Symplocamide A, a potent cytotoxin and chymotrypsin inhibitor from the marine cyanobacterium *Symploca* sp. *J Nat Prod* 71: 22-27.

Lopez-Archilla AI, Moreira D, López-García P, Guerrero C (2004) Phytoplankton diversity and cyanobacterial dominance in a hypereutrophic shallow lake with biologically produced alkaline pH. *Extremophiles* 8(2): 109-115.

Lipok J, Owsiak T, Młynarz P, Forlani G, Kafarski P (2007) Phosphorus NMR as a tool to study mineralization of organophosphonates—The ability of *Spirulina* spp. to degrade glyphosate. *Enzyme Microb Technol* 41(3): 286-291.

Luescher-Mattli M (2003) Algae, a possible source for new drugs in the treatment of HIV and other viral diseases. *Current Medicinal Chemistry-Anti-Infect Agents* 2: 219-225.

Mocan T, Matea CT, Pop T, Mosteanu O, Buzoianu AD, Puia C, Mocan L (2017) Development of nanoparticle-based optical sensors for pathogenic bacterial detection. *J Nanobiotechnol* 15(1): 1-14.

Mulkidjanian AY, Koonin EV, Makarova KS, Mekhedov SL, Sorokin A, Wolf YI, Dufresne A, Partensky F, Burd H, Kaznadzey, D (2006) The cyanobacterial genome core and the origin of photosynthesis. *Proc Natl Acad Sci* 103: 13126-13131.

MubarakAli D, Gopinath V, Rameshbabu N, Thajuddin N (2012). Synthesis and characterization of CdS nanoparticles using C-phycoerythrin from the marine cyanobacteria. *Mater Lett* 74:8-11.

Morsy FM, Nafady NA, Abd-Alla MH, Elhady DA (2014) Green synthesis of silver nanoparticles by water soluble fraction of the extracellular polysaccharides/matrix of the cyanobacterium *Nostoc commune* and its application as a potent fungal surface sterilizing agent of seed crops. *Univ J Microbiol Res* 2(2): 36-43.

MubarakAli D, Arunkumar J, Nag KH, SheikSyedIshack KA, Baldev E, Pandiaraj D, Thajuddin N (2013) Gold nanoparticles from Pro and eukaryotic photosynthetic microorganisms—Comparative studies on synthesis and its application on biolabelling. *Colloids Surf B: Biointerfaces* 103: 166-173.

Masuda Y, Suzuki J, Onda Y, Fujino Y, Yoshida M, Doi T (2014) Total synthesis and conformational analysis of apratoxin C. *J Org Chem* 79: 8000-8009.

Mazard S, Penesyan A, Ostrowski M, Paulsen IT, Egan S (2016) Tiny microbes with a big impact: the role of cyanobacteria and their metabolites in shaping our future. *Mar Drugs* 14(5): 97.

Naidoo D, Roy A, Kar P, Mutanda T, Anandraj A (2020) Cyanobacterial metabolites as promising drug leads against the $M^{pro}$ and $PL^{pro}$ of SARS-CoV-2: An *in-silico* analysis. *J Biomol Struct Dyn* 1-13.

Nanotechnology AOo. National Nanotechnology Strategy Annual Report 2008: 2007-2008.

Nisha R, Kaushik A, Kaushik CP (2007) Effect of indigenous cyanobacterial application on structural stability and productivity of an organically poor semi-arid soil. *Geoderma* 138(1-2): 49-56.

Nozzi NE, Oliver JW, Atsumi S (2013) Cyanobacteria as a platform for biofuel production. *Front Bioeng Biotechnol* 1: 7.

Nunnery JK, Mevers E, Gerwick WH (2010) Biologically active secondary metabolites from marine cyanobacteria. *Curr Opin Biotechnol* 21: 787-793.

Ou Y, Lin L, Yang X, Pan Q, Cheng X (2013) Antidiabetic potential of phycocyanin: Effects on KKAy mice. *Pharm Biol* 51: 539-544.

Oza G, Pandey S, Mewada A, Kalita G, Sharon M, Phata J, Sharon M (2012) Facile biosynthesis of gold nanoparticles exploiting optimum pH and temperature of fresh water algae *Chlorella pyrenoidusa*. *Adv Appl Sci Res* 3(3): 1405-1412.

Pan S, Jeevanandam J, Danquah MK (2019) Benefits of Algal Extracts in Sustainable Agriculture. *Grand Challenges in Algae Biotechnology*: Springer, Cham, pp. 501-534.

Panda D, Deluca K, Williams D, Jordan MA, Wilson L (1998) Antiproliferative mechanism of action of cryptophycin-52: kinetic stabilization of microtubule dynamics by high-affinity binding to microtubule ends. *Cell Biol* 95: 9313-9318.

Pandurangan M, Kim DH (2016) Therapeutic potential of cyanobacteria against streptozotocin-induced diabetic rats. 3 *Biotech* 6: 94.

Panjiar N, Mishra S, Yadav AN, Verma P (2017) Functional foods from cyanobacteria: an emerging source for functional food products of pharmaceutical importance. *Microbial Functional Foods and Nutraceuticals* 21-37.

Pardee KI, Ellis P, Bouthillier M, Towers GH, French CJ (2004). Plant virus inhibitors from marine algae. *Can J Bot* 82(3): 304-309.

Park JH, von Maltzahn G, Zhang L, Schwartz MP, Ruoslahti E, Bhatia SN, Sailor MJ (2008) Magnetic iron oxide nanoworms for tumor targeting and imaging. *Adv Mater* 20(9): 1630-1635.

Parmar A, Singh NK, Pandey A, Gnansounou E, Madamwar D (2011) Cyanobacteria and microalgae: a positive prospect for biofuels. *Bioresour Technol* 102(22): 10163-10172.

Patel V, Berthold D, Puranik P, Gantar M (2015) Screening of cyanobacteria and microalgae for their ability to synthesize silver nanoparticles with antibacterial activity. *Biotechnol Rep (Amst)* 5: 112–119.

Pavela R, Benelli G (2016) Essential oils as ecofriendly biopesticides? Challenges and constraints. *Trends Plant Sci* 21(12): 1000-1007.

Pisciotta JM, Zou Y, Baskakov IV (2010) Light-dependent electrogenic activity of cyanobacteria. *PloS One* 5: 1-10.

Prasanna R, Triveni S, Bidyarani N, Babu S, Yadav K, Adak A, Saxena AK (2014) Evaluating the efficacy of cyanobacterial formulations and biofilmed inoculants for leguminous crops. *Arch Agron Soil Sci* 60(3): 349-366.

Priatni S, Budiwati TA, Ratnaningrum D, Kosasih W, Andryani R, Susanti H, Susilaningsih D (2016) Antidiabetic screening of some Indonesian marine cyanobacteria collection. *Biodiversitas* 17: 642-646.

Price GD, Pengelly JJ, Forster B, Du J, Whitney SM, von Caemmerer S, Evans JR (2013) The cyanobacterial CCM as a source of genes for improving photosynthetic $CO_2$ fixation in crop species. *J Exp Bot* 64(3): 753-768.

Priotto G, Kasparian S, Ngouama D, Ghorashian S, Arnold U, Ghabri S, Karunakara U (2007) Nifurtimox-eflornithine combination therapy for second-stage *Trypanosoma brucei* gambiense sleeping sickness: a randomized clinical trial in Congo. *Clin Infect Dis* 45(11): 1435-1442.

Priya M, Gurung N, Mukherjee K, Bose S (2014) Microalgae in removal of heavy metal and organic pollutants from soil. *Microbial Biodegradation and Bioremediation*: Elsevier, pp. 519-537.

Qamar H, Hussain K, Soni A, Khan A, Hussain T, Chénais B (2021) Cyanobacteria as natural therapeutics and pharmaceutical potential: Role in antitumor activity and as nanovectors. *Molecules* 26(1): 247.

Reinert R, Donald EL, Rosi FX, Watal C, Dowzicky M (2007) Antimicrobial susceptibility among organisms from the Asia/Pacific Rim, Europe and Latin and North America collected as part of TEST and the in vitro activity of tigecycline. *J Antimicrob Chemother* 60: 1018-1029.

Renuka N, Guldhe A, Prasanna R, Singh P, Bux F (2018) Microalgae as multi-functional options in modern agriculture: current trends, prospects and challenges. *Biotechnol Adv* 36(4): 1255-1273.

Roco MC. *The Long View of Nanotechnology Development: The National Nanotechnology Initiative at 10 Years*. Springer; 2011.

Rodríguez AA, Stella AM, Storni MM, Zulpa G, Zaccaro MC (2006). Effects of cyanobacterial extracellular products and gibberellic acid on salinity tolerance in *Oryza sativa* L. *Saline Syst* 2(1): 1-4.

Rodriguez-Garcia A, Hosseini S, Martinez-Chapa SO, Cordell GA (2017) Multi-target activities of selected alkaloids and terpenoids. *Mini-Reviews Org Chem* 14(4): 272-279.

Romay C, Ledon N, Gonzalez R (1999) Phycocyanin extract reduces leukotriene B4 levels in arachidonic acid-induced mouse-ear inflammation test. *J. Pharm. Pharmacol* 51: 641-642.

Ronga D, Biazzi E, Parati K, Carminati D, Carminati E, Tava A (2019) Microalgal biostimulants and biofertilizers in crop productions. *Agronomy* 9(4): 192.

Ruffing AM (2011) Engineered cyanobacteria: teaching an old bug new tricks. *Bioeng Bugs* 2(3): 136-149.

Sakamoto Y, Mori K, Matsuo Y, Mukojima N, Watanabe W, Sobaru N, Chaya M (2017) Breeding of a new potato variety 'Nagasaki Kogane'with high eating quality, high carotenoid content, and resistance to diseases and pests. *Breed Sci* 67(3): 320-326.

Saran T, Sharma G, Kumar M, Ali M (2017) Biosynthesis of copper oxide nanoparticles using cyanobacteria *Spirulina platensis* and its antibacterial activity. *Int J Pharm Sci Res* 8: 3887.

Schaeffer DJ, Krylov VS (2000) Anti-HIV activity of extracts and compounds from algae and cyanobacteria. *Ecotoxicol Environ Saf* 45: 208-227.

Schmid G (2011) *Nanoparticles: From Theory to Application*: John Wiley & Sons.

Senthilkumar R, John S (2008) Hypoglycaemic activity of marine cyanobacteria in alloxan induced diabetic rats. *Pharmacol Online* 2: 704-714.

Shang Y, Hasan M, Ahammed GJ, Li M, Yin H, Zhou J (2019) Applications of nanotechnology in plant growth and crop protection: a review. *Molecules* 24(14): 2558.

Sharma A, Sharma S, Sharma K, Chetri SP, Vashishtha A, Singh P, Agrawal V (2016) Algae as crucial organisms in advancing nanotechnology: a systematic review. *J Appl Phycol* 28(3): 1759-1774.

Shetty K, Pometto A, Paliyath G, Levin RE (Eds.) (2006) *Food Biotechnology* (2nd ed.). Boca Raton: CRC Press.

Simmons TL, Engene N, Ureña LD, Romero LI, Ortega Barría E, Gerwick L, Gerwick WH (2008) Viridamides A and B, lipodepsipeptides with antiprotozoal activity from the marine cyanobacterium *Oscillatoria nigro-viridis*. *J Nat Prod* 71(9): 1544-1550.

Singh G, Babele PK, Shahi SK, Sinha RP, Tyagi MB, Kumar A (2014) Green synthesis of silver nanoparticles using cell extracts of *Anabaena doliolum* and screening of its antibacterial and antitumor activity. *J Microbiol Biotechnol* 24(10): 1354-1367.

Singh H, Du J, Singh P, Yi TH (2018) Extracellular synthesis of silver nanoparticles by *Pseudomonas* sp. THG-LS1.4 and their antimicrobial application. *J Pharm Anal* 8(4): 258-264.

Singh JS, Kumar A, Rai AN, Singh DP (2016) Cyanobacteria: a precious bio-resource in agriculture, ecosystem, and environmental sustainability. *Front Microbiol* 7: 529.

Singh R, Parihar P, Singh M, Bajguz A, Kumar J, Singh S, Prasad SM (2017) Uncovering potential applications of cyanobacteria and algal metabolites in biology, agriculture and medicine: current status and future prospects. *Front Microbiol* 8: 515.

Singh RK, Tiwari, SP, Rai AK and Mohapatra TM (2011) Cyanobacteria: an emerging source for drug discovery. *J. Antibiotics,* 64: 401-412.

Singh S, Kate BN, Banerjee UC (2005) Bioactive compounds from cyanobacteria and microalgae: an overview. *Crit Rev Biotechnol* 25: 73-95.

Singh Y, Kaushal S, Sodhi RS (2020) Biogenic synthesis of silver nanoparticles using cyanobacterium *Leptolyngbya* sp. WUC 59 cell-free extract and their effects on bacterial growth and seed germination. *Nanoscale Advances.* 2(9): 3972-82.

Skulberg OM (2000) Microalgae as a source of bioactive molecules: experience from cyanophyte research. *J Appl Phycol* 12: 341-348.

Sonker AS, Pathak J, Kannaujiya VK, Sinha RP (2017) Characterization and *in vitro* antitumor, antibacterial and antifungal activities of green synthesized silver nanoparticles using cell extract of *Nostoc* sp. strain HKAR-2. *Can J Biotechnol* 1(1): 26.

Tamoi M, Nagaoka M, Miyagawa Y, Shigeoka S (2006) Contribution of fructose-1, 6-bisphosphatase and sedoheptulose-1, 7-bisphosphatase to the photosynthetic rate and carbon flow in the Calvin cycle in transgenic plants. *Plant Cell Physiol* 47(3): 380-390.

Tan LT (2010) Filamentous tropical marine cyanobacteria: A rich source of natural products for anticancer drug discovery. *J Appl Phycol* 22: 659-676.

Thakkar KN, Mhatre SS, Parikh RY (2010) Biological synthesis of metallic nanoparticles. *Nanomedicine* 6(2): 257-262.

Thornburg CC, Cowley ES, Sikorska, J, Shaala LA, Ishmael JE, Youssef DT, McPhail KL (2013) Apratoxin H and apratoxin A sulfoxide from the Red Sea cyanobacterium *Moorea producens*. *J Nat Prod* 76(9): 1781-1788.

Tiwari AK, Tiwari BS (2020) Cyanotherapeutics: an emerging field for future drug discovery. *Appl Phycol* 1(1): 44-57.

Usman M, Farooq M, Wakeel A, Nawaz A, Cheema SA, ur Rehman H, Sanaullah, M (2020) Nanotechnology in agriculture: current status, challenges and future opportunities. *Sci Tot Environ* 721: 137778.

Vadiraja BB, Gaikwad NW, Madyastha KM (1998) Hepatoprotective effect of C-phycocyanin: protection for carbon tetrachloride and R-(+)-pulegone-mediated hepatotoxicty in rats. *Biochem Biophys Res Commun* 249: 428-431.

Vijayakumar S (2012) Potential applications of cyanobacteria in industrial effluents-a review. *J Bioremed Biodeg* 3:1-6.

Vijayan SR, Santhiyagu P, Singamuthu M, Kumari Ahila N, Jayaraman R, Ethiraj K (2014) Synthesis and characterization of silver and gold nanoparticles using aqueous extract of seaweed, *Turbinaria conoides*, and their antimicrofouling activity. *Sci World J* 2014: 938272.

Wood NT (2001). Nodulation by numbers: the role of ethylene in symbiotic nitrogen fixation. *Trends Plant Sci* 6(11): 501-502.

Whitton BA, Potts M (Eds.) (2007) *The Ecology of Cyanobacteria: their Diversity in Time and Space*. Springer Science & Business Media.

Wu M, Okino T, Nogle LM, Marquez BL, Williamson RT, Sitchitta N (2000) Structure, synthesis, and biological properties of kalkitoxin, a novel neurotoxin from the marine cyanobacterium *Lyngbya majuscula. J Am Chem Soc* 122: 12041-12042.

Yasin D, Zafaryab M, Ansari S, Ahmad N, Khan NF, Zaki A, Rizvi MMA, Fatma T (2019) Evaluation of antioxidant and anti-proliferative efficacy of *Nostoc muscorum* NCCU-442. *Biocatal Agric Biotechnol* 17: 284-293.

Yang H, Lee E, Kim H (1997) *Spirulina platensis* inhibits anaphylactic reaction. *Life Sci* 61: 1237-1244.

Zulpa G, Zaccaro MC, Boccazzi F, Parada JL, Storni M (2003) Bioactivity of intra and extracellular substances from cyanobacteria and lactic acid bacteria on "wood blue stain" fungi. *Biol Control* 27(3): 345-348.

## Chapter 6

# Mycosporine-Like Amino Acids: Multipurpose Stress-Relieving Secondary Metabolites in Marine Ecosystems

**J. Devika[1], Anu Gopinath[1,*], Idrees Babu[2] and Rajeshwar P. Sinha[3]**

[1]Department of Aquatic Environment Management,
Kerala University of Fisheries and Ocean Studies, Kochi, Kerala, India
[2]Scientist, Department of Science andTechnology, Union Territory of Lakshadweep, Kavaratti, Lakshadweep, India
[3]Laboratory of Photobiology and Molecular Microbiology,
Centre of Advanced Study in Botany, Institute of Science, Banaras Hindu University, Varanasi, India

## Abstract

Marine organisms especially the seaweeds, dinoflagellates and other microbes such as cyanobacteria encompass a variety of secondary metabolites. Mycosporine-like amino acids (MAAs) are a class of recently reported secondary metabolites with multiple functions including the protection of living cells from UV rays-induced damages. These are low molecular weight, water soluble compounds with amino acid or imino alcohol substituents on an aminocyclohexanone or aminocyclohexanamine ring. The production of MAAs in living cells are found to be in correlation with its UV exposure and their biosynthesis is presumed to occur *via* the Shikimate pathway. Apart from being natural sunscreens, MAAs are also involved in functions like antioxidant activities, providing tolerance to salt and thermal stress and also

[*] Corresponding Author's Email: dranugopinath@gmail.com.

In: Cyanobacteria
Editor: Rajeshwar P. Sinha
ISBN: 979-8-88697-169-9

protection of cells from mutation. The principle role of MAAs in sustaining life in the most biodiverse ecosystems like the coral reefs, seaweeds and the marine microbial community should undergo a comprehensive evaluation as these can be used as the indicators of the vulnerability of these ecosystems in near future. Their species-wise and seasonal variations may also serve as key points to assess the tolerance limit of living cells to various stress factors. This chapter thus focuses on the presence, variation and stress relieving functions including the photoprotective nature of MAAs in marine ecosystems with emphasis to coral dinoflagellates, seaweeds and the marine microbes.

**Keywords:** mycosporine-like amino acids, coral reefs, dinoflagellates, seaweeds, marine ecosystems

## Introduction

Marine organisms are known to have evolved to live in a wide range of habitats and are functionally adapted to extreme conditions, especially those of the low latitudes where they are frequently exposed to high solar radiation in their normal environment (Baker et al. 1980). Variable environmental conditions induce the production of a wide array of secondary metabolites in these organisms as a means to maintain their homeostasis in stressful conditions (Akula and Ravishankar 2011). Certain group of metabolites from marine organisms which functions mainly for absorbing harmful UV radiation were discovered and comprehensively studied since early 1960s. A powerful UV-B absorption agent with an absorption maximum at 305 nm was identified in the gas gland of an epipelagic *Physalia* by Wittenburg in 1960, but the isolated material was properly defined later only. Since then, natural photo protective compounds were at the focus of research and in 1969 the presence of certain UV absorbing compounds in *Acropora* corals and a cyanobacterium from the Great Barrier Reef were detected (Shibata 1969). The isolated compound was a water-soluble material having a broad absorbance maximum ($\lambda_{max}$) at ≈320 nm, named as S-320 and were then characterised as substituted amino acids similar to those isolated from fungal cells by Charles M. Leach (Leach 1965), hence marked as Mycosporine-like amino acids (MAAs). Within the next two decades, around 23 MAAs described in marine organisms, and are assumed to be accumulated as a result of prolonged UV exposure (Jokiel and York 1982).

## Mycosporine-Like Amino Acids (MAAs)

Mycosporine-like amino acids (MAAs) are a class of recently reported secondary metabolites with multiple functions, including the protection of living cells from UV ray-induced damage. These are low molecular weight, water soluble molecules capable of absorbing UV Radiation (UVR) in the wavelength range of 309-365 nm with remarkable stability under fluctuating environmental conditions (Bhatia et al. 2011). They are composed of either an aminocyclohexenone or an aminocyclohexanamine ring, with nitrogen substituents such as mycosporine-glycine, mycosporine-taurine or methylamine. MAAs generally have large molar extinction coefficients ($\xi$= 28,100–50,000 mol$^{-1}$ cm$^{-1}$) underlining their photo-absorptive properties with an absorbance maximum ranging from 268 to 362 nm depending on the type of ring and substituents in the compound (Dunlap and Shick 1998).

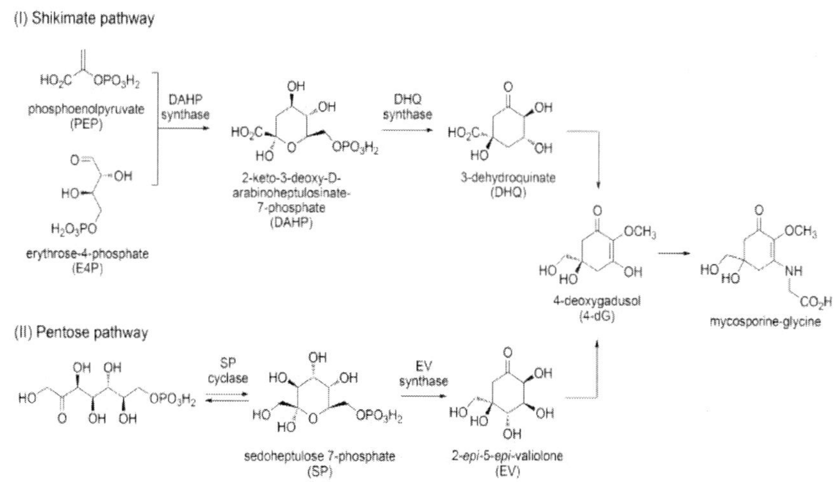


**Figure 1.** Two possible pathways of synthesising mycosporine-glycine (Ref: Shick and Dunlap. 2002).

### Biosynthesis of MAAs

Proper mapping on the biosynthetic pathway for the formation of MAAs is still a matter of research, and the most agreed assumption is about the Shikimate pathway which is the typical route for amino acid synthesis in lower organisms (Bandaranayake 1998).

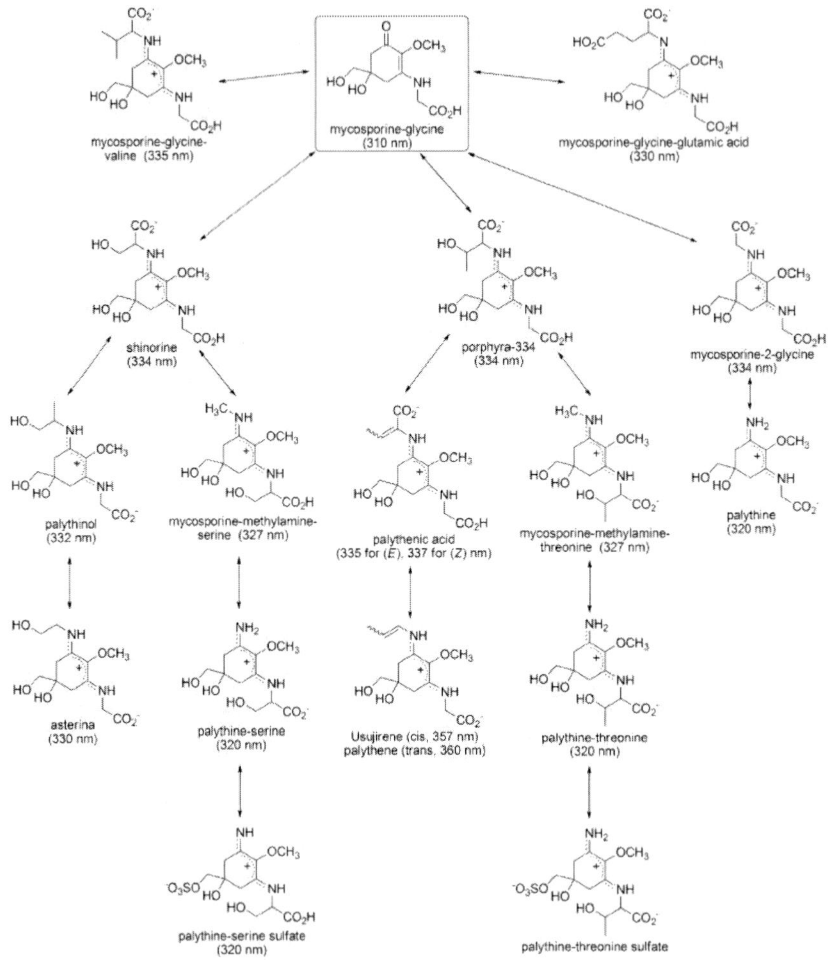


**Figure 2.** Formation of different MAAs through amino- substitution to Mycosporine glycine and chemical conversions (Ref: Wada et al. 2013).

Portwich and Garcia-Pichel (2003) noticed the formation of MAAs in cyanobacteria and their findings are in agreement with the theory that generation of the wide array of mycosporine molecules in nature is attributed to the variations in late biosynthesis processes, which add amino-moieties to the core mycosporine molecule (Portwich and Garcia-Pichel 2003). High concentration of MAAs in lower and primitive organisms again supports their production through Shikimate pathway. In this pathway, it is believed that 4-deoxygadusol (4-dG) an intermediate, is transformed into mycosporine glycine which on subsequent substitution by various kinds of amino acids gives various MAAs. Pentose pathway is another possible means of 4-deoxygadusol (4-dG) intermediate formation.

Mycosporine-glycine-valine, mycosporine-glycine-glutamic acid, shinorine, porphyra-334, and mycosporine-2-glycine are obtained by the substitution of valine, glutamic acid, serine, threonine, and glycine respectively (Wada et al. 2013). Figure 2 depicts the formation various MAAs through subsequent substitution of different amino acid moieties to mycosporine glycine. Absence of Shikimate pathway can be presumed as the key reason for non-production of MAAs in higher animals.

### MAAs - Relevance in Marine Ecosystems

Generally, the marine organisms, cyanobacteria in particular, have been shown to be great suppliers of mycosporine-like amino acids produced via the Shikimate pathway. The role of MAAs as a UV protective filter has been investigated in cyanobacteria (Sinha et al. 2001; Vega et al. 2020), coral-associated dinoflagellates (Dunlap and Chalker 1986; Banaszak et al. 2006), marine macro algae (Sun et al. 2020), and a variety of crustaceans. Besides photoprotection, evidence has emerged in recent years that mycosporines also serve as antioxidant molecules that scavenge toxic oxygen radicals, as compatible solutes that protect cells from salt stress, desiccation, or thermal stress in certain organisms, and as an intracellular nitrogen reservoir (Oren and Cimerman 2007). Thus, these molecules are expected to safeguard the vulnerable marine ecosystems from being destroyed by their various anti-stress functions.

## MAAs in Marine Microorganisms

Marine microbial communities are known to harbour a multitude of metabolites and toxins produced as a response to the environmental condition they are exposed to (Nunnery et al. 2010). Even though the presence of MAAs is more expected in lower organisms, other marine microbes are relatively in the rear-end when compared to the production of MAAs in blue-green algal community.

Cyanobacteria alias blue-green algae are a group of primitive organisms with long evolutionary history and thus might have faced the most deleterious ultraviolet radiation, which lead to an evolution of ultraviolet protecting mycosporine-like amino acids for the survival strategy (Jain et al. 2017). So far, various MAAs such as shinorine, porphyra-334, and mycosporine-glycine have been reported from cyanobacteria and the former one is considered to be most dominant. However, the type of MAA produced is found to vary in different species.

Production of shinorine, a bisubstituted MAA having an absorption maximum at 334 nm was observed in heterocystous $N_2$-fixing cyanobacteria *Anabaena* sp., *Nostoc commune* and *Scytonema* sp. under induction by solar ultraviolet-B (UV-B) radiation points the fact that the organism may protect itself from deleterious short wavelength radiation by their ability to synthesize photoprotective compounds in response to UV-B radiation (Sinha et al. 2001). The bloom-forming freshwater cyanobacterium *Microcystis aeruginosa* was found to produce the MAAs porphyra-334 and shinorine which enabled them to maintain the bloom even at high solar irradiance (Liu et al. 2004).

Other stressful circumstances like hyper-salinity can also be a reason for a hike in MAA production. Oren (1997) detected two MAAs, one with an absorption maximum at 332 nm (Mycosporine-2-glycine), and one at 362 nm in a community of unicellular cyanobacteria inhabiting a gypsum crust developing on the bottom of a hypersaline pond in Eilat, Israel (Oren 1997) Later, Volkmann et al. (2006) identified the unknown MAA as Euhalothece-362 and this turned to be the first elucidation of the MAA Euhalothece-362. *Aphanothece halophytica*, an extremophilic cyanobacterium which lives in high saline and high alkaline condition are studied and the in vivo analysis showed accumulation of the mycosporine-2-glycine but not shinorine or mycosporine-glycine. Also, the accumulation was found to get stimulated more under the stress condition of high salinity than UV-B radiation (Waditee et al. 2011).

Three bloom-forming *Nodularia* species from the Baltic Sea (*N. spumigena, N. baltica* and *N. harveyana*) were reported to produce the MAAs porphyra-334 and shinorine upon UV-B irradiation in an experimental set-up (Sinha et al. 2003). This reflects the stability of MAAs even in extreme circumstances like high UV radiation and hyper salinity due to which MAAs are presumed to be the molecules that safeguarded cyanobacteria in the primitive earth. More instances of MAA production by marine microbes and the particular function identified are discussed in Table 1.

**Table 1.** Various marine microbes producing MAAs and their purpose of production

| ORGANISM | MAA produced | FUNCTION | REFERENCE |
|---|---|---|---|
| *Chlorogloeopsis fritschii* PCC 6912 (Marine cyanobacterium) | Mycosporine-Glycine, Shinorine | UV Protection | Llewellyn et al. 2020 |
| *Gyrodinium dorsum* (Marine dinoflagellate) | Shinorine, Porphyra-334, Palythine, and two unidentified MAAs having $\lambda_{max}$ at 310 and 331nm. | Protection against UV-B-induced damage | Klish et al. 2001 |
| *Nodularia baltica*, *Nodularia harveyana*, *Nodularia spumigena* (Marine cyanobacteria) | Shinorine and Porphyra-334 | UV Protection | Sinha et al. 2003 |
| *Aphanothece halophytica* (Halophilic cyanobacteria) | MAAs having $\lambda_{max}$ at 332 and 362 nm. | | Oren, A. 1997 |
| *Alexandrium excavatum* (Red tide dinoflagellate) | Mycosporine-Glycine, Palythine, Asterina-330, Porphyra-334, Palythenic Acid and Isomeric mixture of Usujirene and Palythene | UV protection | Carreto et al. 1990 |
| *Alexandrium tamarense* (Red tide dinoflagellate) | Porphyra-334 and Palythene | - | Callone et al. 2006 |
| *Scrippsiella sweeneyae* (Marine dinoflagellate) | Shinorine, Palythine, Porphyra-334, and Palythene | UV protection | Taira et al. 2004 |
| *Prorocentrum minimum* (Marine dinoflagellate) | Shinorine, Usujirene, Palythene, Mycosporine-serine-glycine methyl ester and Palythenic acid, Mycosporine-glycine, Palythine, and Porphyra-334. | UV protection | Carreto et al. 2018 |

## MAAs in Marine Macroalgae

Marine macroalgae also called as seaweeds, which are attached to their substrate, are subjected to a variety of periodic pressure in various extent and gradients, to which each species has evolved according to its physiological properties (Bedoux et al. 2020). As a protective measure the seaweeds are known to produce a broad spectrum of secondary metabolites which are now under comprehensive research due to their potent biotechnological applications (Davis and Vasanthi 2011). Red seaweed is now under extensive evaluation for their ability to accumulate natural sun screening compounds along with other pigments.

Hoyer et al. (2001) identified and quantified the distribution pattern of UV-absorbing mycosporine-like amino acids (MAAs) in Antarctic macroalgae and correlated with habitat as well as with the radiation climate under water and found that MAA content was negatively correlated with water depth. In terms of MAA accumulation and induction patterns after exposure to different radiation conditions, this study classified red seaweeds into three different physiological groups, (i) species with no capacity for MAA biosynthesis; (ii) species with a constant MAA composition at high concentration, irrespective of environmental conditions; and (iii) species with a basic MAA concentration adjusted according to environmental conditions.

Major MAAs compounds reported from seaweeds, especially red algae, includes mycosporine-glycine, porphyra-334, asterina-330, shinorine, palythinol, palythine, and palythene (Carreto and Carignan 2011). An evaluation on the production MAAs with seasonal correlation in *Gracilaria vermiculophylla* (Rhodophyta) detected Porphyra-334, Shinorine, Palythine and Asterina-330 where, the highest levels of Porphyra-334 and Shinorine were reached from November to January while Palythine and Asterina-330 from April to August (Barceló-Villalobos et al. 2017). Jofre et al. (2020) scrutinized the content and composition of MAAs in sub-Antarctic intertidal seaweeds *Iridaea tuberculosa, Nothogenia fastigiate*, and *Corallina officinalis* and observed that MAAs varied seasonally, with high concentration recorded in different seasons, depending on the species, i.e., winter (*I. tuberculosa*), spring (*N. fastigiata*), and summer (*C. officinalis*) (Jofre et al. 2020).

**Table 2.** Major MAAs identified from various seaweeds and their functions

| ORGANISM | MAAs IDENTIFIED | FUNCTION | REFERENCE |
|---|---|---|---|
| *Bostrychia* sp. *Caloglossa* sp. *Stictosiphonia* sp. (RED ALGAE) | Mycosporine-Glycine, Shinorine, Porphyra-334, Palythine, Asterina-330, Palythinol, Palythene and an unidentified MAA having $\lambda_{max}$ at 357 nm. | UV protection | Karsten et al. 1998 |
| *Chondrus yendoi* (RED ALGA) | Palythine ($\lambda_{max}$ 320 nm) | UV absorption | Makino1999 |
| *Chondrus crispus* (RED ALGA) | Shinorine ($\lambda_{max}$ 334 nm), Asterina ($\lambda_{max}$ 330 nm), and Palythine ($\lambda_{max}$ 320 nm), | Protection of photosynthetic and growth-related processes from excess UV radiation. | Franklin et al. 1999 |
| *Devaleraea ramentacea* (RED ALGA) Endemic Arctic Macroalga | Mycosporine-glycine, Shinorine, Porphyra-334, Palythine, Asterina-330, Palythinol, and Palythene | As natural UV screening compounds | Karsten et al. 1999 |
| *Gracilaria cornea* (RED ALGA) | Porphyra-334 and Shinorine | Protection from UV and heat stress | Sinha et al. 2000 |
| *Prasiola crispa* ssp. Antarctica (GREEN ALGA) | Unknown MAA with $\lambda_{max}$ at 334 nm | Protection from excess UV radiation | Hoyer et al. 2001 |
| 13 different red algae from intertidal region of Chile | Mycosporine glycine, Shinorine, Porphyra-334, Palythine, Asterina-330, Palythinol | Photoprotection | Huovinen et al. 2004 |
| *Porphyra columbina* (RED ALGA) | Various MAAs | Against UV and Ammonium stress | Peinado et al. 2004 |
| *Porphyra haitanensis* (RED ALGA) | Porphyra-334 and another MAA with peak at 336 nm | Against desiccation stress | Jiang et al. 2008 |
| *Bostrychia scorpioides* (RED ALGA) | Six new MAAs named provisionally as Bostrichine A to Bostrichine F | UV-stress and osmotic stress | Orfanoudaki et al. 2019 |

Apart from the natural occurrence, several species were identified to produce MAAs in reaction to stress given on experimental basis. Also, the differential responses of UV-absorbing compounds concentrations in seaweed are found more associated to the previous light history of the algae depending on their position in the intertidal zone rather than the radiation treatment imposed to the samples (Helbling et al. 2004). Sun et al. (2020) established an open online database to quickly retrieve MAAs in 501 species of marine macroalgae on the basis of a study on MAAs in 572 species of marine macroalgae which belonged to 28 orders, conducted from 1990 to 2019 (Sun et al. 2020). Table 2 gives an account on the major type of MAAs identified from various seaweeds and their functional significance.

## MAAs in Coral Dinoflagellates

Coral bleaching occurs once the homeostatic capacity of the organism fails before the stress factors. In this case, homeostatic capacity refers to the organism's defensive systems in the face of unfavourable situations. When environmental circumstances start to deteriorate, the coral zooxanthellae synthesize MAAs along with other metabolites, as a survival tactic. Around 15 MAAs have been detected from scleractinian corals including the primary MAAs like mycosporine-glycine (maximum $\lambda$ [$\lambda_{max}$] = 310 nm), shinorine ($\lambda_{max}$ = 334 nm), porphyra-334 ($\lambda_{max}$ = 334 nm), and mycosporine-2-glycine ($\lambda_{max}$ = 331 nm) (Rosic and Dove 2011). The presence of mycosporine-Gly, asterina-330, palythinol, and palythine were detected in the colonies of zooxanthellate coral *Acropora microphthalma* (Shick et al. 1995).

Yakovleva and Hidaka (2004) speculated that MAAs are able to protect scleractinian corals from biologically damaging ultraviolet radiation through both the direct sun screening activity of imino-MAAs and the antioxidant properties of oxo-carbonyl-MAAs and suggested that the natural predominance of MAA species with an antioxidant ability may render corals more tolerant to high photosynthetically active and ultraviolet radiation (Yakovleva and Hidaka 2004).

An increase in the concentration of MAAs associated with increased UV flux has been observed directly in marine dinoflagellates. For example, the Great Barrier Reef corals (*Acropora* spp.) at 20 m depth showed significantly lower concentrations of MAAs than those in shallower waters (Dunlap et al. 1986). Shick et al. scrutinized the effects of starvation, ammonium concentration, and photosynthesis on the UV-dependent accumulation of mycosporine-like amino acids (MAAs) in the coral *Stylophora pistillata* and observed that the cultures of zooxanthellae isolated from *S. pistillata* on exposure to UVR showed increased levels of shinorine, and some production of mycosporine-glycine. However, the increased ammonium concentration consistently affected only the accumulation of primary, Symbiodinium-MAAs (mycosporine glycine, shinorine, porphyra-334, and mycosporine-2 glycine) and not secondary MAAs derived from the host's tissues (Shick et al. 2005).

## Functional Versatality of MAAs

Despite the fact that the MAAs are first identified and considered as photo-protective metabolites, further researches manifested their functional versatility before various stress factors facing by the organism. Experimental evidences prevail on their protective accumulation against osmotic stress, thermal stress and also on their antioxidant activities apart from UV protection. However, photo protection is the most discussed function of MAAs.

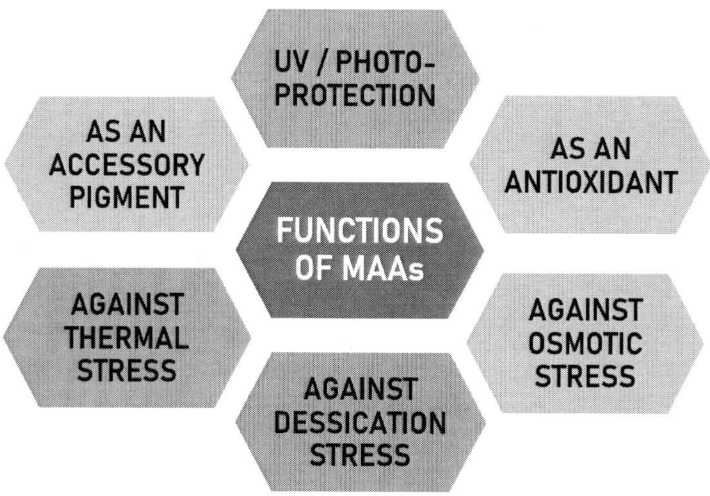


**Figure 3.** Illustration on various functions of MAAs.

## MAAs as Photorotective Molecules

Wide absorption spectra extending from 268 nm to 362 nm and high molar attenuation coefficients are responsible for the photoprotective effect of MAAs against UVR. Their absorption gradient covers the majority of the UVR spectrum (295-400 nm) that reaches the Earth's surface, and their large molar attenuation coefficients show how well MAAs absorb light at this wavelength (Geraldes and Pinto 2021). Several studies on the photodegradation and photophysical characteristics underline their stability to strong radiations and are hence regarded as effective sunscreen compounds.

UV screening properties of MAAs are evaluated mostly in cyanobacteria, dinoflagellates and in higher animals which feeds on MAA producing organisms. High light exposure induced accumulation of MAAs is observed in organisms dwelling extremely different habitats like Antarctic algae *Phaeocystis antarctica* (Riegger and Robinson 1997); red tide dinoflagellate *Gymnodinium sanguineum* (Neale et al. 1998); dinoflagellate *Phaeocystis pouchetii* and the diatom *Guinardia striata* habituating the English Channel (Llewellyn and Harbour 2003) etc. Thus, photo-protection is verified as the prime function of MAAs by analysing their abundance in UV exposed cells.

## MAAs against Osmotic Stress

Exceptionally high concentrations of MAAs in cyanobacteria living in saline and hypersaline environment confirm the ability of the metabolites to protect the organism from osmotic stress they are encountering in their habitat. Enhanced production of MAAs confronting salt stress is reported in halophytic cyanobacteria *Aphanothece halophytica* (Oren 1997); green microalga *Desmodesmus* sp. (Gharib et al. 2020); rice-field cyanobacterium, *Nostoc commune* (Sinha et al. 2003).

A combined exposure to UV-B radiation and osmotic stress in cyanobacterium *Chlorogloeopsis* PCC 6912 was found to synthesize and accumulate two MAAs: mycosporine-glycine and shinorine in which UV-B strongly enhanced the accumulation of shinorine, whereas osmotic stress had a more pronounced effect on mycosporine-glycine (Portwich and Garcia-Pichel 1999). Arrigo and Thomas (2004) reported that MAAs may be involved in the adaptation of sea ice algae to osmotic changes when variations in ambient temperature cause large changes in sea ice brine salinity (Arrigo and Thomas 2004).

Gharib et al. (2020) noticed the green microalga *Desmodesmus* sp. can produce MAA-like compound under hypersaline condition, whereas it was not capable to synthesize MAA in normal condition. In fact, salinity is identified as a stressor which can lead to the induction of MAAs synthesis in this particular microalga and the investigation supports that the production of MAAs helps the organism to survive in harsh saline environments.

## MAA as an Antioxidant

MAAs are reported to show antioxidant properties by scavenging reactive oxygen species such as singlet oxygen, superoxide anions, hydroperoxyl radicals, and hydroxyl radicals.

Dunlap and Yamamoto (1995) suggested mycosporine-glycine as a biological antioxidant in marine organisms from a study in which only mycosporine-glycine was reactive to peroxyl radicals while the imino-mycosporine-like amino acids, shinorine, porphyra-334, palythine, asterina-330 and palythinol, were oxidatively robust (Dunlap and Yamamoto 1995). Mycosporine-glycine was further confirmed as a functional antioxidant in the thermal-stress susceptibility of two scleractinian corals, *Platygyra ryukyuensis* and *Stylophora pistillata* and the study demonstrated its importance in the survival of reef-building corals under thermal stress (Yakovleva et al. 2004). On probing the occurrence and synthesis of UV-absorbing MAAs and characterization of their induction, stability and free radical scavenging capacity in the cyanobacterium *Nostoc* sp. Rastogi et al. (2016) spotted four MAAs: palythine, asterina, porphyra and palythene with significant and dose-dependent *invitro* antioxidant and *in vivo* reactive oxygen scavenging potentials (Rastogi et al. 2016).

Apart from cyanobacteria and dinoflagellates, MAAs accumulated in seaweeds also shows significant anti-oxidant properties. Porphyra-334 and shinorine isolated from *Porphyra rosengurttii*, asterina-330 and palythine from *Gelidium corneum* and shinorine (SH), from *Ahnfeltiopsis devoniensis*, showed excellent anti-oxidant property by scavenging hydroxyl radical in addition to its photo-absorptive nature (De la Coba et al. 2009).

## Other Miscellaneous Functions of MAA

Other than the broadly discussed photoprotective, osmotic stress relief and anti-oxidant functions, several MAAs are reported to accumulate in certain organisms during unfavourable conditions like thermal and desiccation stress also.

Wagner (2001) analysed the concentration of MAAs in shallow water colonies of *Lobophytum compactum* and *Sinularia flexibilis* on correlation with seasonal irradiance levels and observed a positive correlation between MAA level and solar radiation along with temperature. The study inferred that these metabolites are an integral component in the biochemical defence system of soft corals against high irradiance and temperature stress and thus prevents bleaching (Wagner 2001). A possibility on the intracellular nitrogen reserve functions by MAAs was also proposed as the amino acids are furnished with two nitrogen atoms per molecule (Korbee et al. 2006), but still there is a little experimental evidence to support the assumption. Likewise, the suggestion that MAAs can serve as accessory pigments in photosynthesis (Sivalingam et al. 1976) was not evidenced in any further studies and so the claim remains unsubstantiated.

## Conclusion

Climate change is driving changes in the physical and chemical properties of the ocean that have consequences on marine ecosystems. This prevailing climatic instability enticed life beneath the water to evolve and adapt, and thus as adaptive strategy different cellular mechanisms are carried out in living cells resulting in the production of various bioactive molecules (Poloczanska et al. 2016). Ability to tolerate the simultaneous stresses of desiccation, UV irradiation, and oxidation in the cyanobacterium *Nostoc commune* is largely attributed to the protective molecules that the organism is producing and MAAs are detected as a major secondary metabolite in *Nostoc* (Wright et al. 2005).

Traces of MAAs with modifications for specific functioning are detected in higher aquatic organisms also. It is believed that the MAAs produced by algae, fungi, and cyanobacteria are acquired by higher aquatic animals through trophic interactions. Kicklighter et al. (2011) identified three MAAs: asterina-330, aplysiapalythine-A, and aplysiapalythine-B as intraspecific alarm cues for sea hares (*Aplysia californica*) which were present in the algal diet of sea hares and got concentrated in their defensive secretion as well as in their skin (Kicklighter et al. 2011). Different combinations and quantities of MAAs like asterina-330, palythene and mycosporine-N-methylamine serine fund in the mucus secretion of reef fish is again in consonance with the dietary acquirement of MAAs by higher organisms (Eckes et al. 2008). The existence of MAAs in higher animals with proper modification confirms their photoprotective significance in producers, herbivores, and even in carnivores (Chrapusta et al. 2017).

Mycosporine-like amino acids, a group of multifunctional secondary metabolites formed in living cells as a survival strategy are now gaining importance in pharmaceutical and cosmetic industry owing to their anti-stress functions. As the production of MAAs is in proportion to the stress that the organism is facing, their identification and quantification can be taken as bio-tools to assess the tolerance limit of living cells to various stress factors.

## Acknowledgments

The authors are pleased to acknowledge DST- SERB, Govt. of India for the financial support and Faculty of Fisheries Science, Kerala University of Fisheries and Ocean Studies, Kochi for the facilities provided in order to carry out the project. We wish to extend our wholehearted thanks to Prof. R. P. Sinha, editor of the book for providing us a chance to contribute a chapter in the book.

## References

Akula R, Ravishankar GA (2011) Influence of abiotic stress signals on secondary metabolites in plants. *Plant Signal Behav* 6(11): 1720-1731.

Arrigo KR, Thomas DN (2004) Large scale importance of sea ice biology in the Southern Ocean. *Antarctic Sci* 16(4): 471-486.

Baker KS, Smith RC, Green AES (1980) Middle ultraviolet radiation reaching the ocean surface. *Photochem Photobiol* 32(3): 367-374.

Banaszak AT, Santos MGB, LaJeunesse TC, Lesser MP (2006) The distribution of mycosporine-like amino acids (MAAs) and the phylogenetic identity of symbiotic dinoflagellates in cnidarian hosts from the Mexican Caribbean. *J Exp Mar Biol Ecol* 337(2): 131-146.

Bandaranayake W (1998) Mycosporines: are they nature's sunscreens? *Nat Prod Rep* 15(2): 159-172.

Barceló-Villalobos M, Figueroa FL, Korbee N, Álvarez-Gómez F, Abreu MH (2017) Production of mycosporine-like amino acids from *Gracilaria vermiculophylla* (Rhodophyta) cultured through one year in an integrated multi-trophic aquaculture (IMTA) system. *Mar Biotechnol* 19(3): 246-254.

Bedoux G, Pliego-Cortés H, Dufau C, Hardouin K, Boulho R, Freile-Pelegrin Y, Bourgougnon N (2020) Production and properties of mycosporine-like amino acids isolated from seaweeds. *Adv Bot Res* 95: 213-245.

Bhatia S, Garg A, Sharma K, Kumar S, Sharma A, Purohit AP (2011) Mycosporine and mycosporine-like amino acids: A paramount tool against ultra violet irradiation. *Pharmacog Rev* 5(10): 138.

Callone AI, Carignan M, Montoya NG, Carreto JI (2006) Biotransformation of mycosporine like amino acids (MAAs) in the toxic dinoflagellate *Alexandrium tamarense*. *J Photochem Photobiol B: Biol*, 84(3), 204-212.

Carreto JI, Carignan MO (2011) Mycosporine-like amino acids: relevant secondary metabolites. Chemical and ecological aspects. *Mar Drugs* 9(3): 387-446.

Carreto JI, Carignan MO, Daleo G, Marco SD (1990) Occurrence of mycosporine-like amino acids in the red-tide dinoflagellate *Alexandrium excavatum*: UV-photoprotective compounds? *J Plankton Res*, 12(5), 909-921.

Carreto JI, Carignan MO, Montoya NG, Cozzolino E, Akselman R (2018) Mycosporine-like amino acids and xanthophyll-cycle pigments favour a massive spring bloom development of the dinoflagellate *Prorocentrum minimum* in Grande Bay (Argentina), an ozone hole affected area *J Mar Systems*, *178*, 15-28.

Chrapusta E, Kaminski A, Duchnik K, Bober B, Adamski M, Bialczyk J (2017) Mycosporine-like amino acids: Potential health and beauty ingredients. *Mar Drugs* 15(10): 326.

Davis GDJ, Vasanthi AHR (2011) Seaweed metabolite database (SWMD): A database of natural compounds from marine algae. *Bioinfo* 5(8): 361.

De la Coba F, Aguilera J, Figueroa FL, De Gálvez MV, Herrera E (2009) Antioxidant activity of mycosporine-like amino acids isolated from three red macroalgae and one marine lichen. *J Appl Phycol* 21(2): 161-169.

Dunlap WC, Chalker BE (1986) Identification and quantitation of near-UV absorbing compounds (S-320) in a hermatypic scleractinian. *Coral Reefs* 5(3): 155-159.

Dunlap WC, Shick JM (1998) Ultraviolet radiation-absorbing mycosporine-like amino acids in coral reef organisms: a biochemical and environmental perspective. *J Phycol* 34(3): 418-430.

Dunlap WC, Yamamoto Y (1995) Small-molecule antioxidants in marine organisms: antioxidant activity of mycosporine-glycine. *Comp Biochem Physiol Part B: Biochem Mol Biol* 112(1): 105-114.

Eckes MJ, Siebeck UE, Dove S, Grutter AS (2008) Ultraviolet sunscreens in reef fish mucus. *Mar Ecol Prog Ser* 353: 203-211.

Franklin LA, Yakovleva I, Karsten U, Lüning K (1999) Synthesis of mycosporine-like amino acids in *Chondrus crispus* (Florideophyceae) and the consequences for sensitivity to ultraviolet B radiation *J Phycol*, 35(4), 682-693.

Geraldes V, Pinto E (2021) Mycosporine-like amino acids (MAAs): biology, chemistry and identification features. *Pharmaceuticals* 14(1): 63.

Gharib R, Tabarzad M, Hosseinabadi T (2020) Effect of high salinity on mycosporine-like amino acid production in *Desmodesmus* sp. *Trends Pept Protein Sci* 5: 1-6.

Helbling EW, Barbieri ES, Sinha RP, Villafañe VE, Häder DP (2004) Dynamics of potentially protective compounds in Rhodophyta species from Patagonia (Argentina) exposed to solar radiation. *J Photochem Photobiol B: Biol* 75(1-2): 63-71.

Hoyer K, Karsten U, Sawall T, Wiencke C (2001) Photoprotective substances in Antarctic macroalgae and their variation with respect to depth distribution, different tissues and developmental stages. *Mar Ecol Prog Ser* 211: 117-129.

Huovinen P, Gomez I, Figueroa FL, Ulloa N, Morales V, Lovengreen, C (2004) Ultraviolet-absorbing mycosporine-like amino acids in red macroalgae from Chile. *Bot Mar* 47: 21–29.

Jain S, Prajapat G, Abrar M, Ledwani L, Singh A, Agrawal A (2017) Cyanobacteria as efficient producers of mycosporine-like amino acids. *J Basic Microbiol* 57(9): 715-727.

Jiang H, Gao K, Helbling EW (2008) UV-absorbing compounds in *Porphyra haitanensis* (Rhodophyta) with special reference to effects of desiccation. *J ApplPhycol* 20(4): 387-395.

Jofre J, Celis-Plá PS, Figueroa FL, Navarro NP (2020) Seasonal variation of mycosporine-like amino acids in three subantarctic red seaweeds. *Mar Drugs* 18(2): 75.

Jokiel PL, York Jr RH (1982) Solar ultraviolet photobiology of the reef coral *Pocillopora damicornis* and symbiotic zooxanthellae. *Bull Mar Sci* 32(1): 301-315.

Kageyama H, Waditee-Sirisattha R (2018) Mycosporine-like amino acids as multifunctional secondary metabolites in cyanobacteria: From biochemical to application aspects. *Stud Nat Prod Chem* 59: 153-194.

Karsten U, Bischof K, Hanelt D, Tüg H, Wiencke C (1999) The effect of ultraviolet radiation on photosynthesis and ultraviolet-absorbing substances in the endemic Arctic macroalga *Devaleraea ramentacea* (Rhodophyta) *Physiol Plant*, 105(1), 58-66.

Karsten U, Sawall T, Wiencke C (1998) A survey of the distribution of UV-absorbing substances in tropical macroalgae *Phycol Res* 46(4): 271-279.

Kicklighter CE, Kamio M, Nguyen L, Germann MW, Derby CD (2011) Mycosporine-like amino acids are multifunctional molecules in sea hares and their marine community. *Proc Nat Acad Sci* 108(28): 11494-11499.

Klisch M, Sinha RP, Richter PR, Häder D-P (2001) Mycosporine-like amino acids (MAAs) protect against UV-B-induced damage in *Gyrodinium dorsum* Kofoid. *J Plant Physiol* 158(11): 1449-1454.

Korbee N, Figueroa FL, Aguilera J (2006) Acumulación de aminoácidos tipo micosporina (MAAs): biosíntesis, fotocontrol y funciones ecofisiológicas. *Revista Chilena de Historia Natural* 79(1): 119-132.

Leach CM (1965) Ultraviolet-absorbing substances associated with light-induced sporulation in fungi. *Can J Bot* 43(2): 185-200.

Liu Z, Häder DP, Sommaruga R (2004) Occurrence of mycosporine-like amino acids (MAAs) in the bloom-forming cyanobacterium *Microcystis aeruginosa*. *J Plankton Res* 26(8): 963-966.

Llewellyn CA, Greig C, Silkina A, Kultschar B, Hitchings MD, & Farnham, G (2020) Mycosporine-like amino acid and aromatic amino acid transcriptome response to UV and far-red light in the cyanobacterium *Chlorogloeopsis fritschii* PCC 6912 *Sci Reports* 10(1): 1-13.

Llewellyn CA, Harbour DS (2003) A temporal study of mycosporine-like amino acids in surface water phytoplankton from the English Channel and correlation with solar irradiation. *J Mar Biol Assoc Unit Kingd* 83(1): 1-9.

Makino M (1999) The influence of ultraviolet irradiation on marine organisms 3 Effects of ultraviolet absorbing substance, palythine from *Chondrus yendoi*. *Japan J Phycol* (Sorui), 47: 173-177.

Michalek-Wagner K (2001) Seasonal and sex-specific variations in levels of photo-protecting mycosporine-like amino acids (MAAs) in soft corals. *Mar Biol* 139(4): 651-660.

Neale PJ, Banaszak AT, Jarriel CR (1998) Ultraviolet sunscreens in *Gymnodinium sanguineum* (Dinophyceae): mycosporine-like amino acids protect against inhibition of photosynthesis. *J Phycol* 34(6): 928-938.

Nunnery JK, Mevers E, Gerwick WH (2010) Biologically active secondary metabolites from marine cyanobacteria. *Curr Opin Biotechnol* 21(6): 787-793.

Oren A (1997) Mycosporine-like amino acids as osmotic solutes in a community of halophilic cyanobacteria. *Geomicrobiol J* 14(3): 231-240.

Oren A (1997) Mycosporine-like amino acids as osmotic solutes in a community of halophilic cyanobacteria Geomicrobiol J 14(3): 231-240.

Oren A, Gunde-Cimerman N (2007) Mycosporines and mycosporine-like amino acids: UV protectants or multipurpose secondary metabolites? *FEMS Microbiol Lett* 269(1): 1-10.

Orfanoudaki M, Hartmann A, Miladinovic H, Nguyen Ngoc H, Karsten U, & Ganzera M (2019) Bostrychines A–F, six novel mycosporine-like amino-acids and a novel betaine from the red alga *Bostrychia scorpioides* Mar Drugs 17(6): 356.

Peinado NK, Abdala Díaz RT, Figueroa, FL, Helbling EW (2004) Ammonium and UV radiation stimulate the accumulation of mycosporine-like amino acids in *Porphyra columbina* (Rhodophyta) from Patagonia, Argentina. *J Phycol 40*(2): 248-259.

Poloczanska ES, Burrows MT, Brown CJ, García Molinos J, Halpern BS, Hoegh-Guldberg O, Sydeman WJ (2016) Responses of marine organisms to climate change across oceans. *Front Mar Sci* 62.

Portwich A, Garcia-Pichel F (1999) Ultraviolet and osmotic stresses induce and regulate the synthesis of mycosporines in the cyanobacterium *Chlorogloeopsis* PCC 6912. *Arch Microbiol* 172(4): 187-192.

Portwich A, Garcia-Pichel F (2003) Biosynthetic pathway of mycosporines (mycosporine-like amino acids) in the cyanobacterium *Chlorogloeopsis* sp. strain PCC 6912. *Phycologia* 42(4): 384-392.

Rastogi RP, Sonani RR, Madamwar D, Incharoensakdi A (2016) Characterization and antioxidant functions of mycosporine-like amino acids in the cyanobacterium *Nostoc* sp. R76DM. *Algal Res* 16:110-118.

Riegger L, Robinson D (1997) Photoinduction of UV-absorbing compounds in Antarctic diatoms and *Phaeocystis antarctica*. *Mar Ecol Prog Ser* 160: 13-25.

Rosic NN, Dove S (2011) Mycosporine-like amino acids from coral dinoflagellates. *Appl Environ Microbiol* 77(24): 8478-8486.

Shibata K (1969) Pigments and a UV-absorbing substance in corals and a blue-green alga living in the Great Barrier Reef. *Plant Cell Physiol* 10(2): 325-335.

Shick JM, & Dunlap, WC (2002). Mycosporine-like amino acids and related gadusols: biosynthesis, accumulation, and UV-protective functions in aquatic organisms. *Annu Rev Physiol*64(1): 223-262.

Shick JM, Ferrier-Pagès C, Grover R, Allemand D (2005) Effects of starvation, ammonium concentration, and photosynthesis on the UV-dependent accumulation of mycosporine-like amino acids (MAAs) in the coral *Stylophora pistillata. Mar Ecol Prog Ser* 295: 135-156.

Shick JM, Lesser MP, Dunlap WC, Stochaj WR, Chalker BE, Won JW (1995) Depth-dependent responses to solar ultraviolet radiation and oxidative stress in the zooxanthellate coral *Acropora microphthalma. Mar Biol* 122(1): 41-51.

Sinha RP, Ambasht NK, Sinha JP, Häder DP (2003) Wavelength-dependent induction of a mycosporine-like amino acid in a rice-field cyanobacterium, *Nostoc commune*: role of inhibitors and salt stress. *Photochem Photobiol Sci* 2(2): 171-176.

Sinha RP, Ambasht NK, Sinha JP, Klisch M, Häder DP (2003) UV-B-induced synthesis of mycosporine-like amino acids in three strains of *Nodularia* (cyanobacteria). *J Photochem Photobiol B: Biol* 71(1-3): 51-58.

Sinha RP, Häder D-P (2008) UV-protectants in cyanobacteria. *Plant Sci* 174(3): 278-289.

Sinha RP, Klisch M, Gröniger A, & Häder D-P (2000) Mycosporine-like amino acids in the marine red alga *Gracilaria cornea*—effects of UV and heat. Env Exp Bot 43(1): 33-43.

Sinha RP, Klisch M, Helbling EW, Häder D-P (2001) Induction of mycosporine-like amino acids (MAAs) in cyanobacteria by solar ultraviolet-B radiation. *J Photochem Photobiol B: Biol* 60(2-3): 129-135.

Sivalingam PM, Ikawa T, Nisizawa K (1976) Physiological roles of a substance 334 in algae. Bot Mar 19: 9-21.

Stochaj WR, Dunlap WC, Shick JM (1994) Two new UV-absorbing mycosporine-like amino acids from the sea anemone *Anthopleura elegantissima* and the effects of zooxanthellae and spectral irradiance on chemical composition and content. *Mar Biol* 118(1): 149-156.

Sun Y, Zhang N, Zhou J, Dong S, Zhang X, Guo L, Guo G (2020) Distribution, contents, and types of mycosporine-like amino acids (MAAs) in marine macroalgae and a database for MAAs based on these characteristics. *Mar Drugs* 18(1): 43.

Taira H, Aoki S, Yamanoha B, & Taguchi S (2004) Daily variation in cellular content of UV-absorbing compounds mycosporine-like amino acids in the marine dinoflagellate *Scrippsiella sweeneyae. J Photochem Photobiol B: Biol 75*(3): 145-155.

Vega J, Bonomi-Barufi J, Gómez-Pinchetti JL, Figueroa FL (2020) Cyanobacteria and red macroalgae as potential sources of antioxidants and UV radiation-absorbing compounds for cosmeceutical applications. *Mar Drugs* 18(12): 659.

Volkmann M, Gorbushina AA, Kedar L, Oren A (2006) Structure of euhalothece-362, a novel red-shifted mycosporine-like amino acid, from a halophilic cyanobacterium (*Euhalothece* sp.). *FEMS Microbiol Lett* 258(1): 50-54.

Wada N, Sakamoto T, Matsugo S (2013) Multiple roles of photosynthetic and sunscreen pigments in cyanobacteria focusing on the oxidative stress. *Metabolites* 3(2): 463-483.

Waditee-Sirisattha R, Kageyama H, Sopun W, Tanaka Y, Takabe T (2014) Identification and upregulation of biosynthetic genes required for accumulation of mycosporine-2-glycine under salt stress conditions in the halotolerant cyanobacterium *Aphanothece halophytica*. *Appl Environ Microbiol* 80(5): 1763-1769.

Wright DJ, Smith SC, Joardar V, Scherer S, Jervis J, Warren A, Potts M (2005) UV irradiation and desiccation modulate the three-dimensional extracellular matrix of *Nostoc commune* (Cyanobacteria). *J Biol Chem* 280(48): 40271-40281.

Yakovleva I, Bhagooli R, Takemura A, Hidaka M (2004) Differential susceptibility to oxidative stress of two scleractinian corals: antioxidant functioning of mycosporine-glycine. *Compar Biochem Physiol Part B: Biochem Mol Biol* 139(4): 721-730.

Yakovleva I, Hidaka M (2004) Diel fluctuations of mycosporine-like amino acids in shallow-water scleractinian corals. *Mar Biol* 145(5): 863-873.

## Chapter 7

# Natural Compounds Derived from Cyanobacteria: Potential Benefits and Applicability to Skin-Care

**Hakuto Kageyama[1,2,*], Masaki Honda[1], Takashi Hibino[1,2] and Rungaroon Waditee-Sirisattha[3,†]**

[1]Faculty of Science and Technology, Meijo University, Shiogamaguchi, Tenpaku-ku, Nagoya, Aichi, Japan
[2]Graduate School of Environmental and Human Sciences, Meijo University, Shiogamaguchi, Tenpaku-ku, Nagoya, Aichi, Japan
[3]Department of Microbiology, Faculty of Science, Chulalongkorn University, Payathai Road, Pathum-wan, Bangkok, Thailand

## Abstract

Cyanobacteria are photosynthetic prokaryotes that are widely distributed in aquatic and terrestrial environments. They are found to be a promising source of beneficial bioactive compounds. In this review article, we describe our current understanding of cyanobacterial secondary metabolites by focusing on beneficial biological properties for skin care. Mycosporine-like amino acids (MAAs), which are known to be produced in cyanobacteria, show ultraviolet (UV)-screening and potential anti-photoaging properties. Accumulated evidence indicates the potential applications of MAAs not only in physical sunscreens but also in various therapeutic agents for skin protection. In addition, scytonemin, phycobiliproteins, carotenoids, and enzymes derived from cyanobacteria

[*] Corresponding Author's Email: kageyama@meijo-u.ac.jp.
[†] Corresponding Author's Email: Rungaroon.W@chula.ac.th.

In: Cyanobacteria
Editor: Rajeshwar P. Sinha
ISBN: 979-8-88697-169-9

are described and discussed as important molecules from the cosmetic and dermatological points of view.

**Keywords:** cyanobacteria, bioactive compound, skin, mycosporine-like amino acid, scytonemin, antioxidant

## Introduction

The skin, which consists of multiple layers, is the largest organ in the human body. This complicated organ can play many roles. For instance, it acts as a physical and biological barrier to the external environment by covering the internal organs, maintains the body temperature, and prevents water loss (Pangestuti et al. 2021). Exposure to radiation from the sun is one of the major environmental stresses to the skin. Sunlight consists of nonionizing electromagnetic radiation, which includes ultraviolet (UV), visible, and infrared (IR) radiations. UV radiation can be divided into three distinct types: UV-A (315–400 nm), UV-B (280–315 nm), and UV-C (100–280 nm). Because of their reduction through the Earth's atmosphere, the radiation doses of UV-A and UV-B reaching the Earth's surface represent 6.3% and 1.5% of the total solar irradiance, respectively (Frederick et al. 1989). On the other hand, essentially all UV-C radiation is absorbed by $O_2$ and $O_3$ before it reaches the Earth's surface. It is known that both UV-A and UV-B radiations can cause skin damage, skin photoaging, and skin cancer (Gruber et al. 2007; Ikehata 2018; Browne et al. 2014; Oyamada et al. 2008). Therefore, blocking the exposure to UV radiation is important for skin care. For skin care purposes, different civilizations have used various compounds from different natural sources, including minerals and plants, as well as chemically synthesized products (González-Minero and Bravo-Díaz 2018). In particular, there has been a growing demand for natural skin care products because they have been proven to be safer for human utilization than a number of synthetic chemical products (Jesumani et al. 2019).

Cyanobacteria have recently gained much attention as promising microbial chassis for the bioproduction of beneficial natural compounds because of their photoautotrophic properties and relatively higher growth rates (Kageyama and Waditee-Sirisattha 2018a). If sustainable cultivation systems are developed at an industrial scale, atmospheric carbon capture by cyanobacteria and its utilization may play a significant role in alleviating the harmful effects of increased $CO_2$ levels. Cyanobacteria are widely distributed in both aquatic and terrestrial areas, including extreme habitats, such as high light, high or low temperature, desiccation, and high salinity conditions (Wada et al. 2013). Diverse cyanobacterial species are known to exhibit unique survival strategies to thrive under such harsh conditions. For example, several cyanobacterial species can accumulate specific small compounds, including osmolytes, antioxidants, and sunscreen molecules, to combat environmental stresses. These compounds could be considered for cosmetic and pharmaceutical applications. Moreover, certain cyanobacterial species are known to be transformable, and the genomic sequences of various cyanobacterial species are available (Macallef et al. 2015). These facts indicate that genetically engineered cyanobacteria, which can efficiently produce useful compounds, can be generated. Herein, we review potential natural skin care products, including mycosporine-like amino acids (MAAs), scytonemin, and other antioxidants, derived from cyanobacteria. In addition to describing their chemical characteristics, we discuss the current state, advantages, and drawbacks of their applications.

## MAAs

MAAs are a group of low-molecular-weight secondary metabolites known as natural sunscreens. They are believed to serve as potential UV-absorbing compounds (Gao and Garcia-Pichel 2011; Kageyama and Waditee-Sirisattha 2018b). It is known that MAAs are bioproduced in a wide range of organisms, including cyanobacteria, micro- and macro-algae, fungi, actinomyces, corals, and dinoflagellates (Kageyama and Waditee-Sirisattha 2018b).

MAAs are characterized by a cyclohexenone or cyclohexenimine ring as the core chromophore structure that generally binds to amino acid residues or their amino alcohol with further modifications at positions C1 and C3 (Wada et al. 2013; Kageyama and Waditee-Sirisattha 2018b). So far, more than 30 different MAAs have been reported. Depending on their structure, MAAs can be classified in three types: monosubstituted cyclohexanone-type MAAs, disubstituted cyclohexeneimine-type MAAs, and further modified derivatized-type MAAs (Wada et al. 2015).

## Cyanobacterial MAAs

To date, several cyanobacterial species have been reported to be MAA accumulators. We have provided examples of cyanobacterial MAAs in Table 1. Although shinorine is thought to be a common and dominant MAA in cyanobacteria, various cyanobacterial MAAs, including monosubstituted, disubstituted, derivatized, and uncharacterized ones, have been detected.

**Table 1.** Examples of MAAs accumulated in cyanobacteria

| **MAAs** | **Cyanobacterial species** | **Reference** |
|---|---|---|
| Monosubstituted cyclohexanone-type | | |
| Mycosporine-glycine ($\lambda_{max}$ = 310 nm) 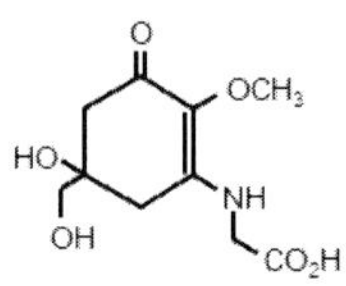  | *Chlorogloeopsis* PCC6912<br>*Anabaena doliolum*<br>*Arthrospira* sp. CU2556 | Portwich and Garcia-Pichel 1999<br>Singh et al. 2008a<br>Rastogi and Incharoensakdi 2014a |
| Mycosporine-GABA ($\lambda_{max}$ = 310 nm) 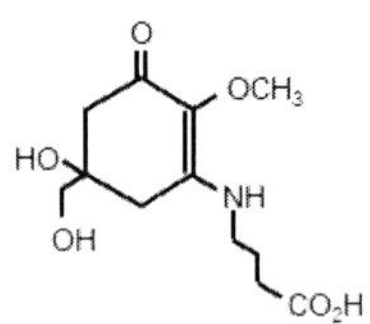  | *Nostoc commune* | Wada et al. 2015 |
| Disubstituted cyclohexeneimine-type | | |
| Shinorine ($\lambda_{max}$ = 334 nm) 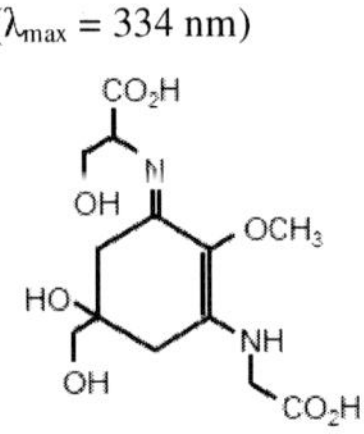  | *N. punctiforme* ATCC29133<br>*Nostoc* sp. strain HKAR-2<br>*Nostoc* sp. strain HKAR-6<br>*N. verrucosum*<br>*N. commune*<br>*Gloeocapsa* sp. CU2556<br>*Microcystis aeruginosa* strain PCC7806<br>*A. doliolum*<br>*A. variabilis* ATCC29413<br>*Chlorogloeopsis* PCC6912<br>*Scytonema* cf. *crispum* | Gao and Garcia-Pichel 2011<br>Richa and Sinha 2015<br>Richa and Sinha 2015<br>Inoue-Sakamoto et al. 2018<br>Singh et al. 2003<br>Rastogi and Incharoensakdi 2014b<br>Hu et al. 2015<br>Singh et al. 2008a<br>Singh et al. 2010<br>Portwich and Garcia-Pichel 1999<br>D'Agostino et al. 2016 |
| Porphyra-334 ($\lambda_{max}$ = 334 nm) | *Nostoc* sp. strain HKAR-2<br>*Nostoc* sp. strain HKAR-6 | Richa and Sinha 2015<br>Richa and Sinha 2015 |

| MAAs | Cyanobacterial species | Reference |
|---|---|---|
| 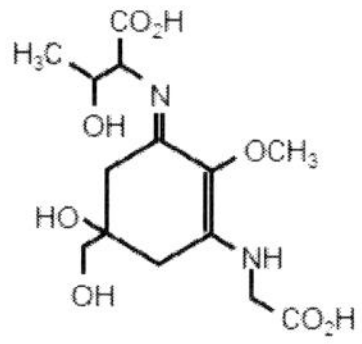 | *N. verrucosum*<br>*A. doliolum* | Inoue-Sakamoto et al. 2018<br>Singh et al. 2008a |
| Mycosporine-2-glycine<br>($\lambda_{max}$ = 331 nm)<br>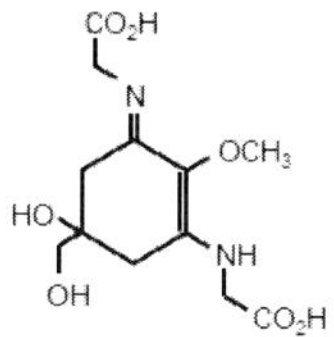 | *Halothece* sp. PCC7418<br>*Euhalothece* sp. | Waditee-Sirisattha et al. 2014<br>Kedar et al. 2002 |
| Palythine<br>($\lambda_{max}$ = 320 nm)<br>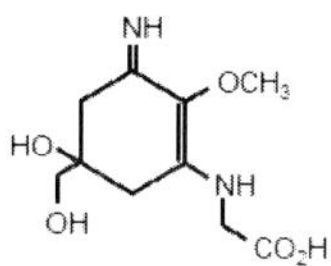 | *Lyngbya* sp. CU2555 | Rastogi and Incharoensakdi 2014c |
| Palythine-serine<br>($\lambda_{max}$ = 320 nm)<br>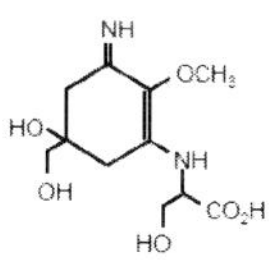 | *A. variabilis* ATCC29413<br>*Scytonema* cf. *crispum* | Singh et al. 2010<br>D'Agostino et al. 2016 |
| Euhalothece-362<br>($\lambda_{max}$ = 362 nm)<br>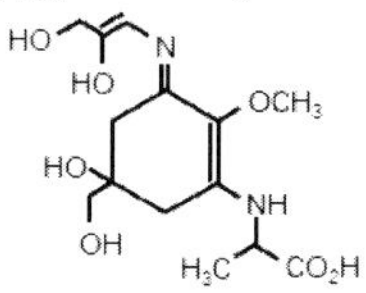 | *Euhalothece* sp. | Kedar et al. 2002 |
| Asterina-330<br>($\lambda_{max}$ = 330 nm)<br>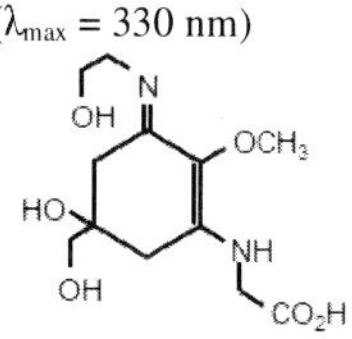 | *Lyngbya* sp. CU2555<br>*Trichodesmium* sp. | Rastogi and Incharoensakdi 2014c<br>Subramaniam et al. 1999 |

**Table 1.** (Continued)

| MAAs | Cyanobacterial species | Reference |
|---|---|---|
| Palythene ($\lambda_{max}$ = 360 nm) 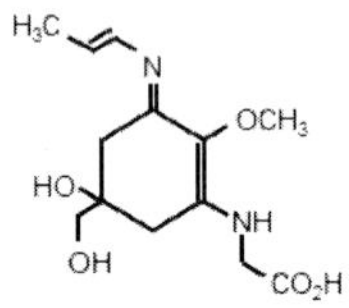 | *Trichodesmium* sp. | Subramaniam et al. 1999 |
| Derivatized-type | | |
| Hexose-bound porphyra-334 (508-DaMAA) ($\lambda_{max}$ = 334 nm) | *N. commune* | Nazifi et al. 2013 |
| Two-hexose-bound palythine-threonine (612-Da MAA) ($\lambda_{max}$ = 322 nm) | *N. commune* | Nazifi et al. 2015 |
| Arabinose-bound porphyra-334 (478-Da MAA) ($\lambda_{max}$ = 335 nm) | *N. commune* | Nazifi et al. 2015; Korteerakul et al. 2020 |
| Pentose-bound shinorine (464-Da MAA) ($\lambda_{max}$ = 332 nm) | *N. commune* | Nazifi et al. 2015; Korteerakul et al. 2020 |
| 880-Da MAA ($\lambda_{max}$ = 331 nm) | *N. commune* | Nazifi et al. 2015 |
| 1050-Da MAA ($\lambda_{max}$ = 332 nm) | *N. commune* | Nazifi et al. 2015 |
| 756-Da MAA ($\lambda_{max}$ = 322 nm) | *N. commune* | Sakamoto et al. 2019 |
| 13-O-(β-galactosyl)-porphyra-334 ($\lambda_{max}$ = 334 nm) | *N. sphaericum* | Ishihara et al. 2017 |
| Hexose-shinorine ($\lambda_{max}$ = 333 nm) | *S.* cf. *crispum* | D'Agostino et al. 2016 |
| Hexose-palythine-serine ($\lambda_{max}$ = 320 nm) | *S.* cf. *crispum* | D'Agostino et al. 2016 |
| Unclassified MAAs | | |
| M-307 ($\lambda_{max}$ = 307 nm) | *Gloeocapsa* sp. CU2556 | Rastogi and Incharoensakdi 2014b |
| M-312 ($\lambda_{max}$ = 312 nm) | *Lyngbya* sp. CU2555 | Rastogi and Incharoensakdi 2014c |

## Biosynthetic Pathway of Cyanobacterial MAAs and Its Regulation

The biosynthetic pathway of cyanobacterial MAAs was intensively elucidated in the 2000s. Finally, biosynthetic gene clusters were identified in *Anabaena variabilis* ATCC29413 and *Nostoc punctiforme* ATCC29133 (Gao and Garcia-Pichel 2011; Balakus and Walsh 2010). Then, their homologous genes in other cyanobacteria, such as *Halothece* sp. PCC7418 and *Cylindrospermum stagnale* PCC7417, were confirmed to produce MAAs (Waditee-Sirisattha et al. 2014; Katoch et al. 2016). In general, it is believed that 4-deoxygadusol (4-DG) acts as a common precursor compound for MAA synthesis. This precursor compound is predicted to be synthesized by two genes encoding 3-dehydroquinate (DHQ) synthase (DHQS) and *O*-methyltransferase (*O*-MT) from DHQ or sedoheptulose-7-phosphate (SHP). DHQ and SHP are intermediates of the shikimate pathway and pentose phosphate pathway, respectively. Next, mycosporine-glycine is produced by the enzyme adenosine triphosphate (ATP)-grasp, which catalyzes the conjugation of 4-DG and glycine. Disubstituted MAAs are further synthesized by the enzyme non-ribosomal peptide synthase (NRPS)-like protein or D-Ala-D-Ala ligase, which catalyzes the conjugation of mycosporine-glycine and an additional amino acid moiety.

It is well known that the bioproduction of MAAs is upregulated by UV radiation exposure. For example, UV-B irradiation was found to induce MAA accumulation in *A. variabilis* ATCC29413, *A. doliolum*, *Anabaena* sp., *N. commune*, and *Scytonema* sp. (Singh et al. 2008a; 2008b; Sinha et al. 2001). In addition to UV radiation exposure, other abiotic stresses could enhance MAA production in cyanobacteria. One such example is salt stress. In *Halothece* sp. PCC7418 and *Chlorogloeopsis* PCC6912, NaCl stress was found to increase intracellular MAA amounts (Portwich and Garcia-Pichel 1999; Waditee-Sirisattha et al. 2014). These observations suggest that MAAs function as osmoprotectants in cyanobacteria. Temperature shift could also be a critical factor for the induction of MAAs. In *Halothece* sp. PCC7418, cold stress strongly induced mycosporine-2-glycine production, whereas heat stress did not have this effect (Patipong et al. 2019). In addition, nutrient availability could affect cyanobacterial MAA production. In *Halothece* sp. PCC7418, the supplementation of nitrate, serine, or glycine strongly induced mycosporine-2-glycine production (Waditee-Sirisattha et al. 2015). In *A. variabilis* ATCC29413, sulfur deficiency induced palythine-serine production (Singh et al. 2010). Further investigations are needed to elucidate the molecular mechanisms underlying the regulation of MAA biosynthesis in the future.

## Use of MAAs for Skin Care

Given their UV-screening properties, MAAs have potential for cosmetic and pharmaceutical applications. In addition to being UV-screening compounds, MAAs have recently been shown to have several biological properties, such as antioxidative and DNA-protective properties. In this subsection, the applicable bioactivities and the industrial production of MAAs are described.

## UV-Screening Properties

MAAs show a maximum absorbance in the UV range of 310 to 362 nm, with high molar extinction coefficients ($\varepsilon$ = 28,100-50,000 $M^{-1}cm^{-1}$) (Gao and Garcia-Pichel 2011). They can convert absorbed UV into harmless energy in the form of heat without the generation of reactive oxygen species (ROS) (Conde et al. 2007). Sunscreen products that contain MAAs have already been commercialized. Some examples are Helioguard 365® (Mibelle Biochemistry, Switzerland) and Helinori® (Biosil Technologies, USA). Both these products contain MAAs extracted from the red alga *Porphyra umbilicalis*. Helioguard 365® contains shinorine and porphyra-334, which are encapsulated in liposomes, whereas Helinori® contains shinorine, porphyra-334, and palythine (Soule and Garcia-Pichel 2013). Helioguard 365® was found to reduce lipid peroxidation and improve the firmness and smoothness of human skin, while Helinori® was found to reduce sunburn. Regarding the use of cyanobacterial MAAs, the Japanese companies Skinxia and Lekarka, for example, deal with a series of cosmetic skin care products containing MAAs extracted from cyanobacteria. In addition, Soule and Garcia-Pichel have discussed the development of a sunscreen pill because marine invertebrates could accumulate MAAs after ingestion from their diet (Soule and Garcia-Pichel 2013). As certain MAA-accumulating cyanobacteria are traditionally edible (Korteerakul et al. 2020), cyanobacteria may be used for producing such functional foods containing MAAs.

## Antioxidative Properties

UV exposure can induce ROS generation, which is one of the major causes of the skin photoaging process. Therefore, ROS-scavenging antioxidant molecules are commonly used in cosmetic formulations to prevent skin aging. To date, many MAAs have been reported to exhibit antioxidative activities. In 2009, de la Coba et al. reported a higher antioxidant activity of mycosporine-glycine isolated from the marine lichen *Lichina pygmaea* at pH 8.5 than of shinorine and porphyra-334 (de la Coba et al. 2009). In that study, superoxide scavenging activities of these MAAs were assessed by inhibiting the autoxidation of pyrogallol. It should be noted that the assessment of the antioxidative activities of MAAs could be affected by variations in the methods used and their parameters. For example, shinorine purified from *Chlamydomonas hedleyi* did not exhibit 1,1-diphenyl-2-picrylhydrazyl (DPPH) free radical scavenging activity (Suh et al. 2014), although several reports have indicated the antioxidative activities of shinorine using other methods (de la Coba et al. 2009; Ngoennet et al. 2018). Moreover, Ngoennet et al. recently reported the antioxidative activity of porphyra-334 extracted from Helioguard 365® using 2,2'-azinobis (3-ethylbenzonthiazoline-6-sulfonic acid) (ABTS) as a substrate (Ngoennet et al. 2018). However, Inoue-Sakamoto et al. reported that porphyra-334 extracted from *N. verrucosum* did not exhibit antioxidative activities using the ABTS method (Inoue-Sakamoto et al. 2018). This discrepancy may have resulted from the differences in the source and purity of shinorine. Ngoennet et al. analyzed the antioxidative activities of high-purity shinorine, porphyra-334, and mycosporine-2-glycine using the ABTS method and reported half-maximal inhibitory concentration ($IC_{50}$) values of 94, 133, and 40 μM, respectively (Ngoennet et al. 2018). The relationship between antioxidative activity and the chemical structure of MAAs would be an interesting topic to be explored in the future.

## Antiglycative Properties

Nonenzymatic protein glycation leads to the generation of advanced glycation-end products (AGEs). The accumulation of AGEs in the human body is believed to be associated with the normal aging process (Münch et al. 1997). To date, several reports have suggested that MAAs act as promising inhibitors against protein glycation. Tarasuntisuk et al. demonstrated that mycosporine-2-glycine purified from a halotolerant cyanobacterium exhibited glycation-dependent crosslinking of hen egg white lysozyme (Tarasuntisuk et al. 2018). Its $IC_{50}$ value of 1.6 mM was smaller than that of the well-known synthetic glycation inhibitor amino guanidine (4.7 mM). Later, Orfanoudaki et al. detected the inhibitory effects of 10 MAAs, namely shinorine, porphyra-334, mycosporine-methylamine-threonine, mycosporine-alanine-glycine, aplysiapalythine A, asterina-330, palythine, bostrychine-D, bostrychine-E, and bostrychine-F, on the glycation of bovine serum albumin (Orfanoudaki et al. 2019). Although the molecular mechanisms underlying the inhibitory activity of MAAs on AGE formation remain largely unknown, of the 10 MAAs examined, shinorine, porphyra-334, mycosporine-alanine-glycine, and bostrychine-D exhibited particularly higher activities.

## Other Bioactivities

MAAs are known to exhibit inhibitory effects on collagenase, which is associated with the skin aging process. Several MAAs have been shown to inhibit collagenase activity *in vitro* (Hartmann et al. 2015; Tarasuntisuk et al. 2018; Orfanoudaki et al. 2019). Because MAAs have been suggested to acted as metal chelators, the metal-chelating properties of MAAs may be involved in the inhibition of collagenase, which could be activated by divalent cations (Tarasuntisuk et al. 2018). The DNA-protecting activities of MAAs are also thought to be important for skin care. As DNA can be damaged by ROS generated through UV exposure, the antioxidative activities of MAAs could protect DNA. In fact, mycosporine-2-glycine was found to prevent DNA damage induced by oxidative stress in the A375 human melanoma cell line (Cheewinthamrongrod et al. 2016). MAAs, including mycosporin-2-glycine, could also exhibit inhibitory effects on lipopolysaccharide (LPS)-induced nitric oxide (NO) production in an *in vitro* model of RAW 264.7 macrophages (Tarasuntisuk et al. 2019). Because NO is an inflammatory signaling molecule, these MAAs could exert anti-inflammatory activities. In addition, Suh et al. found that porphyra-334 exerted protective effects against UV-induced apoptosis through the attenuation of the caspase pathway in HaCaT cells (Sue et al. 2017).

## Industrial Production of MAAs

Given their beneficial bioactivities, as described above, MAAs have recently attracted considerable attention not only in academic research but also at an industrial level in the fields of cosmetology and pharmacology. In fact, several patents related to MAA application have appeared (Kageyama and Waditee-Sirisattha 2018b). For the industrial application of MAAs, their preparation strategy would be very important. From the viewpoint of sustainable production, the cultivation of cyanobacteria is attractive. Using their photosynthetic machinery, useful target compounds could be obtained from cyanobacterial biomass grown utilizing sunlight or artificial light and atmospheric $CO_2$. Cyanobacteria can be cultivated in open, closed, or ultrathin layer systems (Grewe and Pulz 2012). However, to the best of our knowledge, the industrial large-scale cyanobacterial platform for MAA production has not yet been successful for sustainable operation. For the effective bioproduction of MAAs, the optimization of growth conditions for each cyanobacterium is important. Alternatively, cyanobacterial MAA synthesis genes could be introduced into other host cells; this is an interesting topic. So far, it has been reported that the heterologous expression of genes from *Halothece* sp. PCC7418 and *C. stagnale* PCC7417 in *Escherichia coli* resulted in the production of specific MAAs (Katoch et al. 2016; Patipong et al. 2017). The purification process of MAAs from biomass should also be considered. In the case of using cyanobacterial cells, broadly applicable methods for extraction, characterization, and chromatographic purification have been proposed (Volkmann and Gorbushina 2006; Ngoennet et al. 2018). In addition to chromatographic separation, the development of other separation methods, such as adsorption, solid-phase extraction, and crystallization, may be valuable for effective MAA production from biomass in the future.

## Scytonemin

Scytonemin, which is known as a cyanobacterial UV-screening compound, is commonly found in the sheath of many cyanobacterial species (Bultel-Poncé et al. 2004). This pigment is present in the oxidized yellow-brown and reduced red forms (Figure 1). Scytonemin could count for 5% of the total dry weight of cyanobacterial cells (Proteau et al. 1993). A characteristic absorption maximum peak at 386 ± 2 nm and significant peaks at 252 ± 2, 278 ± 2, and

300 ± 2 nm were observed *in vitro* (Rastogi et al. 2015). Scytonemin dissolves in lipids but does not dissolve in water. In addition to the range of spectral UV absorption, this is one of the differences between scytonemin and the other cyanobacterial UV-screening compounds, i.e., MAAs.

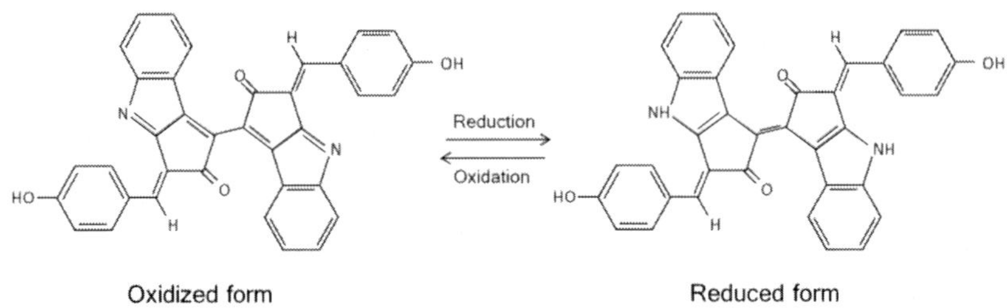


**Figure 1.** Structure of scytonemin.

## Biosynthetic Pathway of Scytonemin and Its Regulation

The bioproduction of scytonemin by cyanobacteria could be induced by UV-A irradiation; however, green or red light were not found to influence scytonemin synthesis (Garcia-Pichel and Castenholz 1991). In addition to UV-A treatment, other abiotic stresses could enhance scytonemin production. Oxidative stress, temperature stress, desiccation stress, and nitrogen limitation stress have been reported to induce scytonemin synthesis in cyanobacteria (Dillon et al. 2002; Fleming et al. 2007; Fleming et al. 2008). The genes associated with the scytonemin biosynthetic pathway have been partially elucidated by characterizing an 18-gene cluster containing NpR1276–NpR1259 in the cyanobacterium *N. punctiforme* ATCC29133 (Soule et al. 2007; Balskus and Walsh 2009; Soule et al. 2009). Scytonemin is believed to be synthesized from two precursors, tryptophan and prephenate, in the cytoplasm followed by the periplasm, and it finally gets located in the extracellular sheath. However, the molecular mechanisms underlying the transportation of scytonemin to the extracellular sheath remain unknown and need further investigation.

## Potential Uses of Scytonemin for Skin Care

Given its specific characteristics, scytonemin has good prospects in cosmetic and biomedical fields. Assessment of the photoprotective efficacy of scytonemin revealed that scytonemin can prevent up to 90% of solar UV radiation from entering cyanobacterial cells (Garcia-Pichel and Castenholz 1991; Derikvand et al. 2017). Besides its UV-screening property, scytonemin possesses an antioxidant property. Rastogi et al. demonstrated that scytonemin exhibited efficient photoprotective ability by detoxifying ROS generated by UV irradiation and by reducing the formation of thymine dimers in *Rivularia* sp. HKAR-4 cells (Rastogi et al. 2013). Moreover, Matsui et al. showed that scytonemin purified from *N. commune* quenched an organic radical *in vitro* and accounted for up to 10% of the total activity of an ethanol extract of *N. commune* (Matsui et al. 2012). Furthermore, Kang et al. found that scytonemin inhibited skin inflammation (Kang et al. 2012). They found that the topical application of scytonemin inhibited 12-*O*-tetradecanoylphorbol-13-acetate (TPA)-induced skin inflammation in mice. They also demonstrated that scytonemin suppressed LPS-induced expression of tumor necrosis factor-α (TNF-α) and inducible nitric oxide synthase (iNOS) in a mouse macrophage cell line. Considering these beneficial effects and its nontoxic properties, scytonemin offers biotechnological advantages for the development of cosmetic products in the industry.

Like MAAs, to the best of our knowledge, a sustainable large-scale industrial platform for scytonemin production is not yet available. To develop this platform, the selection of a cyanobacterial strain for scytonemin production is important. Because the biosynthesis of complete scytonemin by the heterologous expression of the cyanobacterial scytonemin biosynthetic gene cluster in other host microorganisms, such as *E. coli*, has not been successful (Gao et al. 2021), we have no choice but to use scytonemin-producing cyanobacterial strains at present. It should be noted that some cyanobacterial species, such as *Synechocystis* sp. PCC6803 and *Anabaena* sp. PCC7120, do not produce scytonemin. For the extraction of scytonemin from cyanobacterial biomass, organic solvents, such as ethyl acetate and acetone, are often used because of its hydrophobicity (Rastogi et al. 2014b; Pathak et al. 2017; Fuentes-Tristan et al. 2019). The use of large amounts of organic solvents may become an obstacle for scaling up the process. In addition, there is room for improvement of the subsequent chromatographic separation process. Although reversed-phase high-performance liquid chromatography (HPLC) has been used for the separation of scytonemin (Rastogi et al. 2014b; Pathak et al. 2017), other strategies, including normal-phase HPLC and low-pressure liquid chromatography, may be more suitable for its purification.

## Other Pigments

In addition to MAAs and scytonemin, other cyanobacterial pigments, including phycobiliproteins (PBPs) and carotenoids, could be used in skin care products.

### PBPs

PBPs are water-soluble, light-harvesting proteins that commonly exist in cyanobacteria, rhodophytes, cryptomonads, and cyanelles (Bolte et al. 2008). They are covalently made up of two components: phycobilins and apoproteins. PBPs could be classified on the basis of their spectral patterns. The major classes, namely blue-colored phycocyanin (PC), red-colored phycoreythrin (PE), and light blue-colored allophycocyanin (APC), have maximum absorption peaks at 610–620, 555–565, and 650 nm, respectively (Bryant et al. 1979).

Regarding the application of PBPs, it is well known that their colors are utilized as natural dyes in foods. For example, blue-colored PC is known to be important for the food industry in bubble gums, milk products, and jellies (Santiago-Santos et al. 2004). On the other hand, PBPs are also known as water-soluble and strong antioxidants (Hossain et al. 2016). Romay et al. established the antioxidative activity of purified PC, the major PBP in several cyanobacteria (Romay et al. 1998). In addition, they revealed the anti-inflammatory properties of PC by assessing the inhibitory effect of PC on liver microsomal lipid peroxidation and the edema index in glucose oxidase-induced inflammation of mouse paws (Romay et al. 1998). The anti-aging activities of PBPs have also been observed in animals. Sonani et al. found that the supply of PE purified from *Lyngbya* sp. A09DM expanded the lifespan of *Caenorhabditis elegans* (Sonani et al. 2014). They discussed that the observed effect of PE may be triggered in part by its antioxidative property. Thus, the characteristics of PBPs make them attractive as safe natural products not only to the food industry but also to cosmetic and pharmaceutical industries.

## Carotenoids

Carotenoids are organic yellow, orange, or red photosynthetic accessory pigments biosynthesized by a wide variety of organisms, including plants, algae, and cyanobacteria; however, animals are not able to synthesize carotenoids *de novo* (Moeller et al. 2005). These pigments play an important role in photosynthetic organisms; they function alongside chlorophylls to absorb light for photosynthesis and exert photoprotection by quenching ROS from oxidative stress (Tapiero et al. 2004). Carotenoids are used as natural additives in foods to enhance color or confer health benefits. They are also used in cosmetics and in pharmaceutical products (Ryu et al. 2015). Given their health benefits in humans, carotenoids have been extensively explored so far. They can be roughly divided into two types: carotenes and xanthophylls. Carotenes are hydrocarbons, whereas xanthophylls contain oxygen in their structures (Batista et al. 2006). In the cyanobacterial carotenogenesis pathway, all carotenoids are derived from lycopene. Lycopene is produced by a biosynthetic pathway conserved in various cyanobacterial species (Liang et al. 2006). Various carotenoids are then produced from lycopene via enzymatic reactions. The chemical structures of representative cyanobacterial carotenoids are shown in Figure 2. At present, the microalgae *Dunaliella* sp. and *Haematococcus* sp. are well known as producers of commercialized carotenoids. The regulation of carotenoid biosynthesis has been extensively studied. For example, the production of β-carotene and astaxanthin could be induced under high light conditions in *Dunaliella* sp. and *Haematococcus* sp., respectively (Koller et al. 2014; Safafar et al. 2015; Hamed et al. 2016). Moreover, Janchot et al. reported that the intracellular accumulation of carotenoids, including canthaxantin and astaxanthin, was enhanced under salt stress in the green chlorococcal microalga strain TITSR 9500 (Janchot et al. 2019). It is thought that cyanobacterial carotenoid bioproduction could also be influenced by various abiotic stresses. In fact, high light stress was found to alter the accumulation of carotenoids in cyanobacteria (Nonnengiesser et al. 1996; Olaizola et al. 1990; Steiger et al. 1999).

Although the absorption maxima of carotenoids are between 440 and 520 nm, they are involved in light absorption in the UV-A range as well as low-wavelength photosynthetically active radiation (PAR). Therefore, from the viewpoint of skin care, the direct absorption of UV-A light may be responsible for the prevention of sun damage to the skin (Morocho-Jácome et al. 2020). Moreover, evidence indicates that innate resistance of the skin against UV-B-induced erythema (sunburn) is enhanced by the dietary supplementation of β-carotene, lycopene, lutein, astaxanthin, and mixed carotenoids (Baswan et al. 2021). Carotenoids are known as excellent natural antioxidants. Because the antioxidant capacity of the skin is an important defense mechanism against environmental stress, the supplementation of antioxidative carotenoids could reduce skin damage.

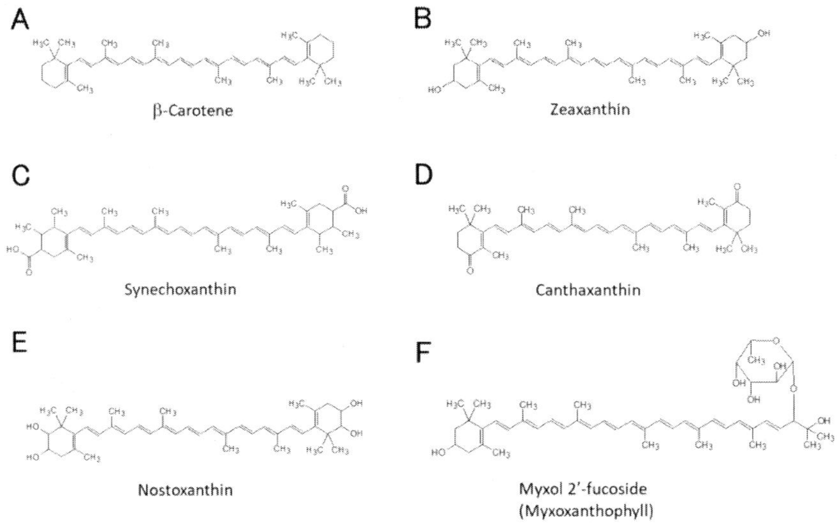


**Figure 2.** Chemical structures of representative cyanobacterial carotenoids.

## Enzymes

In addition to natural pigments, enzymes that possess unique properties are of special interest in the industry, including cosmetic and pharmaceutical fields (Nigam 2013). For skin care, antioxidant enzymes, such as superoxide dismutase (SOD), may help prevent skin aging boosted by ROS. SOD catalyzes the dismutation of the superoxide radical into molecular oxygen and hydrogen peroxide. Another example is protease that catalyzes the cleavage of peptide bonds within proteins. For instance, assessment of the effects of the serine protease trypsin on skin permeability from a dermatological viewpoint revealed that the topical application of serine protease enhanced drug penetration (Germer et al. 2012). All cyanobacterial species possess various proteases, including serine proteases, cysteine proteases, aspartate proteases, threonine proteases, metalloproteases, and other proteases. Interestingly, the distributions of proteases differ depending on species. For example, the numbers of genes encoding putative protease are 58 and 119 in the freshwater cyanobacterium *Synechococcus elongatus* PCC7942 and the halotolerant cyanobacterium *Halothece* sp. PCC7418, respectively (Patipong et al. 2021). In the case of serine proteases, specific enzymatic characteristics of HtrA proteases in response to salt stress have been found in *Halothece* ps. PCC7418 (Patipong et al. 2020). Mining of enzymes that show unique catalytic properties from specific cyanobacteria would be an intriguing topic for future research.

## Conclusion

In this review, we have presented evidence of various benefits of cyanobacterial products, including MAAs, scytonemin, PBPs, carotenoids, and enzymes, focusing on their biological properties in terms of skin care. Although there is still room for consideration for industrial production and commercial product development, cyanobacteria are thought to be promising sources of natural bioactive compounds because of their photoautotrophic and gene-manipulative properties. Alternatively, the bioinformatical use of genetic sources of cyanobacteria will be a good strategy for the bioproduction of natural products using other microbial host cells as well as cyanobacterial cells.

## Acknowledgements

Research for the findings described in this review was supported in part by the Cosmetology Research Foundation (J-18-1) and Hoyu Science Foundation.

## References

Balskus EP, Walsh CT (2009) An enzymatic cyclopentyl[*b*]indole formation involved in scytonemin biosynthesis. *J Am Chem Soc* 131: 14648-14649.

Balakus EP, Walsh CT (2010) The genetic and molecular basis for sunscreen biosynthesis in cyanobacteria. *Science* 329: 1653-1656.

Baswan SM, Klosner AE, Weir C, Salter-Venzon D, Gellenbeck KW, Leverett J, Krutmann J (2021) Role of ingestible carotenoids in skin protection: A review of clinical evidence. *Photodermatol Photoimmunol Photomed* 37: 490-504 .

Batista AP, Raymundo A, Sousa I, Empis J (2006) Rheological characterization of coloured oil-in-water food emulsions with lutein and phycocyanin added to the oil and aqueous phases. *Food Hydrocoll* 20: 44-52.

Bolte K, Kawach O, Prechtl J, Gruenheit N, Nyalwidhe J, Maier UG (2008) Complementation of a phycocyanin-bilin lyase from *Synechocystis* sp. PCC 6803 with a nucleomorph-encoded open reading frame from the cryptophyte *Guillardia theta*. *BMC Plant Biol* 8: 56.

Browne N, Donovan F, Murray P, Saha S (2014) Cyanobacteria as bio-factories for production of UV-screening compounds. *OA Biotechnol* 3: 6.

Bryant DA, Guglielmi G, de Marsac NT, Castets AM, Cohen-Bazire G (1979) The structure of cyanobacterial phycobilisomes: a model. *Arch Microbiol* 123: 113–127.

Bultel-Poncé V, Felix-Theodose F, Sarthou C, Ponge JF, Bodo B (2004) New pigments from the terrestrial cyanobacterium *Scytonema* sp. collected on the Mitaraka inselberg, French Guyana. *J Nat Prod* 67:678-681.

Cheewinthamrongrod V, Kageyama H, Palaga T, Takabe T, Waditee-Sirisattha R (2016) DNA damage protecting and free radical scavenging properties of mycosporine-2-glycine from the Dead Sea cyanobacterium in A375 human melanoma cell lines. *J Photochem Photobiol* 164: 289–295.

Conde FR, Churio MS, Previtali CM (2007) Experimental study of the excited-state properties and photostability of the mycosporine-like amino acid palythine in aqueous solution. *Photochem Photobiol Sci* 6: 669-674.

D'Agostino PM, Javalkote VS, Mazmouz R, Pickford R, Puranik PR, Neilan BA (2016) Comparative profiling and discovery of novel glycosylated mycosporine-like amino acids in two strains of the cyanobacterium *Scytonema* cf. *crispum*. *Appl Environ Microbiol* 82: 5951–5959.

de la Coba F, Aguilera J, Figueroa FL, Gálvez MV, Herrera E (2009) Antioxidant activity of mycosporine-like amino acids isolated from three red macroalgae and one marine lichen. *J Appl Phycol* 21: 161–169.

Derikvand P, Llewellyn CA, Purton S (2017) Cyanobacterial metabolites as a source of sunscreens and moisturizers: a comparison with current synthetic compounds. *Eur J Phycol* 52: 43-56.

Dillon JG, Tatsumi CM, Tandingan PG, Castenholz RW (2002) Effect of environmental factors on the synthesis of scytonemin, a UV-screening pigment, in a cyanobacterium (*Chroococcidiopsis* sp.). *Arch Microbiol* 177: 322-331.

Fleming ED, Castenholz RW (2007) Effects of periodic desiccation on the synthesis of the UV-screening compound, scytonemin, in cyanobacteria. *Environ Microbiol* 9: 1448-1455.

Fleming ED, Castenholz RW (2008) Effects of nitrogen source on the synthesis of the UV-screening compound, scytonemin, in the cyanobacterium *Nostoc punctiforme* PCC 73102. *FEMS Microbiol Ecol* 63: 301-308.

Frederick J, Snell H, Haywood E (1989) Solar ultraviolet radiation at the earth's surface. *J Photochem Photobiol* 50: 443-450.

Fuentes-Tristan S, Parra-Saldivar R, Iqbal HMN, Carrillo-Nieves D (2019) Bioinspired biomolecules: Mycosporine-like amino acids and scytonemin from *Lyngbya* sp. with UV-protection potentialities. *J Photochem Photobiol* 201: 111684.

Gao Q, Garcia-Pichel F (2011) Microbial ultraviolet sunscreens. *Nature Rev Microbiol* 9: 791-802.

Gao X, Jing X, Liu X, Lindblad P (2021) Biotechnological production of the sunscreen pigment scytonemin in cyanobacteria: Progress and strategy. *Mar. Drugs 19*: 129.

Garcia-Pichel F, Castenholz RW (1991) Characterization and biological implications of scytonemin, a cyanobacterial sheath pigment. *J Phycol* 27: 395-409.

Germer G, Ohigashi T, Yuzawa H, Kosugi N, Flesch R, Rancan F, Vogt A, Rühl E (2012) Improved sin prmeability after topical treatment with serine protease: Probing the penetration of rapamycin by scanning transmission X-ray microscopy. *ACS omega*, 6: 12213–12222.

González-Minero FJ, Bravo-Díaz L (2018) The use of plants in skin-care products, cosmetics and fragrances: Past and present. *Cosmetic* 5: 50.

Grewe CB, Pulz O (2012) The Biotechnology of Cyanobacteria. In: Whitton B, Editors. *Ecology of Cyanobacteria II*. Springer, Dordrecht, 707-739.

Gruber F, Peharda V, Kastelan M, Brajac I (2007) Occupational skin diseases caused by UV radiation. *Acta Dermatovenerologica Croat* 15(3): 191–198.

Hamed I (2016) The evolution and versatility of microalgal biotechnology: a review. *Compr Rev Food Sci Food Saf* 15: 1104–1123.

Hartmann A, Gostner J, Fuchs JE, Chaita E, Aligiannis N, Skaltsounis L, Ganzera M (2015) Inhibition of collagenase by mycosporine-like amino acids from marine sources. *Planta Medica* 81: 813–820.

Hossain MF, Ratnayake RR, Meerajini K, Wasantha Kumara KL (2016) Antioxidant properties in some selected cyanobacteria isolated from fresh water bodies of Sri Lanka. *Food Sci Nutr* 4: 753-758.

Hu C, Völler G, Süβmuth R, Dittmann E, Kehr JC (2015) Functional assessment of mycosporine-like amino acids in *Microcystis aeruginosa* strain PCC 7806. *Environ Microbiol* 17: 1548-1559.

Ikehata H (2018) Mechanistic considerations on the wavelength-dependent variations of UVR genotoxicity and mutagenesis in skin: The discrimination of UVA-signature from UV-signature mutation. *Photochem Photobiol Sci* 17: 1861–1871.

Inoue-Sakamoto K, Nazifi E, Ysuji C, Asano T, Nishiuchi T, Matsugo S, Ishihara K, Kanesaki Yu, Yoshikawa H, Sakamoto T (2018) Characterization of mycosporine-like amino acids in the cyanobacterium *Nostoc verrucosum*. *J Gen Appl Microbiol* 64: 201-211.

Ishihara K, Watanabe R, Uchida H, Suzuki T, Yamashita M, Takenaka H, Nazifi E, Matsugo S, Yamaba M, Sakamoto T (2017) Novel glycosylated mycosporine-like amino acid, 13-O-(β-galactosyl)-porphyra-334, from the edible cyanobacterium *Nostoc sphaericum*-protective activity on human keratinocytes from UV light. *J Photochem Photobiol* 172: 102–108.

Janchot K, Rauytanapanit M, Honda M, Hibino T, Sirisattha S, Praneenararat T, Kageyama H, Waditee-Sirisattha R (2019) Effects of potassium chloride-induced stress on the carotenoids canthaxanthin, astaxanthin, and lpid accumulations in the green Chlorococcal microalga strain TISTR 9500. *J Eu Microbiol* 66: 778-787.

Jesumani V, Du H, Aslam M, Pei P, Huang N (2019) Potential use of seaweed bioactive compounds in skincare—A review. *Mar. Drugs* 17: 688.

Kageyama H, Waditee-Sirisattha R (2018a) Cyanobacterial UV sunscreen: Biosynthesis, regulation, and application. In: Rastogi RP, Editors. *Sunscreens Source, Formulations, Efficacy and Recommendations*. Nova Science Publishers: New York, USA 1–28.

Kageyama H, Waditee-Sirisattha R (2018b) Mycosporine-like amino acids as multifunctional secondary metabolites in cyanobacteria: From biochemical to application aspects. In: Atta-ur-Rahman, Editors. *Stud Nat Prod Chem* 59: 153–194.

Kang MR, Jo SA, Lee H, Yoon YD, Kwon JH, Yang JW, Choi BJ, Park KH, Lee MY, Lee CW, Lee KR, Kang JS (2020) Inhibition of skin inflammation by scytonemin, an ultraviolet sunscreen pigment. *Mar. Drugs* 18: 300.

Katoch M, Mazmouz R, Chau R, Pearson LA, Pickford R, Neilan BA (2016) Heterologous production of cyanobacterial mycosporine-like amino acids mycosporine-ornithine and mycosporine-lysine in *Escherichia coli*. *Appl Environ Microbiol* 82: 6167–6173.

Kedar L, Kashman Y, Oren A (2002) Mycosporine-2-glycine is the major mycosporine-like amino acid in a unicellular cyanobacterium (*Euhalothece* sp.) isolated from a gypsum crust in a hypersaline saltern pond. *FEMS Microbiol Lett* 208: 233-237.

Koller M, Muhr A, Braunegg G (2014) Microalgae as versatile cellular factories for valued products. *Algal Res* 6: 52–63.

Korteerakul C, Honda M, Siripat M, Hibino T, Waditee-Sirisattha R, Kageyama H (2020) Antioxidative and antiglycative properties of mycosporine-like amino acids-containing aqueous extracts derived from edible terrestrial cyanobacteria. *J Nutr Sci Vitaminol* 66: 339-346.

Liang C, Zhao F, Wei W, Wen Z, Qin S (2006) Carotenoid biosynthesis in cyanobacteria: structural and evolutionary scenarios based on comparative genomics. *Int J Biol Sci* 2: 197-207.

Macallef ML, D'Agostino PM, Al-Sinawa B, Neilan BA, Moffitt MC (2015) Exploring cyanobacterial genomes for natural product biosynthesis pathway. *Mar Genomics* 21: 1-12.

Matsui K, Nazifi E, Hirai Y, Wada N, Matsugo S, Sakamoto T (2012) The cyanobacterial UV-absorbing pigment scytonemin displays radical-scavenging activity. *J Gen Appl Microbiol* 58:137-44.

Moeller RE, Gilroy S, Williamson CE, Grad G (2005) Dietary acquisition of photoprotective compounds (mycosporine-like amino acids, carotenoids) and acclimation to ultraviolet radiation in a fresh water coperod. *Limnol Oceanogr* 50: 427-439.

Morocho-Jácome AL, Ruscinc N, Martinez RM, de Carvalho J, Santos de Almeida T, Rosado C, Costa JG, Velasco M, Baby AR (2020) (Bio)Technological aspects of microalgae pigments for cosmetics. *Appl Microbiol Biotechnol* 104: 9513–9522.

Münch G, Thome J, Foley P, Schinzel R, Riederer P (1997) Advanced glycation end products in ageing and Alzheimer's disease. *Brain Res Rev* 23: 134-143.

Pangestuti R, Shin KH, Kim SK (2021) Anti-photoaging and potential skin health benefits of seaweeds. *Mar. Drugs* 19: 172.

Nazifi E, Wada N, Yamaba M, Asano T, Nishiuchi T, Matsugo S, Sakamoto T (2013) Glycosylated porphyra-334 and palythine-threonine from the terrestrial cyanobacterium *Nostoc commune. Mar. Drugs* 11: 3124-3154.

Nazifi E, Wada N, Asano T, Nishiuchi T, Iwamuro Y, Chinsks S, Matsugo S, Sakamoto T (2015) Characterization of the chemical diversity of glycosylated mycosporine-like amino acids the terrestrial cyanobacterium *Nostoc commune. J Photochem Photobiol* 142: 154-168.

Ngoennet S, Nishikawa Y, Hibino T, Waditee-Sirisattha R, Kageyama H (2018) A method for the isolation and characterization of mycosporine-like amino acids from cyanobacteria. *Methods Protoc* 1: 46.

Nigam PS (2013) Microbial enzymes with special characteristics for biotechnological applications. *Biomolecules* 3:597-611.

Nonnengiesser K, Schuster A, Koenig F (1996) Carotenoids and reaction center II-D1 protein in light regulation of the photosynthetic apparatus in *Aphanocapsa. Plant Biol* 109: 115-124.

Olaizola M, Duerr EO (1990) Effects of light intensity and quality on the growth rate and photosynthetic pigment content of *Spirulina platensis. J Appl Phycol* 2: 97-104.

Orfanoudaki M, Hartmann A, Alilou M, Gelbrich T, Planchenault P, Derbré S, Schinkovitz A, Richomme P, Hensel A, Ganzera M (2019) Absolute configuration of mycosporine-like amino acids, their wound healing properties and in vitro anti-aging effects. *Mar. Drugs* 18: 35.

Oyamada C, Kaneniwa M, Ebitani K, Murata M, Ishihara K (2008) Mycosporine-like amino acids extracted from scallop (Patinopecten yessoensis) ovaries: UV protection and growth stimulation activities on human cells. *Mar. Biotechnol* 10: 141–150.

Pangestuti R, Shin KH, Kim SK (2021) Anti-photoaging and potential skin health benefits of seaweeds. *Mar. Drugs* 19(3): 172.

Pathak J, Sonker AS, Kannaujiya VK, Singh V, Ahmed H, Sinha RP (2017) Screening and partial purification of photoprotective pigment scytonemin from cyanobacterial crusts dwelling on the historical monuments in and around Varanasi, India. *Microbiol Res* 8: 4-12.

Patipong T, Hibino T, Waditee-Sirisattha R, Kageyam H (2017) Efficient bioproduction of mycosporine-2-glycine, which functions as potential osmoprotectant, using *Escherichia coli* cells. *Nat Prod Commun* 12: 1593–1594.

Patipong T, Hibino T, Waditee-Sirisattha R, Kageyama H (2019) Induction of antioxidative activity and antioxidant molecules in the halotolerant cyanobacterium *Halothece* sp. PCC7418 by temperature shift. *Nat Prod Commun* 7: 1-6.

Patipong T, Hibino T, Kageyama H, Waditee-Sirisattha R (2020) The evolutionarily conserved HtrA is associated with stress tolerance and protein homeostasis in the halotolerant cyanobacterium *Halothece* sp. PCC7418. *Extremophiles* 24: 377-389.

Patipong T, Kageyama H, Waditee-Sirisattha R (2021) Insights into the phylogeny and transcriptional response of serine proteases in a halotolerant cyanobacterium *Halothece* sp. PCC7418. *Plant Signal Behav*, (in press).

Portwich A, Garcia-Pichel F (1999) Ultraviolet and osmotic stresses induce and regulate the synthesis of mycosporines in the cyanobacterium *Chlorogloeopsis* PCC 6912. *Arch Microbiol* 172: 187–192.

Proteau PJ, Gerwick WH, Garcia-Pichel F, Castenholz R (1993) The structure of scytonemin, an ultraviolet sunscreen pigment from the sheaths of cyanobacteria. *Experientia* 49: 825-9.

Rastogi RP, Sinha RP, Incharoensakdi A (2013) Partial characterization, UV-induction and photoprotective function of sunscreen pigment, scytonemin from *Rivularia* sp. HKAR-4. *Chemosphere* 93: 1874-8.

Rastogi RP, Incharoensakdi A (2014a) Analysis of UV-absorbing photoprotectant mycosporine-like amino acid (MAA) in the cyanobacterium *Arthrospira* sp. CU2556. *Photochem Photobiol Sci 13*: 1016-1024.

Rastogi RP, Incharoensakdi A (2014b) UV radiation-induced biosynthesis, stability and antioxidant activity of mycosporine-like amino acids (MAAs) in a unicellular cyanobacterium *Gloeocapsa* sp. CU2556. *J Photochem Photobiol* 130: 287-292.

Rastogi RP, Incharoensakdi A (2014c) Characterization of UV-screening compounds, mycosporine-like amino acids, and scytonemin in the cyanobacterium *Lyngbya* sp. CU2555. *FEMS Microbiol Lett* 87: 244-256.

Rastogi RP, Sonani RR, Madamwar D (2015) Cyanobacterial sunscreen scytonemin: Role in photoprotection and biomedical research. *Appl Biochem Biotechnol* 176: 1551-1563.

Richa, Sinha RP (2015) Biochemical characterization of sunscreening mycosporine-like amino acids from two *Nostoc* species inhabiting diverse habitats. *Protoplasma 252*: 199-208.

Romay C, Armesto J, Remirez D, González R, Ledon N, García I (1998) Antioxidant and anti-inflammatory properties of C-phycocyanin from blue-green algae. *Inflamm Res* 47: 36-41.

Ryu BM, Himaya SWA, Kim S-K (2015) Applications of microalgae-derived active ingredients as cosmeceuticals. In: Kim S-W, Editors. *Handbook of marine microalgae: biotechnology advances,* Academic Press: Massachusetts, USA, 309–315.

Safafar H, Van Wagenen J, Møller P, Jacobsen C (2015) Carotenoids, phenolic compounds and tocopherols contribute to the antioxidative properties of some microalgae species grown on industrial wastewater. *Mar. Drugs* 13: 7339–7356.

Sakamoto T, Hashimoto A, Yamaba M, Wada N, Yoshida T, Matsugo S, Sakamoto-Inoue K, Nishiuchi T, Matsugo S (2019) Four chemotypes of the terrestrial cyanobacterium *Nostoc commune* characterized by differences in the mycosporine-like amino acids. *Phycol Res* 67: 3-11

Santiago-Santos MC, Ponce-Noyola T, Olvera-Ramírez R, Ortega-López J, Cañizares-Villanueva RO (2004) Extraction and purification of phycocyanin from *Calothrix* sp. *Process Biochem* 39: 2047-2052.

Singh SP, Ambasht NK, Sinha JP, Häder DP (2003) Wavelength-dependent induction of a mycosporine-like amino acid in a rice-field cyanobacterium, *Nostoc commune*: role of inhibitors and salt stress. *Photochem Photobiol Sci* 2: 171-176.

Singh SP, Sinha RP, Kilsch N, Häder DP (2008a) Mycosporine-like amono acids (MAAs) profile of a rice-field cyanobacterium *Anabaena doliolum* as influenced by PAR and UVR. *Planta* 229: 225-233.

Singh SP, Klisch M, Sinha RP, Häder DP (2008b) Effects of abiotic stressors on synthesis of the mycosporine-like amino acid shinorine in the cyanobacterium *Anabaena variabilis* PCC 7937. *Photochem Photobiol* 84: 1500–1505.

Singh SP, Kilsch M, Sinha RP, Häder DP (2010) Sulfer deficiency changes mycosporine-like amino acid (MAA) composition of *Anabaena variabilis* PCC 7937: A possible role of Sulphur in MAA bioconversion. *Photochem Photobiol* 86: 862-870.

Sinha RP, Klisch M, Helbling EW, Häder D (2001) Induction of mycosporine-like amino acids (MAAs) in cyanobacteria by solar ultraviolet-B radiation. *J Photochem Photobiol* 60: 129–135.

Sonani RR, Singh NK, Kumar J, Thakar D, Madamwar D (2014) Concurrent purification and antioxidant activity of phycobiliproteins from *Lyngbya* sp. A09DM: An antioxidant and anti-aging potential of phycoerythrin in *Caenorhabditis elegans*. *Process Biochem* 49: 1757-1766.

Soule T, Stout V, Swingley WD, Meeks JC, Garcia-Pichel F (2007) Molecular genetics and genomic analysis of scytonemin biosynthesis in *Nostoc punctiforme* ATCC 29133. *J Bacteriol* 189: 4465-4472.

Soule T, Garcia-Pichel F, Stout V (2009) Gene expression patterns associated with the biosynthesis of the sunscreen scytonemin in *Nostoc punctiforme* ATCC 29133 in response to UVA radiation. *J Bacteriol* 191: 4639-4646.

Soule T, Garcia-Pichel F (2013) Ultraviolet photoprotective compounds from cyanobacteria in biomedical applications. In: Sharma NK, Rai AK, Stal LJ, Editors. *Cyanobacteria: An Economic Perspective*, Wiley: New York, USA, 119-143.

Steiger S, Schäfer L, Sandmann G (1999) High-light-dependent upregulation of carotenoids and their antioxidative properties in the cyanobacterium *Synechocystis* PCC 6803. *J Photochem Photobiol* 52: 14-18.

Subramaniam A, Carpenter EJ, Karentz D, Falkowski PG (1999) Biooptical properties of the marine diazotrophic cyanobacteria *Trichodesmium* spp. I. Absorption and photosynthetic action spectra. *Limnol Oceanogr* 44: 608-617.

Suh SS, Hwang J, Park M, Seo HH, Kim HS, Lee JH, Moh SH, Lee TK (2014) Anti-inflammation activities of mycosporine-like amino acids (MAAs) in response to UV radiation suggest potential anti-skin aging activity. *Mar Drugs* 12:5174-5187.

Suh SS, Oh SK, Lee SG, Kim IC, Kim S (2017) Porphyra-334, a mycosporine-like amino acid, attenuates UV-induced apoptosis in HaCaT cells. *Acta Pharm* 67:257-264.

Tapiero H, Townsend D, Tew K (2004) The role of carotenoids in the prevention of human pathologies. *Biomed Pharmacother* 58: 100-110.

Tarasuntisuk S, Patipong T, Hibino T, Waditee-Sirisattha R, Kageyama H (2018) Inhibitory effects of mycosporine-2-glycine isolated from a halotolerant cyanobacterium on protein glycation and collagenase activity. *Lett Appl Microbiol* 67: 314-320.

Tarasuntisuk S, Palaga T, Kageyama H, Waditee-Sirisattha R (2019) Mycosporine-2-glycine exerts anti-inflammatory and antioxidant effects in lipopolysaccharide (LPS)-stimulated RAW 264.7 macrophages. *Arch Biochem Biophy* 662: 33-39.

Volkmann M, Gorbushina AA (2006) A broadly applicable method for extraction and characterization of mycosporines and mycosporine-like amino acids of terrestrial, marine and freshwater origin. *FEMS Microbiol Lett* 255: 286–295.

Wada N, Sakamoto T, Matsugo S (2013) Multiple roles of photosynthetic and sunscreen pigments in cyanobacteria focusing on the oxidative stress. *Metabolites* 20: 463-483.

Wada N, Sakamoto T, Matsugo S (2015) Mycosporine-like amino acids and their derivatives as natural antioxidants. *Antioxidants* 4: 603-646.

Waditee-Sirisattha R, Kageyama H, Sopun W, Tanaka Y, Takabe T (2014) Identification and upregulation of biosynthetic genes required for accumulation of mycosporine-2-glycine under salt stress conditions in the halotolerant cyanobacterium *Aphanothece halophytica*. *Appl Environ Microbiol* 80: 1763-1769.

Waditee-Sirisattha R, Kageyama H, Fukaya M, Rai V, Takabe T (2015) Nitrate and amino acid availability affects glycine betaine and mycosporine-2-glycine in response to changes of salinity in a halotolerant cyanobacterium *Aphanothece halophytica*. *FEMS Microbiol Lett* 362: fnv198.

# Chapter 8

# Recent Advances in Cyanobacterial Photopigments and Their Role in Sustainable Development

**Ashish P. Singh, Amit Gupta, Jyoti Jaiswal, Prashant R. Singh, Niharika Sahu and Rajeshwar P. Sinha***

Laboratory of Photobiology and Molecular Microbiology, Centre of Advanced Study in Botany, Institute of Science, Banaras Hindu University, Varanasi, India

## Abstract

Cyanobacteria are photosynthetic, oxygenic prokaryotes that first arose during the Precambrian period (2.8-3.5 billion years ago) and set the stage for the evolution of aerobic life today. As photosynthetic organisms, cyanobacteria contain a light-harvesting complex that absorbs light energy. These harvesting complexes are split into three chemical groups: chlorophylls, carotenoids, and phycobiliproteins.

The principal pigments for light absorption in cyanobacteria are phycobiliproteins, whereas carotenoids play an important role in non-photochemical quenching (NPQ) as a defense against saturating light and a reactive oxygen species quencher (ROS). For photosynthesis, phototrophic organisms must generate not just carotenoids but also chlorophylls. Chl *a* is required for practically all oxygenic photosynthetic organisms, although Chl *d* and Chl *f* have recently gained popularity due to their unique ability to absorb light in the far-red region.

Interestingly, in recent years, scientists and industrialists throughout the world have been working to produce natural products that are both

* Corresponding Author's Email: rpsinhabhu@gmail.com.

In: Cyanobacteria
Editor: Rajeshwar P. Sinha
ISBN: 979-8-88697-169-9

sustainable for the future and less detrimental to human civilization. Cyanobacterial photopigments have been significant in the development of a wide range of biotechnological applications, including food, cosmetics, medicines, agriculture, bioremediation, bioenergy, nutraceuticals, and biofuels. The goal of this chapter is to go through many types of cyanobacterial photopigments as well as the most important properties for biotechnological applications, with an emphasis on their bioactivities. Finally, photopigments are briefly articulated in terms of their significance in sustainable development, and their future potential for creating a decent, sustainable world.

**Keywords:** cyanobacteria, photopigments, phycobiliproteins, biotechnological applications, role in sustainable development

## Introduction

Cyanobacteria are unique photosynthetic organisms that are found in practically all natural and man-made environments. They have tiny unicellular, filamentous, or colonial cells that are occasionally big enough to be seen by the naked eye, particularly during blooms. They are the oldest known fossils, with formations such as stromatolites dating back over 3.5 billion years, and they are one of the biggest and most significant groups of bacteria on the planet. Cyanobacteria synthesize several types of photopigments for photosynthesis, protection, and survival in diverse types of environments. The most significant ones are determined to be phycobiliproteins, carotenoids, and chlorophylls. Photopigments are coloured fluorescent compounds that allow light energy to be captured in auxiliary photosynthetic complexes. Photopigments contain a number of structural and physicochemical properties that have been retained and have been demonstrated to exhibit a variety of biological activities, including antioxidant, antibacterial, and antitumor properties, making them a promising target for biotechnological applications in fields such as bioenergy, biomedicine, and scientific research. Cyanobacteria and micro and macro algae are the primary suppliers of these photopigments (i.e., phycobiliproteins, carotenoids, and chlorophylls). They have drawn the interest of several businesses, including as food, biomedical, and cosmetics, due to their unique biological activity. This is why a large number of patents relating to the production, extraction, purification, and application of photopigments as cosmetics, biopharmaceuticals, or diagnostic applications have been filed, with the goal of reducing environmental impact in macroalgae natural prairies and increasing cyanobacteria culture time or productivity to meet the markets and applications that require large amounts of these molecules. In this chapter, we summarize the main structural characteristics of photopigments, their biosynthesis and biotechnological applications. We also address current trends and future perspectives of the photopigment market.

## Structure

Phycobilisomes (PBS) are a large light-harvesting accessory complex made up of vividly colouredbiliproteins (BPs). Structurally, PBSs are supramolecular complexes with a wide molecular mass range (5-30 MDa) and they are integrally attached to a protein scaffold that contains a fluorescent cofactor (PBPs). PBPs are divided into three classes based on colour and absorption properties: phycocyanin (PC), phycoerythrin (PE), and allophycocyanin (APC). Furthermore, each PBP subunit is made up of a heterodimer and subunits made up of regular tetrapyrrole chromophores (one to four chains with the A, B, C, and D rings) covalently bonded to various cysteine residues in biliproteins (Sidler 1994) (Figure 1).

The open-chain tetrapyrrole chromophores will collect photonic energy from the sun's spectrum and transmit it to the photosystem for use in photosynthesis. In terms of energy transmission, it might begin at the terminal PE and go to the PC, then to the APC, and lastly to the photosystem for photosynthesis (Johnson et al. 2014). PE is also found near the distal end of the PB array and is involved in the response to variations in photosynthetically active radiation. A variety of colourless linker polypeptides act as structural components in the production and stability of PBPs (Ana et al. 2009).

PBPs are found in a hexameric form $(\alpha\beta)_6$, that form a flattened disc (~11 nm diameter) with a central hollow. Two trimers are placed face to face with an interior cavity in the hexameric PBP complexes. The hollow, which has a diameter of around 3.5 nm, is thought to be dominated by linker polypeptides produced from PBP crystals (Stadnichuk et al. 2015). This heterodimer of $\alpha$ and $\beta$ monomer is regarded as the fundamental building component of PBPs' monomeric structure, which then assembles into trimeric discs $(\alpha\beta)_3$. According to certain experimental results, trimetric structures of biliproteins are further developed by chromophore attachment and form a $(\alpha\beta)_6$ hexamer with the rod and unpigmented linker polypeptides (Table 1) (Adir et al. 2005).

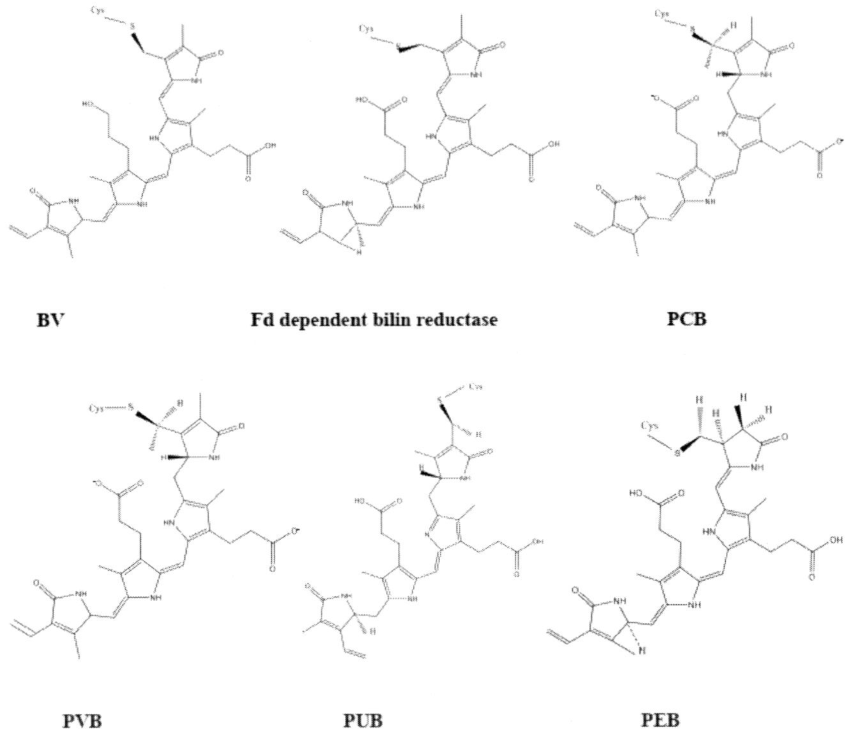


Figure 1.Structure of chromophores in phycobiliproteins. BV (biliverdin), DBV (Fd-dependent bilin reductase), PCB (phycocyanobilin), PVB (phycoviolobilin), PUB (phycourobilin), and PEB (phycoerythrobilin) (Modified from Kannaujiya et al. 2017).

**Table 1.** Chromophore binding sites in phycobiliproteins (Adopted from Kannaujiya et al. 2017)

| **Types of biliproteins** | **Chromophore binding sites on cysteine residue** | | | | | |
|---|---|---|---|---|---|---|
| | **β-50/60** | **β-84*** | **β-155** | **α-75** | **α-84** | **α-140** |
| Phycocyanin# | | PCB | PCB | | PUB | |
| C-Phycocyanin | | PCB | PCB | | PCB | |
| R-Phycocyanin | | PCB | PEB | | PUB | |
| R-Phycoerythrin | | PCB | PEB | | PEB | |
| C-Phycoerythrin | PEB | PEB | PEB | | PEB | PEB |
| Allophycocyanin | | PCB | | | PCB | |
| Phycoerythrocyanin | | PCB | PCB | | PVB | |
| CU-Phycoerythrin (PEI)* | PUB | PEB | PEB | | PEB | PUB |
| CU-Phycoerythrin (PEI)* | PUB | PEB | PEB | PUB | PEB | PEB |

**Table 2.** Types and characteristics of linker polypeptides (Adopted from Kannaujiya et al. 2017)

| Gene | Protein | Amino acids | Symbol | MW (kDa) | pI | Annotation |
|---|---|---|---|---|---|---|
| *cpcC* | CpcC | 219–291 | PC LR | 24.8–32.6 | 9.5–9.6 | PC-associated linker |
| *cpcD* | CpcD | 70–87 | PC LR | 7.8–9.9 | 9.8–10.5 | Rod capping linker |
| *cpcG* | CpcG | 231–279 | LRC | 26.8–31.9 | 9.3–9.6 | Rod–core linker |
| *cpcH* | CpcH | 271–273 | LRC | 30.4–30.8 | 8.8–9.7 | Rod–core linker |
| *cpcI* | CpcI | 288 | LRC | 32.7 | 8.9 | Rod–core linker |
| *cpeC* | CpeC | 285–294 | PE LR | 31.8–33.1 | 9.6 | PE-associated linker |
| *cpeD* | CpeD | 249–255 | PE LR | 27.9–28.4 | 8.2–8.6 | PE-associated linker |
| *cpeE* | CpeE | 244–254 | PE LR | 27.1–28.4 | 9.7 | PE-associated linker |
| *pecC* | PecC | 278–279 | PEC LR | 31.3–31.5 | 9.6–9.7 | PEC-associated linker |
| *apcC* | ApcC | 66–69 | LC | 7.7–7.8 | 10.9–11.4 | APC-associated linker |
| *apcE* | ApcE | 683–1155 | LCM | 76.5–129.8 | 9.5–9.7 | Core–membrane linker |

Four trimers of APC biliproteins combine to form core cylinders. Following that, two to five core cylinders unit into the core substructure, aided by linker polypeptides, (Table 2) to form a fully functioning structure. Finally, the rods and core cylinders link up to form a whole PBS. PBPs are made up of three parts: two subunits ($\alpha$ and $\beta$) and a third component ($\gamma$), which is a linker peptide present in phycoerythrin. The gamma ($\gamma$) component isn't found everywhere; it's only seen in red algae and the rod subunit of PE in some marine cyanobacteria (*Synechococcus* sp.) (Six et al. 2005). Each PBP (Cyanophyta) consists of a heteromonomer with two subunits $\alpha$ (12–20 kDa) and $\beta$ (15–22 kDa) (Kannaujiya and Sinha 2016), whereas R-PE is a trimeric complex with three subunits $\alpha$ (18–20 kDa), $\beta$ (19–21 kDa), and $\gamma$ (30 kDa).

Colored phycobiliproteins (PBPs) which absorb and divert light energy towards the photosynthetic reaction centre and non-pigmented linker proteins which arrange the PBPs and modify their light absorption property make up these PBPs. On the basis of spectrophotometric property, phycobiliproteins are divided into four forms such as phycoerythrin (PE; $\lambda_{max}$ = 540–570 nm), phycocyanin (PC; $\lambda_{max}$ = 610–625 nm), allophycocyanin (APC; $\lambda_{max}$ = 650–660 nm), and phycoerythrocyanin (PEC; $\lambda$max = 560–600 nm), (Bermejo et al. 2003) (Table 3). The covalently bonded prosthetic group is mostly responsible for the PBPs' coloration. These prosthetic groups are phycobilins, which are A, B, C, and D rings open-chain tetrapyrrole chromophores. Red colored Phycoerythrobilin (PEB), blue colored phycocyanobilin (PCB), yellow-colored phycourobilin (PUB), and purple-colored phycobiliviolin (PVB) are all present.

**Table 3.** Spectroscopic and fluorescence properties of phycobiliproteins (Adopted from Bermejo et al. 2003)

| Phycobiliproteins | Absorption maxima (nm) | Fluorescence emission maxima (nm) | Absorption color (visible spectrum) | Fluorescence color (UV-radiation) | Distribution |
|---|---|---|---|---|---|
| C-Phycocyanin (C-PC) | 610–620 | 630–650 | Purple, violet | Strong red | Cyanobacteria |
| Phycoerythrocyanin | 570–590 | 600–610 | Purple, light red | Orange | Cyanobacteria |
| C-Phycoerythrin (C-PE) | 540–565 | 568–575 | Pink, red | Yellow orange | Cyanobacteria |
| Allophycocyanin (APC) | 650–655 | 660–665 | Violet, gray | Weak red | Red algae/ cyanobacteria |
| R-Phycocyanin (R-PC) | 617–620 | 635–638 | Purple, violet | Strong red | Red algae/ cyanobacteria |
| R-Phycoerythrin (R-PE) | 562–565 | 570–575 | Red | Yellow orange | Red algae |

## Biosynthesis Mechanism

A two- to three times increase in heme pigment initiates the biosynthesis of PBPs and tetrapyrrole chromophores (Stadnichuk et al. 2015). Heme oxygenase (HOs) which converts heme to biliverdin IXα (BV) catalyses the biosynthesis of PBPs from heme (Figure 2) (Dammeyer et al. 2008). Seven reducing electrons and three molecules of molecular oxygen are required by oxygenase to complete the heme catalytic cycle, which results in the conversion of the BV IXα chromophore. Biliverdin reductase is basically found in a many marine and freshwater cyanobacteria. These enzymes help into catalysis of the reduction of methine bridge between BV Ixα's rings B and C, resulting in conversion to bilirubin IXα (BR IXα).

PcyA, which is involved in PCB production, is found in the majority of cyanobacteria. PcyAbiliproteins have a well-known structural arrangement that includes an integration of α-helix and β-strand sheet. The presence of extremely conserved residues such as $His^{74}$, $His^{88}$, and $Asp^{105}$ in the α-helix sandwich fold provides polar interactions in substrate BV IXα in PC conversion, according to neutron crystallography studies (Unno et al. 2015). In contrast to that PcyA, PebS uses independently active FDBRs and four electrons for the conversion of BV to PEB, with intermediate 15,16-DHBV (Figure 3) (Martiny et al. 2006). PebS plays a key role in the control of the alternative route for PEB production (Dammeyer et al. 2008). PVB and PUB, which are alternatively coupled with native PBPs, are generated for some of the intermediate components of the harvesting complex. Furthermore, the synthesis of PVB begins with the catalytic enzymes PecE/PecF lyases isomerizing the cysteine-84 residue of PEC in *Mastigocladus laminosus* (Böhm et al. 2007). However, the precise role of the BR IXα in cyanobacteria has yet to be determined. During photosynthesis, BR IXα is not found in any known light sensors or light-harvesting protein complexes. In photosynthetic cyanobacteria, particular genes known as the FDBR family genes (ferredoxin-dependent bilin reduction) are required for the conversion of bilin to phycobilin.

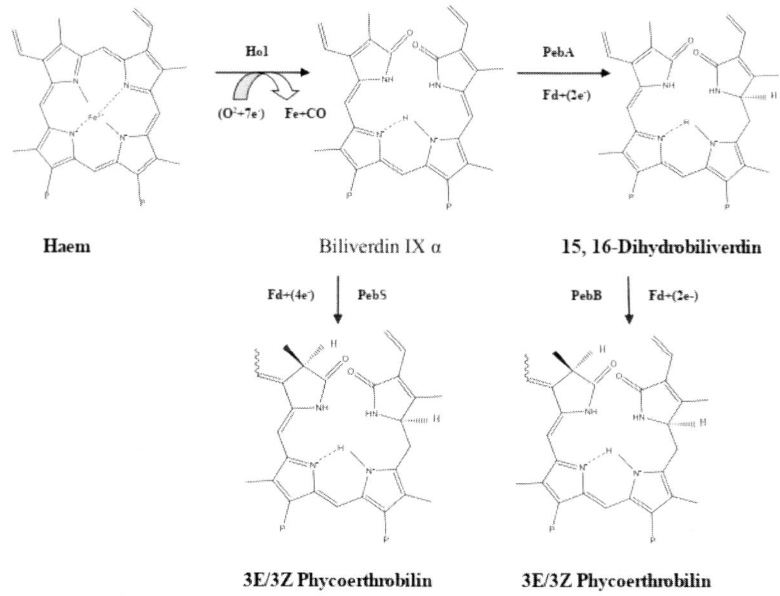


**Figure 2**. Biosynthesis of chromophore of PE by catalytic reduction of electron of biliverdin IXα in the presence of PebA, PebB, and PebS (Adopted from Kannaujiya et al. 2017).

In the current scenario, a group of FDBR family genes have been recognised, which includes HO (heme oxygenase), pcyA (phycocyanobilin: ferredoxin oxidoreductase), pebA (15,16-dihydrobiliverdin: ferredoxin oxidoreductase), pebB (phycoerythrobilin: ferredoxin oxidoreductase) (Dammeyer et al. 2008; Scheer et al. 2013).

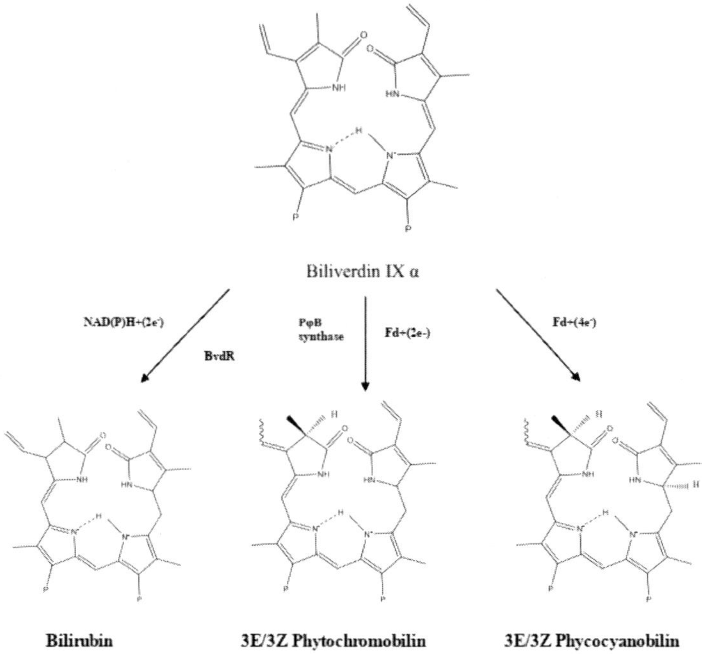


**Figure 3**. PC Chromophore synthesized by PcyA, PΦB synthase via the intermediate 181,182-DHBV from same precursor (Adopted from Kannaujiya et al. 2017).

## Applications

The beautifully coloured light-harvesting proteins known as phycobiliproteins (PBPs) are responsible for the natural hues of cyanobacteria, cryptomonads, and red algae (Glazer 1981). The current use of synthetic hues for food colouring contains hazardous substances that are potentially detrimental to the human body and have a key part in the development of cancer. They are widely employed in the production of food, drinks, dairy products, ice cream, and confectionery, among other things. PBPs have unique fluorescence qualities that are extremely valuable for biomedical, pharmacological, therapeutic applications, and molecular biology, in addition to their capabilities as natural colourants (Eriksen 2008). Currently, only a few species, such as *Spirulina sp.* and *Porphyridium sp.*, are used to produce PBPs. PBPs are the subject of hundreds of patents in a variety of sectors, including 55 patents for PBP manufacture, 236 patents for fluorescence uses, and 30 patents pertaining to the clinical value of PBPs (Sekar and Chandramohan 2008).

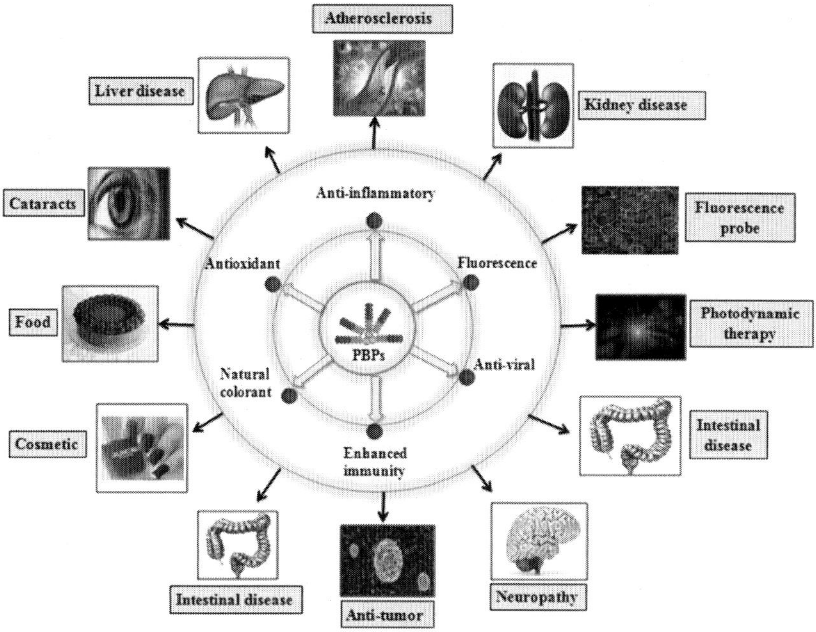


**Figure 4**.Commercial and Pharmaceutical applications of phycobiliproteins (PBPs).

## Nutraceuticals and Natural Food Colorants

The Food and Drug Administration (FDA) in the United States controls the identification and modification of nutritional value in the development of successful food products for health promotion and disease vaccination. Many functional meals have now been produced by a branch of science and technology that offer significant health advantages and enhanced quality of life for people. DIC International (previously Dainippon Ink and Chemicals) is one of the largest enterprises involved in the development of food products from *Spirulina* sp., with an annual gross production of over 300 t and rising (DIC 2009). Earthrise® Nutritionals, a division of DIC, has the world's biggest *Spirulina* farm, which spans 444,000 $m^2$ and is located in California, USA. Its *Spirulina*-based pills are sold in over 20 different countries (Earthrise Nutritionals LLC 2004).

PE is obtained from red algae such as *Porphyridium* sp. It is utilised in colourful confectionery and gelatin desserts (Dufosse et al. 2005). PE's dye qualities include strong yellow fluorescence, which is useful for special effects in low-light situations when exposed to ultraviolet (UV) radiation. A vast variety of food products that glow in the dark are created. When exposed to UV radiation at an acidic pH of 5–6, a variety of foods, including lollipops, soft drinks, dry sugar-drop sweets, and alcoholic beverages, glow. For colour stability, the fluorescent characteristic of PC has been retained in 30 percent alcohol (Dufosse et al. 2005). The cyanobacterium *Nostoc* sp. strain HKAR-2, which was isolated from a hot spring, generates a PC that is more heat stable and pH and light sensitive (Kannaujiya and Sinha 2016).

Colored items, like drinks, demands pH stability and light to avoid deterioration. In the confectionery industry, the use of PC colourant in jelly gums and candy manufacture has risen. The appearance of PC's blue hue is denatured in hard candy; however, the presence of a gelatin matrix can keep the blue colour in jelly gum (Jespersen et al. 2005). When exposed to extreme sunlight for 24 hours, the jelly gum cores discolour. PE is highly sensitive to light and the oxygen activity of food items, necessitating the use of long-term preservatives to keep it unspoiled and safe. In the form of antioxidant health foods, unpurified forms of PC extracted from a crude extract of *Arthrospira platensis* are mostly accepted (Bermejo et al. 2008).

PC, which is made from the dried biomass of *A. platensis*, has recently received more attention as a nutraceutical. PC production has risen to a total market value of US$10–50 million per year, with a value-added coloured food product value of roughly US$450 per kilogramme (Leema et al. 2010). In order to optimise large-scale manufacture of PBPs for this application, extensive understanding of the colorant's stability is required for proper application of any natural colourant in food. Soft drinks (e.g., Pepsi® and Bacardi Breezer®) include the blue hue of PC, which lasts for roughly a month at room temperature. PC, a natural blue colourant, is often employed in dry sweets where sugar is present. This makes it ideal for floral cake decorations, as it keeps its colour for up to a year. Long-term thermal stability of PC has been documented with the addition of food-grade preservatives such as sodium chloride, citric acid, and sucrose (Chaiklahan et al. 2012). Benzoic acid is very good preservation agent for PBPs in food items maintained at 4, 25, and 40°C, (Kannaujiya and Sinha 2016) probably because it functions as an antioxidant and has antibacterial action that limits microbe development.

## Pharmaceuticals and Fluorescence Probes

### *Pharmaceutical Agents*

Green approaches to the manufacturing of antibiotics, pharmaceuticals, and fluorescent tags have sparked interest in industrial production in recent decades (Sekar and Chandramohan 2008). In terms of anti-inflammatory, antioxidant, anti-aging, anticancer, neuroprotective, and hepatoprotective actions, PC has a significant advantage over the pharmacological capabilities of other substances. PC's antioxidant capabilities include scavenging of hydroxyl, peroxide, alkoxyl, nitrogen-containing radicals, and superoxide as well as a possible function in lipid peroxidation inhibition (Romay et al. 2003). PC derived from *Aphanizomenonflos-aquae* has high antioxidative characteristics and can effectively guard against oxidative damage in a variety of situations (Benedetti et al. 2004).

Similarly, PC plays an important function in preventing peroxyl radical-induced oxidative hemolysis of red blood cells. In plasma samples subjected to pro-oxidant synthetic compounds like cupric chloride, PC has been demonstrated to drastically minimise lipid peroxidation. Many oxidation-related clinical illnesses might be prevented by these PC-induced antioxidative processes. During inflammation, PC can prevent the formation of edoema and the excessive production of myeloperoxide (MPO), prostaglandin E2 (PGE2), histamine, and leukotriene B4 (LTB4) in tissue (Richa et al. 2011). Furthermore, in human leukaemia K562 cell lines, PC from *Spirulina platensis* has strong inhibitory effects on cancer growth and development in terms of dosage dependency and time duration. PC also lowers cholesterol levels, which may aid in the prevention of heart disease (Nagaoka et al. 2005). PC has been proven to lower the levels of enzymes like alanine aminotransferase (ALT), malondialdehyde (MDA), and aspartate aminotransferase (AST) in the blood serum.PC also affects the thyroxine hormone (T3) by increasing serum nitrite levels. R-PE, which is derived from red algae, is another promising antioxidant for photodynamic treatment and cancer prevention. APC, in addition to PC and PE, is important in the suppression of enterovirus 71 in host cells (Shih et al. 2003).

### *Fluorescence Agents*

PBPs autofluorescence makes them ideal for a variety of experimental techniques that need fluorescent probes. After dissociation from the photosystem, the individual subunits PE, PC, and APC are extremely fluorescent. Natural fluorescent proteins are currently in high demand in the market due to their numerous uses. PBPs have a high molar extinction coefficient, high excitation/emission, a big wavelength shift, and a high fluorescence quantum yield, among other spectrum features. As a result, PBPs outperform traditional fluorescent agents. Fluorescence microscopy, flow cytofluorimetry, fluorescence activated cell sorting (FACS), single-cell analysis, and immunoassays are only a few of the key cell biology procedures that employ PBPs as the primary fluorescent agents (Eriksen 2008).

PE fluorescence is notable for the large shift in emission wavelengths (580 nm) following excitation at 488 nm. In molecular biology, PE is utilized as the second best coloured fluorescent product for fluorescein-labeled antibodies (Sekar and Chandramohan2008). In the near future, the proliferation of these luminous compounds might lead to diagnostic tools for cancer, the human immunodeficiency virus (HIV), and other devastating illnesses. The combination of acceptor cyanine dyes Cy5/Cy7 (indodicarbocyanine/ indotricarbocyanine) and donor PE has recently been used to create multicolor fluorescence studies. The same combination can change the emission wavelength from red to deep red, allowing for better colour examination of any cells (Waggoner 2006).

PE's broad fluorescence has also been exploited in the construction of Affymetrix DNA microarray chips. In biotin-labeled DNA/protein probes for array detection, the combined PE–streptavidin shows a high signal (Benvin et al. 2007). PBP conjugation with immunoglobulins, protein A, biotin, and avidin has been described for FACS's crucial dual colour single-cell analysis (Oi et al. 1982).

## Cosmetics

PBPs' commercial uses have grown as a result of modern research and development (Richa et al. 2011). Furthermore, PBPs have played an important role in the long-term development of cosmetics. In Japan, PC and PE are sold as cosmetic colouring agents (Prasanna et al. 2007). Food rules in Japan require the use of food colours without the inclusion of synthetic preservatives. Lina-Blue is a DIC Corporation-produced blue commercial dye that may be used in natural cosmetics and food items. Currently, a number of enterprises are developing unique and environmentally friendly PC-based items such as bath powder, whitening cream, soap, body lotion, lipstick, eyeliner, and other cosmetics. In the near future, the PC/PE-based cosmetics industry may reach substantial economic growth, allowing for the long-term creation of healthy cosmetics.

## Bio-Medical Application

**Table 4.** Biological effects of PBPs and their proposed applications. Studies reported between 2019 and 2021 (Adopted from Dagnino-Leone et al. 2022)

| PBP | Biological activity | Source of PBP/ Biological Effect | Proposed Application | Reference |
|---|---|---|---|---|
| PC | Antioxidant | *Arthrospira maxima* / PCB prevents alterations in oxidative stress markers, antioxidant enzymes, and caspase 9 activities. | Nephro-protective action on acute kidney injurycaused by mercury. | Garcia-Pliego et al. 2021 |
| | | *Spirulina* sp. / Attenuate radiation-induced oxidative stress damage in liver by activating Nrf2/ HO-1 signaling pathway and reduce DNA damage. | Protective effect on hepatic damage induced by X-ray. | Liu et al. 2020 |
| | | *Spirulina* sp. / Lengthening delivery of PC, encapsulated in poly (ethylene glycol)-b- (poly (l-glutamic acid)-g-polyethylenimine nano-carrier, through the abdominal subcutaneous injection in rats. | Attenuation of hepatic ischemia/reperfusioninduced pancreatic islet injury. | Tong et al. 2019 |
| | | *Phormidium versicolor* / Prevents cadmium-induced elevation of ALAT, ASAT and bilirubin levels in rats. Enhance the levels of antioxidant enzymes. | Prevention action against hepatotoxicity causedby cadmium | Gammoudi et al. 2019 |
| | Anti-inflammatory | *Spirulina* sp. / Inhibits of albumin denaturation, anti-proteinase, hypotonicity-induced haemolysis and anti-lipoxygenase activities. | Potential drug development. | Prabakaran et al. 2020 |
| | | *Spirulina platensis* / Inhibits COX-2 expression during the radiation therapy of the colon cancer cell lines and in normal colonic cells. | Countermeasure reduction by C-PC treatment during colonic cancer radiation therapy. | Kefayat et al. 2019 |
| | | *Spirulina platensis* / Attenuates the pulmonary fibrosis by inhibition of the production of interleukin-1 beta, tumor necrosis factor-a, and lipopolysaccharide. Increases the intestinal bacterial diversity, richness, and reduces pro-inflammatory bacterias. | Idiopathic pulmonary fibrosis treatment | Xie et al. 2019 |

**Table 4.** (Continued)

| PBP | Biological activity | Source of PBP/ Biological Effect | Proposed Application | Reference |
|---|---|---|---|---|
| PC | Anti-metabolic diseases | *Arthrospira platensis* / APC and C-PC tryptic peptides inhibit enzyme DPPIV activity in vitro, therefore, acting as hypoglycemic peptides. | Nutraceuticals orfunctional foods for the management of type 2diabetes. | Li et al. 2020 |
| | | *Spirulina platensis* / Reduce lipid accumulation in the steatosis L02 cells and in the liver of NASH mice. Improve the antioxidant capacity of liver by activating AMPK pathway of hepatocytes. | Nutraceuticals and therapeutics of non-alcoholic fatty liver disease. | Ma et al. 2020 |
| | | *Spirulina platensis* / Inhibits a-amylase and b-glucosidase enzymes activity. | Potential drug development for diabetes type 2 treatment | Prabakaran et al. 2020 |
| | Anti-cancer | *Limnothrix* sp. / Regulates both anti- and pro-apoptotic genes by increasing levels of Bax, Apaf-1 and activates caspase-8, caspase-9, and caspase-3. Decreasing expression of Bcl-2, Mcl-1, and survivin. In combination with Topotecan, increase expression of the death receptor FAS and cleaved PARP. | Prostate cancer cells (LNCaP) treatment with reduction of conventional chemotherapeutic agents such as Topotecan. | Kaur et al. 2020 |
| | | *Spirulina platensis* / Anti-proliferative effect against HepG-2 cell lines. | Potential drug development for cancer treatment. | Prabakaran et al. 2020 |
| | | *Spirulina* sp. / Ameliorates the radiation-induced intestinal injury and regulates the effect on gut microbiota by TLR4/Myd88/NF-jB pathway | Countermeasure for the radiation-induced gastrointestinal syndrome. | Lu et al. 2020 |
| PC | Anti-cancer | *Spirulina* sp. / Induce apoptosis, suppressed the growth of NSCLC cells by down regulation of TIRAP/NF-jB activity. | *Non-small cell lung cancer treatment.* | Hao et al. 2019a |
| | | Spirulina platensis / Anti-proliferative and anti-migratory function by the reduction of RIPK1/NF-jB activity in the NSCLC cells. | *Non-small cell lung cancer treatment.* | Hao et al. 2019b |
| | | *Spirulina platensis* LEB 52 / In multiple drug resistance phenotype of erythroleukemic cells, C-PC modulates the expression of COX2 and ABCB1 for the K562-Lucena cells in a ROS-dependent manner and the expression of ALOX5 for the FEPS cells in a ROS-independent manner. | *Erythroleukemic cells multiple drug resistance (MDR) phenotype treatment.* | Fernandes e Silva et al. 2019 |
| | | Galdieriasulphuraria / Aqueous extract containing C-PC has antioxidant activity and exert cytotoxic activity in the human adenocarcinoma A549 cells. | *Anticancer candidate for therapeutic treatments.* | Bottone et al. 2019 |

| PBP | Biological activity | Source of PBP/ Biological Effect | Proposed Application | Reference |
|---|---|---|---|---|
| | Anti-neuro-degenerative | *Spirulina platensis* / PCB reduces brain injury in PC12 neuronal cells after endothelin-1- induced focal cerebral ischaemia in Wistar rats. PCB restored the myelin basic protein and CNPase enzyme expression levels in ischaemic rats. *Spirulina* sp. / Prevents streptozotocin-induced increase activity of hippocampal cholinesterases, BAX, and the levels of BCL-2 and ChAT. | PCB as a therapeutic pharmacological alternative for ischaemic stroke patients.<br><br>Therapeutic agent in managing Alzheimer's Disease | Pavón-Fuentes et al. 2020<br><br>Agrawal et al. 2020 |
| | Anti-microorganism | *Spirulina* sp. / PCB inhibits the main protease and papain-like proteas activities from coronaviruses (CoV) in vitro. In silico, showed inhibitory activity using other phycobilins such as PUB, PEB, and PVB.<br>*Spirulina* sp. / Bacterial abundance and diversity increase in mice intestinal. Decrease intestinal permeability and increases barrier function.<br>*Spirulina* sp. / Reduce the culture viability of *E. faecalis* in human premolar teeth by increasing the bacterial internal ROS. | PCB as broad-spectrum antiviral drugs against coronaviruses.<br><br>Basis for a positive modulation of intestinal microbes by treatment with PC.<br>Adjunct therapy in endodontic infection treatment. | Pendyala et al. 2021<br><br>Xie et al. 2019<br><br>Pourhajibagher et al. 2019 |
| APC | Antioxidant | *Phormidium* sp. / APC increase anti-aging effect in worm model by improving physiological markers. Moderates the expression of human Ab1–42. | The use of APC as drug development for agerelated disorders. | Chaubey et al. 2020 |
| | Anti-inflammatory | *Pyropia pseudolinearis* / Peptides produced from the water-soluble proteins (riched in PE, PC, APC, and Rubisco) had inhibitory effect on th angiotensin-I-converting enzyme. | Prevention of cardiovascular diseases and high blood pressure. | Kumagai et al. 2021 |

**Table 4.** (Continued)

| PBP | Biological activity | Source of PBP/ Biological Effect | Proposed Application | Reference |
|---|---|---|---|---|
| PE | Antioxidant | Phormidium persicinum / C-PE reduce expression of nephrin and podocin in a mouse model of HgCl2-induced acute kidney injury. Reduced activation of IRE1a pathway and avoids caspase mediated cell death. | *Prevention of HgCl2-induced acute kidney injury.* | Blas-Valdivia et al. 2021 |
| | Anti-inflammatory | Colaconema sp. / Enhance prophenoloxidase and phagocytosis activity but inhibits ROS in hemocytes from whiteleg shrimp. In vivo, PE induce immune genes and enhance the innate immune parameters. | *Potential immunomodulator in shrimp culture to prevent the Vibrio parahaemolyticus or white spot syndrome virus.* | Lee et al. 2021 |
| | Anti-metabolic diseases | *Porphyridium purpureum* / Peptide from PE reduce the foam cell formation, intracellular lipid accumulation and the secretion of TNF- and IL-6 in RAW 264.7 cell line by targeting the scavenging receptors CD36, SRA1, and Map Kinase p38. | Prevention atherogenic foam cell formation. | Kavitha et al. 2019 |
| | Anti-cancer | Inhibits growth and induce apoptosis of SKOV-3 cells. The ROS/JNK/Bcl-2 signaling pathway, upregulation of JNK, GADD45A, RAD23, and downregulation of XBP1 and OS9 are critical in PE -induced apoptosis in this cell. | PE as a potential treatment for human ovarian cancer. | Ying et al. 2021 |
| | | R-PE has modulated the innate and adaptive immune systems through TLR4/NF-jB mediated $CD4^+$ T cell activation and differentiation in a hydrocortisone immune-suppressive model. | Immuno-modulator during long-term high-dos glucocorticoids therapy for autoimmune diseases and cancer. | Wang et al. 2020 |
| | Antineurodegenerative | *Porphyra haitanensis* / C-PE inhibits the precursor protein of BACE1, therefore, reduce accumulation of amyloid-b precursor protein. | Therapeutic drug for complement Alzheimer's disease treatment. | Chaubey et al. 2019 |
| | Antimicroorganism | Recombinant PE / R-PE increase the abundance of beneficial bacteria in the intestinal flora and reduce the detrimental bacteria from H22- bearing mice. | Gut microbiota modulation for prevention of resistance diseases and nutritional in animals. | Qi et al. 2019 |

## Anticancer

Because of its high nutraceutical value, high-stroke shift, protein nature, and intense UV fluorescence, PC is widely utilised as a blue colourant in food and cosmetics. PC possesses anticancer properties in a range of cancer cell types, including colon cancer, lung cancer, bone marrow cancer, and breast cancer (Table 4) (Gardeva et al. 2014). Although PC isolated from *Spirulina platensis* has anticancer action, its large molecular weight and complicated structure may make molecular mechanism identification difficult. Prior to its use as a medication, column chromatography and enzymatic hydrolysis were used to degrade PC into a low molecular weight peptide, which was then assessed for tumour inhibitory efficacy on HeLa and 293T tumour cell, with disintegrated PC showing a significantly greater response than non-disintegrated one (Wang et al. 2012). The lung cancer cell line SPC-A-1 growth response was similarly impacted by the component of α and β subunits of PC (Sun et al. 2010). In compared to β subunits, subunits have distinct impacts on tumour progression. PC is a highly active component that interacts with tubulin proteins and metabolic enzymes like glyceraldehyde-3-phosphate dehydrogenase (GAPDH), causing caspase-3 and caspase-9 factor to activate, resulting in $G_0/G_1$ cell cycle arrest, apoptosis, and tumour growth suppression. (Liu et al. 2016).

## Anti-Inflammatory

PC is a dietary protein that reduces inflammatory cytokines such as interleukin-6 (IL-6), NO, MMP-3, TNF, and sulfated glycosaminoglycans in osteoarthritis patients. In a mouse ear inflammation test, PC inhibits cyclooxygenase-2 activity and stimulates the production of leukotriene B4 (Reddy et al. 2000).The catalytic control of nitric oxide synthase (NOS) produces nitric oxide (NO), which is vital in regulating the healthy and pathological inflammatory response of cells (Table 4) (Moncada et al. 1991). The presence of a large quantity of NO induced the production of proinflammatory cytokines, lipopolysaccharide (LPS), and free radicals which is a key mediator in the pathophysiology of inflammatory disorders (Szabo and Thiemermann 1995). PC can control excessive levels of PGE2 and NO by down regulating COX-2 and iNOS expression, which reduces the generation of TNF and neutrophil infiltration to inflammatory sites (Shih et al. 2009).

## Antioxidant

PC has recently been used in a number of biotechnology and pharmaceutical goods. PC is also used as a hepatoprotective, neuroprotective, and antioxidative agent in the treatment of a variety of disorders (Romay et al. 2003). PC can scavenge alkoxyl, peroxyl, hydroxyl, hypochlorous acid (HOCl), and peroxynitrite ($ONOO^-$) and hence performs an important antioxidant function in the body (Bermejo et al. 2008). Free radicals are extremely harmful since they may infiltrate cells and cause a variety of illnesses, including atherosclerosis, inflammation, reperfusion damage, cancer, and others (Gupta et al. 2011). PC is a neuroprotective agent with a lot of power. They, together with N-acetylcysteine (a neuroprotective medication), demonstrate protective properties against tributyltin chloride-induced neurotoxicity (Mitra et al. 2015).

By suppressing oxidative stress, mitochondrial damage, and lipid metabolism, PC has been shown to reduce the risk of cardiovascular disease (CVD). PC significantly decreases oxidative stress-induced inflammatory damage in atherosclerosis and promotes COX-2 production, which improves antioxidant enzymes in the body and controls blood cholesterol levels for optimal heart function (Riss et al. 2007).As a result, PC can slow the progression of atherosclerosis by increasing $CD_{59}$ expression and decreasing muscle cell proliferation, blood fat, and endothelial cell apoptosis (Li et al. 2013). Cataract is an age-related oxidative stress disease. In rats, PC can modify antioxidative levels, lowering the risk of cataract. PC regulates the transcription of lens crystalline, redox, and expression mRNA for the apoptotic cascade pathway, all of which are important in lens transparency maintenance (Kumari et al. 2015).

## EmergingTechnology

Photodynamic therapy (PDT) is a new noninvasive way for treating life-threatening disorders such as psoriasis, cancer, viral, and bacterial infections (Bharathiraja et al. 2016). This treatment offers a number of benefits over traditional chemotherapy, targeted surgery, and the use of antibiotics to overcome drug resistance mechanisms. PDT is a cancer cell killing method that involves the use of strong sun radiation and photosensitizers to produce free radicals. Many photosensitizers, such as chlorophyll-derived protoporphyrin IX and chlorine 6 have been identified; however their hydrophobic nature causes them to clump easily in physiological solution, reducing the efficiency of the PDT process (Ormond and Freeman 2013). Because of its high-water solubility, nontoxicity, and immune-modulating capabilities, PC offers numerous benefits over chlorophyll-derived photosensitizers. PC is more quickly digested by regular cells than malignant cells; therefore it might be employed in PDT to eradicate cancer while causing minimal damage to healthy cells (Bharathiraja et al. 2016). This new technique has been utilised to treat a variety of conditions, including myocardial infarction, tumour removal, traumatic brain injuries, and other cancer-related illnesses (Avci et al. 2013).

## Carotenoids

The major pigments for light absorption in cyanobacteria are phycobiliproteins, whereas carotenoids play an important role in non-photochemical quenching (NPQ) as a barrier against saturating light and a quencher of reactive oxygen species (ROS) (Masojídek et al. 2013). When it comes to carotenoids, these terpenoids are required for the survival of all photosynthetic organisms and are found throughout them. Carotenoids are a broad group of pigments (about 600) that may be split into two categories: carotenes (such as α- and β-carotene) and xanthophylls (oxygenated derivatives of carotenes) such as zeaxanthin and echinenone. Carotenoids are of tremendous interest to the industry because of their bioactive potential as antioxidants, anti-inflammatory agents, and anti-tumor agents, among other things (Pagels et al. 2020). Between 2016 and 2019, the market for carotenoids topped $1.5 billion (USD), with beta-carotene, astaxanthin, and lutein accounting for 60% of the market (Markets and Market 2020). In the microalgae *Dunaliella salina* (β-carotene) and *Haematococcus pluvialis* (astaxanthin), as well as the vascular plant *Tagetes erecta*, the bioprocess of these carotenoids is well known (lutein). Despite cyanobacteria's promise as pigment producers, the sole large-scale commercial use of these organisms is the manufacture of phycocyanin by *Arthrospira platensis*. The explanation for this is because cyanobacteria have far less carotenoids than microalgae or vascular plants (Hirschberg and Chamovitz 1994). On the other hand, tactics for using cyanobacterial carotenoids have grown in popularity in recent years, signalling the possibility of a future cyanobacterial bloom in this sector. This chapter will go through the many types of cyanobacterial carotenoids as well as the most important properties for biotechnological use, with an emphasis on their bioactivities. It is also urged that their manufacturing, extraction, and purification procedures be optimised. Finally, the field's economic issues and future prospects are briefly examined.

## Structure

Carotenoids are tetraterpenoids, which means that they are made up of a C40 hydrocarbon chain with eight isoprenoids and a succession of double-bound conjugations. They may be divided into carotenes and xanthophylls, which are oxygenated derivatives of carotenes and are either linear or cyclic molecules having one or two rings at their extremities. Furthermore, carotenoid modification can result in glycosylated carotenoids like myxoxanthophyll or even apocarotenoids (shorter chain carotenoids) (Hirschberg and Chamovitz 1994).

## Synthesis

Carotenogenesis is a complicated but well-understood metabolic route that allows cyanobacteria to manufacture a wide range of terpenoids. The geranylgeranyl pyrophosphate precursor is used to make these chemicals (Sugiyama et al. 2020). Carotenoids are synthesised by a set of genes that code for synthases, desaturases, cyclases, and hydroxylases. The existence of the mentioned genes was investigated in different cyanobacterial species by Liang (Table 5) (Liang et al. 2006), who discovered that the majority of the genes are transversal, while particular genes can be found as a replacement or redundant to other genes. The biosynthetic route begins with two molecules of geranyl geranyl pyrophosphate being condensed into phytoene by phytoene synthase (CrtB); from there, phytoene may be converted to ζ-carotene (CrtP), which can then be converted to lycopene by carotene desaturase (CrtQ). A phytoene desaturase (CrtI) may convert phytoene directly to lycopene in a few species, including *Anabaena* and *Nostoc* (Liang et al. 2006). They also generate the standard phytoene desaturase (CrtP). The major primary carotenoids, α- and β-carotene, are generated from lycopene. α-carotene is produced by a lycopene cyclase (CrtL or CruA) in a direct conversion, whereas β-carotene is produced by a lycopene cyclase (CrtL or CruA) in a two-step process (Liang et al. 2006; Sugiyama et al. 2020). Several xanthophylls can be made from the three carotenoids.

**Table 5.** Carotenogenesis with genes and enzymes, whose functions are confirmed in phototrophic microalgae/cyanobacteria. (Adopted from Takaichi 2007)

| Gene | Enzyme | Species |
|---|---|---|
| *crtE, ggps* | Geranyl geranyl pyrophosphate synthase | *Thermosynechococcus elongates* BP-1, *Porphyra umbilicalis* UTEX LB-2591 |
| *crtB, pys, psy* | Phytoene synthase | *Synechocystis* sp. PCC 6803, *Euglena gracilis* |
| *crtI* | Phytoene desaturase (bacterial type) | *Gloeobacter violaceus* PCC 7421 |
| *crtP, pds* | Phytoene desaturase (plant type) | *Arthrospira platensis* NIES-39, *Synechococcus elongatus* PCC 7942, *Synechocystis* sp. PCC 6803. |
| *crtQ, zds* | ζ-Carotene desaturase | *Anabaena* sp. PCC 7120, *Synechocystis* sp. PCC 6803 |
| *Z-ISO* | ζ-Carotene isomerase | *Arthrospira platensis* NIES-39, *Euglena gracilis* |
| *crtH, crtISO* | ζ-Carotene isomerase | *Synechocystis* sp. PCC 6803 |
| *crtL, crtL-b, lcy-b* | Lycopene β-cyclase | *Synechococcus elongatus* PCC 7942 |
| *cruA* | Lycopene β-cyclase | *Arthrospira platensis* NIES-39, *Synechococcus* sp. PCC 7002, *Synechocystis* sp. PCC 6803 |
| *cruP* | Lycopene β-cyclase | *Synechococcus* sp. PCC 7002 |
| *crtL-e, lcy-e* | Lycopene ε-cyclase | *Prochlorococcus marinus* MED4 |
| *crtR* | β-Carotene hydroxylase | *Anabaena variabilis* ATCC 29413 |
| *crtG* | β-Carotene 2-hydroxylase | *Thermosynechococcus elongates* BP-1 |
| *crtO* | β-Carotene ketolase | *Anabaena* sp. PCC 7120, *Synechocystis* sp. PCC 6803 |
| *crtW, bkt* | β-Carotene ketolase | *Anabaena* sp. PCC 7120, *Nostoc punctiforme* PCC 73102 |

A myxoxanthophyll molecule is generated from γ-carotene via hydroxylation by γ-carotene hydroxylase (CruF) and the addition of a glycoside group by a glycosyl transferase (CruG). Myxoxanthophyll is a cyanobacteria-only yellow glycoside terpenoid that is necessary for cell wall formation and thylakoid organisation. A hydroxylase (CrtR) may produce lutein from α-carotene (Duan et al. 2012). A ketolase (CrtO or CrtW) can convert β-carotene to an echinenone, which can then be converted to zeaxanthin by a hydroxylase (CrtR). A ketolase (CrtO or CrtW) may make canthaxanthin from echinenone, while a hydroxylase (CrtR) can make hydroxyechinenone, the major ketocarotenoid in the OCP. The violaxanthin cycle is a reversible activity of epoxidase and de-epoxidase that may produce both antheraxanthin and violaxanthin from zeaxanthin. A hydroxylase (CrtG) can produce nostoxanthin from zeaxanthin, while a neoxanthin synthase

(NSY) can produce neoxanthin from violaxanthin (Sugiyama et al. 2020). Furthermore, by introducing a CrtR or CrtW gene into synthetic cyanobacteria, astaxanthin may be generated. Canthaxanthin is converted to astaxanthin by a hydroxylase (CrtR), while zeaxanthin is converted to astaxanthin by a ketolase (CrtW) (Diao et al. 2020). Despite the fact that carotenogenesis genes are found in most photosynthetic organisms, their expression and subsequent carotenoid synthesis are restricted to some cyanobacteria and are influenced by growth circumstances (Pagels et al. 2020). Taking use of such regulations to improve the production of particular or total carotenoids is critical in a biotechnological approach.

## Orange Carotenoid Protein (OCP)

Carotenoids can be found in cyanobacteria's reaction centre, with chlorophyll molecules, or in a protein complex containing a single carotenoid molecule, which is a crucial component for photoprotection of the OCP (Zhang et al. 2014). The OCP is a 35-kDa water-soluble protein that contains a single molecule of hydroxyechinenone in an inactive (orange) form that is activated by blue-green light and transformed to an active (red) form. Despite its importance for cell defence, the OCP accounts for just 1% of total carotenoids. The OCP is made up of two domains linked together by a flexible linker. The first is an all-helical N-terminal domain, which is unique to cyanobacteria, and the second is a α/β-fold C-terminal domain, which is found in all kingdoms (Kirilovsky and Kerfeld 2016). The OCP is encoded by the *slr*1963 gene, which is constitutively expressed, while stress circumstances like strong illumination or salt stress can cause a brief acclimation and increase in the gene's transcription (Kirilovsky and Kerfeld 2016). In summary, the OCP only connects to the phycobilisome in its red form (activated). The phycobilisome may absorb and send all of the energy to the photosystem under darkness or low-light settings. In saturation light (strong blue-green or white), OCP converts to an active state and binds to the phycobilisome for energy dissipation, allowing the photosystem to absorb light without being saturated. Other proteins are involved in the process besides phycobiliproteins and OCP, such as the fluorescence recovery protein (FRP), which is responsible for inactivating OCP following the NPQ process (Kirilovsky and Kerfeld 2016).

## Applications of Carotenoids

Carotenoids pigments are found in higher plants, microalgae and cyanobacteria from both marine and freshwater environments. They cannot be synthesised by animals or humans; thus, they must be obtained through food. The centre component of the molecule, which provides structure, chemical reactivity, and light-absorbing capabilities, is made up of a lengthy chain of conjugated double bonds.Carotenoids colour, which ranges from light yellow to red, and antioxidant activities are heavily influenced by these conjugated double bonds and other functional groups. The six major types of food additives are taste enhancers, preservatives, stabilisers, emulsifiers, antioxidants, and colouring agents. Despite the fact that carotenoids are rather unstable due to their vulnerability to light, heat, oxygen, autoxidation, and isomerization, certain commercially manufactured carotenoids (e.g., β-carotene) are acceptable colour additives. Some naturally occurring carotenoids (such as β-carotene and astaxanthin) are commonly utilised, whereas others (such as lutein, canthaxanthin, zeaxanthin, and lycopene) are not. The worldwide carotenoid industry is projected to be worth more than \$1 billion each year and is steadily increasing.

Furthermore, carotenoids operate as effective radical quenchers, protecting other compounds from undesired oxidation events. Free radicals and oxidising reagents may easily attack the electron-rich polyene chain; therefore, it contributes electrons and hydrogen atoms to neutralise them. More than 750 carotenoids have been identified. Around 60 of them found in food have an impact on human nutrition. Carotenoids play an essential role in the worldwide food business as food additives. Carotenoids are the most popular food colouring agents in Europe. The global market for carotenoids reached \$1.5 billion in 2017 and is expected to reach \$2.0 billion by 2022, growing at a CAGR of 5.7 percent from 2017 to 2022 (Markets and Markets 2021).

**The Most Important** Carotenoids Used in Food, Feed and Cosmetics Industries

### *Astaxanthin*

Astaxanthin is an orange-red hydrophobic xanthophyll (ketocarotenoid) produced from β-carotene (3, 3'-dihydroxy-,-carotene-4, 4'-dione). The molecule has 13 conjugated double bonds in total, and its keto and hydroxy groups at the 4,4'- and 3,3'-ionone ring locations, respectively, have the greatest antioxidant action yet seen (Liu and Osawa 2007). Astaxanthin is one of the most economically lucrative carotenoids on the market, with an annual sales volume anticipated to rise. Together, astaxanthin and β-carotene account for over half of the carotenoid market. Because natural astaxanthin is relatively costly, the synthetic pigment generated from petrochemical precursors now dominates the worldwide market.

Astaxanthin (E161j) is a food dye in the European Union and the United States. Natural astaxanthin is considered safe by the FDA, and astaxanthin and astaxanthin dimethyl disuccinate, have been approved as food colorings and additives for specific uses in animal and feeds (e.g., carp, salmonids, red sea bream, shrimps, ornamental fish such as cichlids, danios, koi, goldfish, gouramis and tetras; however, it shall not exceed 80 mg/kg). It is now utilised on a scale of one hundred tonnes (Code of Federal Regulations 2019). According to Commission Regulation 1288/2004 (EC), additive E161z is astaxanthin generated by *P. rhodozyma* and can be used in salmon and trout feed at a maximum dosage of 100 mg/kg of the entire feed, whereas E161j is astaxanthin dimethyl succinate with the same purpose. *Paracoccus carotinifaciens* may also generate astaxanthin, which can be supplied to feed at a maximum dosage of 100 mg/kg (Commission Regulation 721/2008).

It also has additional physiological advantages in aquaculture, including increased survival, growth performance, reproductive capacity, stress tolerance, disease resistance, and immune-related gene expression (Lim et al. 2018). To our knowledge, no significant detrimental effects of dietary astaxanthin consumption on animal health have been recorded, whether natural or synthetic, and multiple studies have demonstrated its safety in humans (Fassett and Coombes 2011).

### *β Carotene*

The unoxidized category of carotenoids includes β-carotene, an orange pigment. The presence of β-ionone rings at both ends of the molecule distinguishes it from other compounds. β-carotene is a lipophilic carotenoid due to its strong conjugation and absence of functional groups (Torregrosa-Crespo et al. 2018). The European Commission has accepted carotenes as Group II food additives and food colours permitted at quantum satis. β-carotene has the E number E160 and is used as a food colouring additive (https://webgate.ec.europa.eu/foods_ system/main/). Due to its colourant characteristics and bioactivity, it is a high-valued food additive with significant commercial relevance. β-carotene is utilised as an orange-red colour in a variety of culinary items, including non-alcoholic drinks with a tropical fruit flavour, cheese, pastry, and ice cream (Bogacz-Radomska and Harasym 2018).Synthetic β-carotene is currently the most popular product on the market, whereas β-carotene generated by algae is of little importance (Zakynthinos and Varzakas 2016). The microalgae *Chlorella zofingiensis, Dunaliella sp.,* and *Spirulina (Arthrospira)* sp. are all natural sources of β-carotene.

β-carotene is a precursor to vitamin A, which is produced from carotenoids using the enzyme β-carotene 15,15'-monooxygenase. β-carotene has the highest bioactivity as a vitamin A precursor among these carotenoids (Bogacz-Radomska and Harasym 2018). The consumption of carotenoids has been linked to a lower incidence of colon cancer (Ramadas et al. 2010). β-Carotene was shown to have anti-angiogenic properties, lowering the number of tumor-directed capillaries, suppressing endothelial cell proliferation, migration, and tube formation, as well as influencing serum cytokine levels and blocking transcription factor activation and nuclear translocation (Guruvayoorappan and Kuttan 2007).

Several studies have found that carotenoids lessen the incidence of type 2 diabetes. Higher levels of β-carotene in the blood are linked to reduced insulin resistance (Sluijs et al. 2015). As a powerful antioxidant, β-carotene can inhibit the development of free radicals, which can cause skin cells to age prematurely (Lee et al. 2004). Because β-carotene is lipophilic, it can accumulate in the body. Carotenosis is a benign disorder in which the skin becomes orange if too much β-carotene is consumed. Chronic supplementation with high dosages of β-carotene has been linked to an increased risk of lung cancer in cigarette smokers (Tanvetyanon and Bepler 2008).

### *Lutein*

Lutein (β, ε-carotene-3,3'-diol) is a xanthophyll that has only 10 conjugated double bonds, giving it a brighter yellowish-green colour than other carotenoids (Torregrosa-Crespo et al. 2018). Lutein, along with astaxanthin and canthaxanthin, is one of the industrially important xanthophylls (Breithaupt 2007). The use of xanthophylls, particularly lutein, in human and animal feed is gaining popularity (Breithaupt 2007). The European Commission has approved lutein as a food additive in Group III, Food colours with a combined maximum limit, under the E number E161b (https://webgate.ec.europa.eu/foods system/main/).

Lutein cannot be synthesised because of the long synthesis process (Breithaupt 2007). This carotenoid is present in the majority of green plants, such as spinach and kale, mostly in the leaves, but also in the petals of flowers and yellow carrots (Breithaupt 2007; Torregrosa-Crespo et al. 2018). Microalgae, which contain 0.5–1.2 percent lutein, are another natural source of lutein (Chen et al. 2018). Lutein and astaxanthin are produced by a variety of microalgae (*Chlorella* sp., *Coelastrella* sp., *Parachlorella kessleri, Scenedesmus* sp. *and Vischeria stellata*).

The two primary components of the macular pigment of the retina are lutein and zeaxanthin, and maintaining their levels in the lens and retina of the eye is considered to be vital in safeguarding eyesight in older individuals (Torregrosa-Crespo et al. 2018). Lutein and zeaxanthin consumption has been linked to the prevention of age-related macular degeneration (AMD) (Olmedilla et al. 2001). Additionally, dietary supplements of lutein alone or in combination with other nutrients can help people with AMD improve their vision. Lutein is found in large quantities in human serum and breastfeeding. Infant formula supplemented with lutein and β-carotene were proven to be safe, well-tolerated, and beneficial to newborns' physical development (Kon et al. 2014). Furthermore, lutein has been proven to protect the epidermal and dermal layers of skin from UV-induced oxidative damage as an antioxidant (Lee et al. 2004).

### *Lycopene*

Lycopene (ψ, ψ-carotene) is a carotenoid with a highly unsaturated hydrocarbon chain made up of 11 conjugated and 2 unconjugated double bonds and no β-ionone ring at the end. It can occur in both cis- and trans-isomeric forms due to the presence of double bonds. However, it is mostly found in the all-trans isomeric form in nature. Under thermal processing and storage settings, lycopene is extremely stable (Torregrosa-Crespo et al. 2018). Interfering with phases of carotenoid biosynthesis in microalgae has been attempted in order to boost lycopene production. The creation of the linear lycopene molecule, which is a substrate for the enzyme's lycopene beta and epsilon cyclases, which are responsible for attaching beta and epsilon rings to the two ends of lycopene, is a key step in the carotenoid pathway leading to the development of α- and β-carotenes (Yildirim et al. 2017).

Lycopene is well-known for its ability to scavenge singlet oxygen. Lycopene was discovered to have considerably stronger anti-cancer potential than most other carotenoids, decreasing tumour cell proliferation in a variety of tumour cell lines (Schwarz et al. 2008). Lycopene also has preventive and/or therapeutic benefits in a variety of central nervous system illnesses, including Alzheimer's disease, Parkinson's disease, Huntington's disease, cerebral ischemia, epilepsy, and depression (Chen et al. 2019). Antioxidant micronutrients like lycopene have been shown to minimise skin roughness. Another research found that those with greater antioxidant levels, such as lycopene, had fewer furrows and wrinkles than people with lower antioxidant levels (Darvin et al. 2008).

***Zeaxanthin***

One of the most prevalent carotenoid alcohols found in nature is zeaxanthin (β, β-carotene-3,3'-diol). It's a polyene-like molecule with nine conjugated double and single bonds that alternate. An ionone ring, to which a hydroxyl group is attached, is attached to each end of the carbon backbone (Sajilata et al. 2008). The only difference between zeaxanthin and lutein is the placement of a double bond: in zeaxanthin, all double bonds are conjugated. E161h is the E number assigned to it as a food colourant (Torregrosa-Crespo et al. 2018). Only two carotenoids are produced by the microalga *Porphyridium cruentum*: zeaxanthin and β-carotene (97.4 percent and 2.6 percent of total carotenoids, respectively). Furthermore, only zeaxanthin and β-carotene are produced by the blue-green algae *Cyanophoraparadoxa* and *Glaucocystis nostochinearum*. When prawns are fed a *Spirulina*-supplemented diet, zeaxanthin, one of the principal carotenoids generated by the blue-green alga *Spirulina*, is swiftly transformed to astaxanthin, boosting the carapace colour (Liao et al. 1993).

*Spirulina* also boosts the yellowness and redness of broiled chicken owing to zeaxanthin buildup in the meat. *Neospongio coccum*, a green alga, is one of the FDA-approved alga strains that are widely accepted as safe for feeding to chickens to improve yellow colouring (Sajilata et al. 2008). Furthermore, zeaxanthin is one of the key components of the retina's macular pigment, and it works in tandem with lutein to help prevent AMD. The following two pigments may help protect against cataract development and age-related increases in lens density (Sajilata et al. 2008).

## Chlorophylls

Chlorophylls (Chl) are the most prevalent natural pigment in microalgae and Cyanobacteria, where they acquire energy for metabolism and reproduction through oxygenic photosynthesis. Chls is found in the chloroplast of microalgae (eukaryotes), but in the photosynthetic lamellae of Cyanobacteria (prokaryotes). A big aromatic tetrapyrrole macrocycle (light absorption and redox chemistry), a central Mg ion (excited state lifespan maximisation), and a hydrocarbon tail make up chlorophyll (anchoring in thylakoids). The Endosymbiosis Theory proposes that Cyanobacteria ancestor was engulfed by a eukaryotic host cell and progressively changed into organelles, which were then distributed to other eukaryotes through successive rounds of endosymbiosis.Chls *a, b, c, d*, and *f* were spread across microalgae and Cyanobacteria by evolution. Experiments with leaves and Chlorella vulgaris dating back to the 1940s were used to develop the routes for Chl *a* production. Later, it was shown that two genetically and biochemically distinct mechanisms for chlorophyll a production coexist, one light-dependent and the other light-independent. Far-red light photoacclimation, in which 8 percent of Chl *a* is replaced by Chl *f*, allows cyanobacteria to increase their light absorption range for oxygenic photosynthesis up to 800 nm, allowing them to receive 33 percent more photons than organisms that do not have Chl *f*.

## Chlorophyll *d*

Chlorophyll *d* was discovered as a small, green, Mg-containing pigment in Rhodophyta in 1943 (Manning and Strain 1943). Chl *d* was formerly thought to be an artefact of algal ageing or possibly created during the pigment extraction process due to its structural similarities to an oxidation derivative of Chl *a*. Chl *d*'s natural origin was established in the late 1990s when it was discovered in the oxygenic photosynthetic prokaryote organism *Acaryochloris marina*, a cyanobacterium, in which Chl *d* is an important primary photosynthetic pigment.According to Scheer et al. (2013), Chl *d* may replace Chl *a* in a photochemical function in PSI and PSII in *A. marina*, however Chl *d* would be relevant in near-infrared light (700 750 nm). According to Chen (2014), Chl *d* produces 95 percent of the photosynthetic pigments in *A. marina*, while Chl *a* produce just 5%. Chl *a* is expected to prevail under other oxygenic photosynthetic circumstances. According to Larkum (2016), bacteriochlorophylls are active for anoxygenic photosynthesis in the near infrared light spectrum.

## Chlorophyll f

Chlorophyll *f* is the most recent chlorophyll to be identified. It absorbs farred light, especially in the 700–800 nm range of the light spectrum. This pigment was initially discovered in cyanobacteria that occupy habitats with high levels of far-red and near-infrared light, such as caverns (cavernous cyanobacteria), soils, and plant-shaded places. (Ho et al. 2017). They flourish in such conditions thanks to a process known as far-red light photoacclimation (FaRLiP), in which around 8% of the Chl *a* molecule in the photosystems is replaced by Chl *f* and a little quantity of Chl *d* is created (about 1%). (Kurashov et al. 2019). The presence of chl*f* allows cyanobacteria to broaden their light absorption range for oxygenic photosynthesis to include near-infrared light and wavelengths up to 800 nm (Ho et al. 2016).This permits them to absorb 33 percent more photons than creatures that lack Chl *f* and are only able to absorb visible light. PSII, PSI, and phycobilisomes are all engaged in the FaRLiP process. (Ho et al. 2017).

## Chlorophylls as Food Additive

With the growing popularity of natural food additives, chlorophylls and their derived compounds originating from biological sources have become a reality that has been gaining traction and popularity as alternatives to artificial chemical additions (Gárcia et al. 2017). These new regulations are based on the unfavourable and potentially toxicological consequences of artificial additives, which can cause allergic responses, hyperactivity, and even unpleasant taste. Food additives, according to the Codex Alimentarius (FAO/WHO 2019), are any substance that is intentionally added in the preparation, manufacture, processing, treatment, packaging, packing, transport, and holding of food to perform a technological function and is not normally consumed as a food by itself (including sensorial).

According to their application in the food sector, additives are divided into 25 categories, each of which contains roughly 230 distinct chemicals. Chlorophylls and their derivative chemicals (chlorophyllins) are classed among the several types of food additives and are commonly employed as colour additives (Table 6) (Martins et al. 2019). Thus, these compounds are responsible for diverse bluish-greenish colours and may give, intensify, or rejuvenate food colour if it has been lost through technical processing (Sawicki et al. 2019).In light of these considerations, natural and commercial chlorophylls, such as copper and sodium chlorophyllins, have been extensively studied for physiologically beneficial health activities, including antioxidant, anti-inflammatory, antimutagenic, and antibacterial capabilities, as shown in Table 7 (Ferruzzi et al. 2002).The global economy is evolving, with a greater emphasis on the extraction of natural goods and biorefinery products, resulting in a biologically based economy, where microalgae biomass has a high value and may be used in a variety of applications, mostly in the food business (Barsanti and Gualtieri 2018). Consumer interest in and pursuit of healthy foods, such as food items, functional components, or dietary supplements that help health in some manner, such as preventing, treating, or curing specific diseases, has grown (Coopers 2009). As a result, using microalgae biomass and investigating derived metabolites like chlorophylls is a novel approach to the creation of functional meals (Gouveia et al. 2008).

The global market for chlorophylls is segmented into North America, Latin America, Asia-Pacific, Western Europe, Eastern Europe, Japan, and the Middle East and Africa, according to Transparency Market Research (2017) (MEA). It is predicted to grow, particularly in China, and is currently doing so in Latin America due to the strong demand for natural dyes. Because of the scarcity of raw resources in the MEA, global adoption of chlorophyll and chlorophyllin is lower than in any other area in the world. The industrial production of microalgae-based intracellular products has a financial potential to relinquish market share in the bioactive compounds segment. As a result, the microalgae biotechnology sector has expanded rapidly in a short period of time. The microalgae biomass industry now generates around 9000 t of dry matter per year, with a global market value of US$ 2.8 billion in 2011. With a compounded annual growth rate (CAGR) of 6.7 percent, his market is expected to be worth US$ 3.32 billion by 2022 (García et al. 2017). With the rising need for food colouring, estimates of global production figures vary substantially. The worldwide food colouring market was valued at US$ 3.71 billion in 2017 and is expected to reach US$ 5.12 billion by 2023, increasing at a CAGR of 5.7 percent. According to estimates, chlorophyll production in Europe, notably the United Kingdom, is worth over US$ 5.5 million per year, accounting for up to a third of global output. According to Markets and Markets statistics, Europe was the market leader in food colouring in 2018, a market driven by customer demand for safe goods and the need for the pursuit of health-food natural colours (Jacob-Lopes et al. 2019).

**Table 6.** Cyanobacteria and Microalgae-based products that reached colouring green natural (Adopted from Fernandes et al. 2020)

| SpeciesName | Product | Application | Colouringcharacteristics |
|---|---|---|---|
| *Isochrysis galbana* | Biscuits | Natural green dye and food supplement | High stability of colour |
| *Spirulina maxima* | Pasta | Increase of quality parameters and natural green dye | Stable colouring after cooking (103°C) |
| *Arthrospira platensis* | Pasta | Natural green dye and nutraceutical potential | Stable colour during cooking |
| *Spirulina platensis* | Pasta | Natural green dye and food supplement | Stable colour during cooking |
| *Spirulina platensis* | Kiwi fruit pastille | Natural green dye | Instability in colour over high temperatures (90ºC) |
| *Arthrospira platensis* | Cookies | Natural green dye and food supplement | Stable colour (100ºC–120ºC) |
| *Isochrysisgalbana* | Chewing gum | Natural green dye | Stable and effective colour |

**Table 7.** Bioactive properties of different types of chlorophylls and their catabolites (Adopted from Fernandes et al. 2020)

| Chlorophylls/ catabolites | Bioactive properties |
|---|---|
| Chlorophyll *a* | Antimutagenic, chemopreventive, antioxidant, chemoprevention of aflatoxin B1 (AFB1), preventsDNA-mutagen intercalation, treatment of chronic ulcers and impetigo contagiosa |
| Pheophytin *a* | Antimutagenic, chemopreventive, anti-inflammatory, suppressor activity against genotoxins, inhibition of edema formation, inhibition of hepatitis C virus (HCV) proteins (NS3 protease) |
| Pheophorbide *a* | Antioxidant, chemopreventive, antigenotoxic, antimutagenic activity |
| Chlorophyll *b* | Antimutagenic, antioxidant,treatment of chronic ulcers |
| Pheophytin *b* | Antimutagenic, anti-inflammatory, antioxidant, HuH-7 human hepatocellular carcinoma cells, inhibition of edema formation |
| Pheophorbide *b* | Antioxidante, HuH-7 human hepatocelular carcinoma cells |

## Conclusion

Cyanobacterial photopigments have been widely explored in recent decades in order to unravel the mystery of cyanobacteria's brilliant hues. In several nations, photopigments are now widely used in the manufacture of foods and nutraceuticals. Certain Cyanobacterial photopigments offer unique qualities such as high quantum yields and pH constancy, allowing them to be used in the creation of more stable fluorescent products for biotechnological, pharmaceutical, and medicinal applications.

## Acknowledgments

Ashish P. Singh (NTA Ref. No. 191620014505) and Jyoti Jaiswal (926/CSIR-UGC-JRF DEC, 2018) are grateful to the University Grants Commission (UGC), New Delhi, India, for the financial support in the form of junior research fellowship. Amit Gupta (09/013(0912)/2019-EMR-I), Prashant R. Singh (09/013(0795)/2018-EMR-I) are thankful to Council of Scientific and Industrial Research, New Delhi for the financial support in the form of senior research fellowship and Niharika Sahu (09/013(0927)/2020-EMR-1) thankful to Council of Scientific and Industrial Research, New Delhi for the financial support in the form of junior research fellowship

## References

Adir N (2005) Elucidation of the molecular structures of components of the phycobilisome: reconstructing a giant. *Photosynth Res* 85:15-32.

Agrawal M, Perumal Y, Bansal S, Arora S, Chopra K (2020) Phycocyanin alleviates ICV-STZ induced cognitive and molecular deficits via PI3-Kinase dependent pathway. *Food ChemToxicol* 145:111684.

Ana AA, Ghada A, Egbert JB (2009) Structural organization of phycobilisomes from *Synechocystis* sp. strain PCC6803 and their interaction with the membrane. *Biochim Biophys Acta* 1787:272-279.

Avci P, Gupta A, Sadasivam M, Vecchio D, Pam Z, Pam N, Hamblin MR (2013) Low-level laser (light) therapy (LLLT) in skin:stimulating, healing, restoring. *SeminCutan Med Surg* 32:41-52.

Barsanti L, Gualtieri P (2018) Is exploitation of microalgae economically and energetically sustainable? *Algal Res* 31:107-115.

Benedetti S, Benvenuti F, Pagliarani S, Francogli S, Scoglio S, Canestrari F (2004) Antioxidant properties of a novel phycocyanin extract from the blue-green alga *Aphanizomenon flos-aquae*. *Life sci* 75(19):2353-62.

Benvin AL, Creeger Y, Fisher GW, Ballou B, Waggoner AS, Armitage BA (2007) Fluorescent DNA nanotags:supramolecular fluorescent labels based on intercalating dye arrays assembled on nano structured DNA templates. *J Am Chem Soc* 129(7):2025-2034.

Bermejo P, Piñero E, Villar ÁM (2008) Iron-chelatingability and antioxidant properties of phycocyanin isolated from a proteanextract of *Spirulina platensis*. *Food Chem* 110:436-445.

Bermejo R, Acien FG, Ibanez MJ, Fernandez JM, Molina E, Alvarez-Pez JM (2003) Preparative purification of B-phycoerythrin from the microalga *Porphyridium cruentum*by expanded-bed adsorption chromatography. *J Chromatogr B* 790:317-325.

Bharathiraja S, Seo H, Manivasagan P, Moorthy MS, Park S, Oh J (2016) In vitro photodynamic effect of phycocyanin against breast cancer cells. *Mol* 21(11):1470.

Blas-Valdivia V, Rojas-Franco P, Serrano-Contreras JI, Sfriso AA, Garcia-Hernandez C, Franco-Colín M, Cano-Europa E (2021) C-phycoerythrin from *Phormidium persicinum* prevents acute kidney injury by attenuating oxidative and endoplasmic reticulum stress. *Mar Drugs* 19(11):589.

Bogacz-Radomska L, Harasym J (2018) β-carotene—Properties and production methods. *Food QualSaf* 2:69-74.

Böhm SB, Endres S, Scheer H, Zhao KH (2007) Biliprotein chromophore attachment:chaperone-like function of the PecEsubunit of alpha-phycoerythrocyanin lyase. *J Biol Chem* 282:25357-25366.

Bottone C, Camerlingo R, Miceli R, Salbitani G, Sessa G, Pirozzi G, Carfagna S (2019) Antioxidant and anti-proliferative properties of extracts from heterotrophic cultures of *Galdieria sulphuraria*. *Nat Prod Res* 33(11):1659-1663.

Breithaupt DE (2007). Modern application of xanthophylls in animal feeding—A review. *Trends Food Sci Technol* 18:501-506.

Chaiklahan R, Chirasuwana N, Bunnag B (2012) Stability of phycocyanin extracted from *Spirulina* sp.: influence of temperature, pH and preservatives. *Process Biochem* 47:659-664.

Chaubey MG, Patel SN, Rastogi RP, Madamwar D, Singh NK (2020) Cyanobacterial pigment protein allophycocyanin exhibits longevity and reduces β-mediated paralysis in *C. elegans*: complicity of FOXO and NRF2 ortholog DAF-16 and SKN-1. *3 Biotech* 10(8):1-1.

Chaubey MG, Patel SN, Rastogi RP, Srivastava PL, Singh AK, Madamwar D, Singh NK (2019) Therapeutic potential of cyanobacterial pigment protein phycoerythrin: In silico and in vitro study of BACE1 interaction and in vivo Aβ reduction. *Int J Biol Macromol* 134:368-378.

Chen CY, Lu IC, Nagarajan D, Chang CH, Ng IS, Lee DJ, Chang JS (2018) A highly efficient two-stage cultivation strategy for lutein production using heterotrophic culture of *Chlorella sorokiniana* MB-1-M12. *Bioresour Technol* 253:141-147.

Chen D, Huang C, Chen Z (2019) A review for the pharmacologicaleffect of lycopene in central nervous system disorders. *Biomed Pharmacother* 111:791-801.

Chen M (2014) Chlorophyll modifications and their spectral extension in oxygenic photosynthesis. *Annu Rev Biochem* 83:317-340.

Coopers PW (2009) Leveraging growth in the emerging functional foods industry: Trends and market opportunities. *J Funct Foods* 1-22.

Dagnino-Leone J, Figueroa CP, Castañeda ML, Youlton AD, Vallejos-Almirall A, Agurto-Muñoz A, Pérez JP, Agurto-Muñoz C (2022) Phycobiliproteins: Structural aspects, functional characteristics, and biotechnological perspectives. *Comput Struct Biotechnol J*.

Dammeyer T, Bagby SC, Sullivan MB, Chisholm SW, Frankenberg-Dinkel N (2008) Efficient phage-mediated pigment biosynthesis in oceaniccyanobacteria. *Curr Biol* 18:442-448.

Darvin M, Patzelt A, Gehse S, Schanzer S, Benderoth C, Sterry W, Lademann J (2008) Cutaneous concentration of lycopene correlates significantly with the roughness of the skin. *Eur J Pharm Biopharm* 69(3):943-947.

Diao J, Song X, Zhang L, Cui J, Chen L, Zhang W (2020) Tailoring cyanobacteria as a new platform for highly efficient synthesis of astaxanthin. *Metab Eng* 61:275-287.

DIC (2009) *Private communication*. Tokyo:s.n. 2009.

Duan HK, Yan ZH, Qi DD, Li WL, Hua XJ, Liu YX, Xin DE (2012) Comparative study on the expression of genesinvolved in carotenoid and ABA biosynthetic pathway in response to salt stress in tomato. *J Integr Agric* 11(7):1093-1102.

Dufosse L, Galaup P, Yaron A, Arad SM, Blanc P, Murthy KNC, Ravishankar GA (2005) Microorganisms and microalgae as sources of pigments for fooduse: a scientificoddity or an industrial reality? *Trends Food Sci Technol* 16:389-406.

Earthrise Nutritionals LLC (2004) *Earthrise the company. Earthrise Nutritionals* [Online]. http://www.earthrise.com/company.asp?page=page5.html.

Eriksen NT (2008) Production of phycocyanin—a pigment with applications in biology, biotechnology, foods and medicine. *Appl Microbiol Biotechnol* 80:1–14.

Fassett RG, Coombes JS (2011) Astaxanthin:a potential therapeutic agent in cardiovascular disease. *Mar Drugs* 9(3):447-465.

FDA (U.S. Food and Drug Administration) (2019) *CFR—Code of federal regulations title 21. Part 73—Listing of color additives exempt from certification.* Retrieved April fromhttp://www.accessdata.fda.gov/scripts/cdrh/cfdocs/cfcfr/CFRSearch.cfm?fr=73. 1075.

Fernandes AS, Nass PP, Oliveira Á, Zepka LQ (2020) Chlorophylls as Food Additives. In *Pigments from Microalgae Handbook:* Springer Cham, pp. 391-420.

Fernandes e Silva E, Figueira FS, Lettnin AP, Salgado MT, Lopes AC, Rehbein F, Kalil SJ, Votto AP (2019) Modulation of reactiveoxygenlevels and gene expression in sensitive and resistant tumoral cells by C-phyocyanin. *Mol Biol Rep* 46(1):1349-56.

Ferruzzi MG, Failla ML, Schwartz SJ (2002) Sodium copperchlorophyllin: in vitro digestive stability and accumulation by Caco-2 human intestinal cells. *J Agric Food Chem*50(7):2173-2179.

Gammoudi S, Athmouni K, Nasri A, Diwani N, Grati I, Belhaj D, Bouaziz-Ketata H, Fki L, El Feki A, Ayadi H (2019) Optimization, isolation, characterization and hepatoprotectiveeffect of a novel pigment-proteincomplex (phycocyanin) producingmicroalga:*Phormidium versicolor* NCC-466 usingresponse surface methodology. *Int J Biol Macromol* 137:647-656.

García JL, De Vicente M, Galán B (2017) Microalgae, old sustainable food and fashion nutraceuticals. *Microb Biotechnol* 10(5):1017-1024.

Garcia-Pliego E, Franco-Colin M, Rojas-Franco P, Blas-Valdivia V, Serrano-Contreras JI, Pentón-Rol G, Cano-Europa E (2021) Phycocyanobilinis the molecule responsible for the nephroprotective action of phycocyanin in acute kidney injury caused by mercury. *Food Funct* 12(7):2985-2994.

Gardeva E, Toshkova E, Yossifova L, Minkova K, Ivanova N, Gigova L (2014) Antitumoractivity of C-phycocyanin from *Arthronema africanum* (Cyanophyceae). *Braz Arch Biol Technol* 57:675-684.

Glazer AN (1981) Photosynthetic accessory proteins with bilin prosthetic groups. In: Hatch MD, Boardman NK (eds). *The Biochem of Plants*: Academic New York, pp 51-96.

Gouveia L, Coutinho C, Mendonça E, Batista AP, Sousa I, Bandarra NM, Raymundo A (2008) Functional biscuits with PUFA-ω3 from *Isochrysis galbana*. *J Sci Food Agric* 88(5):891-896.

Gupta M, Dwivedi UN, Khandelwal S (2011) C-phycocyanin: an effective protective agent against thymicatrophy by tributyltin. *Toxicol Lett* 204:2-11.

Guruvayoorappan C, Kuttan G (2007) β-carotene inhibits tumor-specific angiogenesis by altering the cytokine profile and inhibits the nuclear translocation of transcription factors in B16F-10 melanomacells. *Integr Cancer Ther* 6(3):258-270.

Hao S, Li S, Wang J, Yan Y, Ai X, Zhang J, Ren Y, Wu T, Liu L, Wang C (2019a) Phycocyanin exerts anti-proliferative effects through down-regulating TIRAP/NF-κBactivity in human non-small cell lung cancer cells. *Cells* 8(6):588.

Hao S, Li S, Wang J, Zhao L, Yan Y, Wu T, Zhang J, Wang C (2019b) C-phycocyanin suppresses the in vitro proliferation and migration of non-small-celllung cancer cells through reduction of RIPK1/NF-κB Activity. *Mar Drugs* 17(6):362.

Hirschberg J, Chamovitz D (1994) Carotenoids in cyanobacteria. In *The molecular biology of cyanobacteria:*Springer, Dordrecht pp. 559-579.

Ho MY, Shen G, Canniffe DP, Zhao C, Bryant DA (2016) Light-dependent chlorophyll *f* synthase is a highly divergent paralog of PsbA of photosystem II. *Science* 353(6302): aaf9178.

Ho MY, Soulier NT, Canniffe DP, Shen G, Bryant DA (2017) Light regulation of pigment and photosystem biosynthesis in cyanobacteria. *Curr Opin Plant Biol* 37:24-33.

Jacob-Lopes E, Maroneze MM, Deprá MC, Sartori RB, Dias RR, Zepka LQ (2019) Bioactive food compounds frommicroalgae: An innovative framework on industrial biorefineries. *CurOpin Food Sci* 25:1-7.

Jespersen L, Strømdahl LD, Olsen K, Skibsted LH (2005) Heat and light stability of three natural blue colorants for use in confectionery and beverages. *Eur Food Res Technol* 220:261-266.

Johnson EM, Kumar K, Das D (2014) Physicochemical parameters optimization, and purification of phycobiliproteins from the isolated *Nostoc* sp. *Bioresour Technol* 166:541-547.

Kannaujiya VK, Rahman A, Adinath, Sundaram S, Sinha RP (2016) Structural and functional dynamics of tyrosine aminoacid in phycocyanin of hot-spring cyanobacteria: a possible pathway for internal energy transfer. *Gene Rep* 5:83-91.

Kannaujiya VK, Sinha RP (2016) Thermokinetic stability of phycocyanin and phycoerythrin in food grade preservatives. *J Appl Phycol* 28:1063-1070.

Kannaujiya VK, Sundaram S, Sinha RP (2017) Phycobiliproteins:recent developments and future applications. Springer Singapore.

Kaur P, Dhandayuthapani S, Venkatesan T, Gantor M, Rathinavelu A (2020) Molecular mechanism of C-phycocyanin induced apoptosis in LNCaPcells. *Bioorg Med Chem* 28(3):115272.

Kavitha MD, Gouda KG, Aditya Rao SJ, Shilpa TS, Shetty NP, Sarada R (2019) Atheroprotectiveeffect of novel peptides from*Porphyridium purpureum* in RAW 264.7 macrophage cell line and its molecular docking study. *Biotechnol Lett* 41(1):91-106.

Kefayat A, Ghahremani F, Safavi A, Hajiaghababa A, Moshtaghian J (2019) c-phycocyanin:anatural product with radio sensitizing property for enhancement of colon cancer radiation therapy efficacy through inhibition of COX-2 expression. *Sci Rep* 9(1):1-3.

Kirilovsky D, Kerfeld CA (2016) Cyanobacterial photoprotection by the orange carotenoidprotein. *Nat Plants* 2(12):1-7.

Kon IY, Gmoshinskaya MV, Safronova AI, Alarcon P, Vandenplas Y (2014) Growth and tolerance assessment of a lutein-fortified infant formula. *Pediatr Gastroenterol Hepatol Nutr* 17(2):104-111.

Kumagai Y, Toji K, Katsukura S, Morikawa R, Uji T, Yasui H, Shimizu T, Kishimura H (2021) Characterization of ACE inhibitory peptides prepared from *Pyropia pseudolinearis* protein. *Mar Drugs* 19(4):200.

Kumari RP, Ramkumar S, Thankappan B, Natarajaseenivasan K, Balaji S, Anbarasu K (2015) Transcriptional regulation of crystallin, redox, and apoptotic genes by C-Phycocyanin in the selenite-induced cataractogenic rat model. *Mol Vis* 21:26.

Kurashov V, Ho MY, Shen G, Piedl K, Laremore TN, Bryant DA, Golbeck JH (2019) Energy transfer from chlorophyll *f* to the trapping center in naturally occurring and engineered Photosystem I complexes. *Photosynth Res* 141(2):151-63.

Larkum AW (2016) Photosynthesis and light harvesting in algae. In *The physiology of microalgae:*Springer Cham,pp. 67-87.

Lee EH, Faulhaber D, Hanson KM, Ding W, Peters S, Kodali S, Granstein RD (2004) Dietaryluteinreduces ultraviolet radiation-induced inflammation and immuno-suppression. *J Investig Dermatol* 122(2):510-7.

Lee PT, Huang J, Huang CY, Liu ZX, Yeh HY, Huang HT, Chen LL, Nan FH, Lee MC (2021) Phycoerythrin from *Colaconema* sp. has immunostimulatory effects on the white leg shrimp *Litopenaeus vannamei* and increases resistance to *Vibrio parahaemolyticus* and white spot syndrome virus. *Animals* 11(8):2371.

Leema JT, Kirubagaran R, Vinithkumar NV, Dheenan PS, Karthikayulu S (2010) High value pigment production from *Arthrospira (Spirulina) platensis* cultured in seawater. *Bioresour Technol* 101:9221-9227.

Li B, Chu XM, Xu YJ, Yang F, Lv CY, Nie SM (2013) CD59 underlines the anti atherosclerotic effects of C-phycocyanin on mice. *Bio Med Res Int* 2013:729413.

Li Y, Aiello G, Bollati C, Bartolomei M, Arnoldi A, Lammi C (2020) Phycobiliproteins from *Arthrospira Platensis (Spirulina)*: A new source of peptides withdipeptidyl peptidase-IV Inhibitory activity. *Nutrients* 12(3):794.

Liang C, Zhao F, Wei W, Wen Z, Qin S (2006) Carotenoid biosynthesis in cyanobacteria: structural and evolutionary scenarios based on comparative genomics. *Int J Biol Sci* 2(4):197-207.

Liao WL, Okada S, Matsui T (1993) Yamaguchi K. Pigmentation of cultured black tigerprawn [Penaeus monodon] by feedingwith a *Spirulina*-supplemented diet. *Bull Japan Soc Sci Fish* 59:165-169.

Lim KC, Yusoff FM, Shariff M, Kamarudin MS (2018) Astaxanthin as feed supplement in aquatic animals. *Rev Aquac* 10(3):738-73.

Liu Q, Huang Y, Zhang R, Cai T, Yu C (2016) Medical application of *Spirulina platensis*derived C-phycocyanin. *Evid Based Complement Alternat Med* 2016:7803846.

Liu Q, Li W, Qin S (2020) Therapeutic effect of phycocyanin on acute liver oxidative damage caused by X-ray. *Biomed Pharmacother* 130:1105-1153.

Liu X, Osawa T (2007) Cis astaxanthin and especially 9-cis astaxanthin exhibits a higher antioxidant activity in vitro compared to the all-trans isomer. *Biochem Biophys Res Commun* 357(1):187-193.

Lu L, Li W, Sun C, Kang S, Li J, Luo X, Su Q, Liu B, Qin S (2020) Phycocyanin ameliorates radiation induced acute intestinal toxicity by regulating the effect of the gut microbiota on the TLR4/myd88/NFκB pathway. *J Parenter Enter Nutr* 44(7):1308-1317.

Ma P, Huang R, Jiang J, Ding Y, Li T, Ou Y (2020) Potential use of C-phycocyanin in non-alcoholic fatty liver disease. *Biochem Biophys Res Commun* 526(4):906-912.

Manning WM, Strain HH (1943) Chlorophyll d, a green pigment of redalgae. *J Biol Chem* 151(1):1-9.

Markets and Markets. CarotenoidsMarket by Type (Astaxanthin, Beta-Carotene, Lutein, Lycopene, Canthaxanthin, and Zeaxanthin), Application (Feed, Food &Beverages, Dietary Supplements, Cosmetics, and Pharmaceuticals), Source, Formulation, and Region—Global Forecast to 2026 (2020)(accessed on 14 May 2021).

Martins FC, Sentanin MA, De Souza D (2019) Analytical methods in food additives determination: Compounds with functional applications. *Food chem*272:732-750.

Martiny AC, Coleman ML, Chisholm SW (2006) Phosphate acquisition genes in *Prochlorococcus* ecotypes: evidence for genomewide adaptation. *Proc Natl Acad Sci* USA 103:12552–12557.

Masojídek J, Torzillo G, Koblížek M, Masojídek J, Torzillo G, Koblížek M (2013) Photosynthesis in Microalgae. In *Handbook of Microalgal Culture* Richmond A Ed.; John Wiley& Sons: Hoboken NJ USA pp. 21–36.

Mitra S, Siddiqui WA, Khandelwal S (2015) C-phycocyanin protects against acute tributyl tin chloride neurotoxicity by modulating glial cell activity along with its anti-oxidant and anti-inflammatory property:a comparative efficacy evaluation with N-acetyl cysteine in adult rat brain. *Chem Biol Interact* 238:138-150.

Moncada S, Palmer RMJ, Higgs EA (1991) Nitricoxide:physiology, pathophysiology, and pharmacology. *Pharmacol Rev* 43:109-142.

Nagaoka S, Shimizu K, Kaneko H, Shibayama F, Morikawa K, Kanamaru Y, Otsuka A, Hirahashi T, Kato T (2005) A novelprotein C-phycocyaninplays a crucial role in the hypocholesterolemic action of *Spirulina platensis* concentrate in rats. *J Nutr* 135(10):2425-2430.

Oi VT, Glazer AN, Stryer L (1982) Fluorescent phycobiliprotein conjugates for analyses of cellsand molecules. *J Cell Biol* 93:981-86.

Olmedilla B, Granado F, Blanco I, Vaquero M, Cajigal C (2001) Lutein in patients withcataracts and age related macular degeneration:a long term supplementation study. *J Sci Food Agric* 81(9):904-909.

Ormond AB, Freeman HS (2013) Dyesensitizers for photodynamic therapy. *Mater* 6:817-840.

Pagels F, Guedes AC, Amaro HM, Kijjoa A, Vasconcelos V (2019) Phycobiliproteins from cyanobacteria: Chemistry and biotechnological applications. *Biotechnol Adv* 37(3):422-43.

Pagels F, Salvaterra D, Amaro HM, Guedes AC (2020) Pigments from microalgae. In *Handbook of Microalgae-Based Processes and Products*: Academic Press, pp. 465-492.

Pavón Fuentes N, Marín Prida J, Llópiz Arzuaga A, Falcón Cama V, Campos Mojena R, Cervantes Llanos M, Piniella Matamoros B, Pentón Arias E, Pentón Rol G (2020) Phycocyanobilin reduces brain injury after endothelin 1 induced focal cerebral ischaemia. *Clin Exp Pharmacol Physiol* 47(3):383-92.

Pendyala B, Patras A, Dash C (2021) Phycobilins as Potent Food Bioactive Broad-Spectrum Inhibitors against Proteases of SARS-CoV-2 and Other Coronaviruses: A Preliminary Study. *Front Microbiol* 1412:1399.

Pourhajibagher M, Chiniforush N, Bahador A (2019) Antimicrobial action of photoactivated C-Phycocyanin against Enterococcus faecalis biofilms:Attenuation of quorum-sensing system. *Photodiagnosis Photodyn Ther* 28:286-91.

Prabakaran G, Sampathkumar P, Kavisri M, Moovendhan M (2020) Extraction and characterization of phycocyanin from *Spirulina platensis* and evaluation of its anticancer, anti diabetic and anti- inflammatory effect. *Int J Biol Macromol* 153:256-63.

Prasanna R, Sood A, Suresh A, Kaushik BD (2007) Potentials and applications of algal pigments in biology and industry. *Acta Bot Hung* 49:131-156.

Qi H, Liu Y, Qi X, Liang H, Chen H, Jiang P, Wang D (2019) Dietary recombinant phycoerythrin modulates the gut microbiota of H22 tumor-bearing mice. *Mar Drugs* 17(12):665.

Ramadas A, Kandiah M, Jabbar F, Zarida H (2010) Dietary risk factors for colorectal adenomatouspolyps: a mini review. *J Sci Technol* 18(2):321-49.

Reddy CM, Bhat VB, Kiranmai G, Reddy MN, Reddanna P, Madyastha KM (2000) Selective inhibition of cyclooxygenase-2 by C-phycocyanin, a biliprotein from *Spirulina platensis*. *Biochem Biophys Res Commun* 277:599-603.

Richa, Kannaujiya VK, Kesheri M, Singh G, Sinha RP (2011) Biotechnological potentials of phycobiliproteins. *Int J Pharm Bio Sci* 2:446-454.

Riss J, Décordé K, Sutra T, Delage M, Baccou JC, Jouy N, Brune JP, Oréal H, Cristol JP, Rouanet JM (2007) Phycobiliprotein C-phycocyanin from *Spirulina platensis* is powerfully responsible for reducing oxidative stress and NADPH oxidase expression induced by an atherogenic diet in hamsters. *J Agric Food Chem* 55(19):7962-7967.

Romay CH, Gonzalez R, Ledon N, Remirez D, Rimbau V (2003) C-phycocyanin: a biliprotein with antioxidant, anti-inflammatory and neuroprotective effects. *Curr Protein Pept Sci* 4(3):207-216.

Sajilata MG, Singhal RS, Kamat MY (2008) The carotenoid pigment zeaxanthin—a review. *Compr Rev Food Sci Food Saf* 7(1):29-49.

Sawicki A, Willows RD, Chen M (2019) Spectral signatures of five hydroxymethyl chlorophyll a derivative chemically derived from chlorophyll b or chlorophyll *f*. *Photosynth Res* 140(1):115-127.

Scheer H (2013) An overview of chlorophylls and bacteriochlorophylls:Biochemistry, biophysics, functions and applications. In B Grimm, RJ Porra, W Rüdiger, H Scheer (Eds.), *Chlorophylls and bacteriochlorophylls, biochemistry, biophysics, functions and applications:*Dordrecht: Springer, pp. 1-19.

Schwarz S, Obermüller-Jevic UC, Hellmis E, Koch W, Jacobi G, Biesalski HK (2008) Lycopene inhibits disease progression in patients withbenign prostate hyperplasia. *J Nutr* 138(1):49-53.

Sekar S, Chandramohan M (2008) Phycobiliproteins as a commodity: trends in applied research, patents and commercialization. *J Appl Physiol* 20:113-136.

Shih CM, Cheng SN, Wong CS, Kuo YL, Chou TC (2009) Anti inflammatory and anti hyperalgesic activity of C phycocyanin. *Anesth Analg* 108:1303-1310.

Shih SR, Tsai KN, Li YS, Chueh CC, Chan EC (2003) Inhibition of enterovirus 71 induced apoptosis by allophycocyanin isolated from a blue green alga *Spirulina platensis*. *J Med Virol* 70(1):119-125.

Sidler WA (1994) Phycobilisome and phycobiliprotein structures. In: Bryant DA (ed) *The molecular biology of cyanobacteria*: Kluwer Academic Publication, The Netherlands, pp 139-216.

Six C, Thomas JC, Thion L, Lemoine Y, Zal F, Partensky F (2005) Two novel phycoerythrin associated linker proteins in the marine cyanobacterium *Synechococcus* sp. strain WH8102. *J Bacteriol* 187:1685-1694.

Sluijs I, Cadier E, Beulens JW, Spijkerman AM, Van der Schouw YT (2015) Dietaryintake of carotenoids and risk of type 2 diabetes. *Nutr Metab Cardiovasc Dis* 25(4):376-381.

Stadnichuk IN, Krasilnikov PM, Zlenko DV (2015) Cyanobacterial phycobilisomes and phycobiliproteins. *Microbiology* 84:101-111.

Sugiyama K, Takaichi S (2020) Carotenogenesis in cyanobacteria:CruA/CruP-type and CrtL-type lycopene cyclases. *J Gen Appl Microbiol* 66:53-58.

Sun GY, Liang H, Xu QY (2010) Study on antitumor activity of phycocyanin and its antioxidant function. *Prog Modern Biomed* 10:243-245.

Szabo C, Thiemermann C (1995) Regulation of the expression of the inducibleisoform of nitric oxide synthase. *Adv Pharmacol* 34:113-153.

Takaichi S, Mochimaru M (2007) Carotenoids and carotenogenesis in cyanobacteria: unique ketocarotenoids and carotenoid glycosides. *Cell Mol Life Sci* 64(19):2607-19.

Tanvetyanon T, Bepler G (2008) Beta carotene in multivitamins and the possible risk of lung cancer amongsmokers versus former smokers:ameta-analysis and evaluation of national brands. *Cancer* 113(1):150-157.

Tong F, Tang X, Liu D (2019) Phycocyanin/PEG-b-(PG-g-PEI) attenuated hepatic ischemia/reperfusion-induced pancreatic islet injury and enlarged islet functionality. *Int J Nanomed* 14:339.

Torregrosa-Crespo J, Montero Z, Fuentes JL, Reig García-Galbis M, Garbayo I, Vílchez C, Martínez-Espinosa RM (2018) Exploring the valuable carotenoids for the large-scale production by marine microorganisms. *Mar Drugs* 16(6):203.

Transparency Market Research (2017) *Copper Complexes of Chlorophyll and Chlorophyllins Market—Global Industry Analysis, Size, Share, Growth, Trends, and Forecast 2017–2027*. Retrieved April 12, 2019.

Unno M, Ishikawa-Suto K, Kusaka K, Tamada T, Hagiwara Y, Sugishima M, Wada K, Yamada T, Tomoyori K, Hosoya T, Tanaka I, Niimura N, Kuroki R, Inaka K, Ishihara M, Fukuyama K (2015) Insights into the proton transfer mechanism of a bilin reductase PcyA following neutron crystallography. *J Am Chem Soc* 137:5452-5460.

Waggoner A (2006) Fluorescent labels for proteomics and genomics. *Curr Opin Chem Biol* 10:62-66.

Wang C, Shen Z, Li L, Li Y, Zhao H, Jiang X (2020) Immunomodulatory activity of R-phycoerythrin from *Porphyra haitanensis* via TLR4/NF-κB-dependent immunocyte differentiation. *Food Funct* 11(3):2173-2185.

Wang XQ, Deng W, Yang JF, Mao YC, Shi ZM (2012) Isolation and purification and cytotoxic activity of enzymatic hydrolysis peptide of phycocyanin. *Food Sci* 1:136-140.

Xie Y, Li W, Lu C, Zhu L, Qin S, Du Z (2019) The effects of phycocyanin on bleomycin-induced pulmonary fibrosis and the intestinal microbiota in C57BL/6 mice. *Appl Microbiol Biotechnol* 103(20):8559-69.

Yildirim A, AKGÜN İH, Dalay M (2017) Converted carotenoid production in *Dunaliella salina* by using cyclization inhibitors 2-methylimidazole and 3-amino-1, 2, 4-triazole. *Turk J Biol* 41(1):213-9.

Ying J, Tang Z, Zhao G, Li X, Pan R, Lin S, Lou Y, Wang W, Li P, Bao Q, Yan C (2021) Transcriptionomic study on apoptosis of SKOV-3 cells induced by phycoerythrin from *Gracilaria lemaneiformis*. *Anticancer Agents Med Chem* 21(10):1240-9.

Zakynthinos G, Varzakas T (2016) Carotenoids:From plants to foodindustry. *CurrResNutr Food Sci*4(Special Issue):38.

Zhang H, Liu H, Niedzwiedzki DM, Prado M, Jiang J, Gross ML, Blankenship RE (2014) Molecular mechanism of photoactivation and structural location of the cyanobacterial orange carotenoid protein. *Biochem* 53(1):13-9.

Chapter 9

# Cyanobacteria for Novel Drug Development: A Computational Analysis

**Niharika Sahu, Amit Gupta, Varsha K. Singh and Rajeshwar P. Sinha***

Laboratory of Photobiology and Molecular Microbiology,
Centre of Advanced Study in Botany, Institute of Science,
Banaras Hindu University, Varanasi, India

## Abstract

Cyanobacteria are ubiquitous organisms and a source of several unexplored natural bioactive compounds which help them to survive under various harsh conditions. An exciting and wide range of biological activities such as antibacterial, antifungal, antiviral, antiprotozoal, and molluscicidal activities have been shown by the cyanobacterial bioactive compounds. With the above characteristics, the cyanobacterial metabolites are now acting as leading compounds for drug discovery against various contagious and non-contagious diseases. Nowadays, the interest of researchers is growing in natural compounds which can be good replacements for harmful synthetic chemicals for new drug development. The drug development process is a time consuming and costly process which includes various steps such as collection, analysis, and various phases of clinical trials. However, in the current scenario, drug reformulation through bioinformatics study is gaining interest as it provides rapid and efficient results for drug development with dynamic medical needs. Various *in silico* techniques such as virtual screening, and molecular docking studies recognise various lead compounds for drug repurposing. Apart from this, various predictive models such as QSAR

* Corresponding Author's Email: rpsinhabhu@gmail.com.

In: Cyanobacteria
Editor: Rajeshwar P. Sinha
ISBN: 979-8-88697-169-9

model and dynamic simulation model provide a basis to know the activity and modifications needed in lead compounds to improve their activity. Drug potential, inhibitory effect and drug-likeness of any presumed compound can be predicted through computer simulation. In this chapter, brief information has been given about various computational studies which can be used for *in silico* drug reformulation and the role of cyanobacteria in drug development against various diseases through computational studies.

**Keywords:** cyanobacterial bioactive compounds, drug development, *in silico* study

## Introduction

As per the evidence from the 3.5 million years old fossilized record, cyanobacteria are ancient organisms. Populations of cyanobacteria are distributed in every possible habitat from terrestrial to fresh, brackish, marine, and wastewater habitats. They are capable of colonizing in a wide range of conditions like extreme temperature, light intensities, extreme salinities, pH factors etc. (Pathak et al. 2021). Cyanobacteria synthesize a wide variety of natural metabolites which help them to resist stressful and extreme environmental conditions. Cyanobacterial secondary metabolites (CSM) range from low molecular weight compounds such as mycosporine-like amino acids to more complex compounds such as cyanotoxin (Rastogi and Sinha 2009). Recently bioactive compounds like apratoxins, lyngbyabellin, and curacin A are reported in cyanobacteria. All these bioactive compounds are now emerging as potent agents in the pharmaceutical, biomedical industries and provide great opportunities in new drug discovery. However, the development of new drugs is a time consuming and lengthy process. Apart from this, major reason behind unsuccessful drugs is certain factors like lack of effectiveness and side effects. Recently the use of computational methods for molecular modelling and drug design is high rated research by pharmacologists and researchers related to pharmaceuticals industries. This method gains popularity because of its high efficiency and less time consumption properties. Molecular biology, biochemistry and nanotechnology are some common fields using computational skills for new drug development.

Identification of natural products against molecular targets of human disease through computational studies is a great area of research in recent decades. With the help of *in silico* studies, one can obtain an earlier prediction of success regarding different medicinal compounds, and be able to understand the side effects during new drug developments. In the process of drug discoveries (1) Virtual screening & de novo design, (2) *in silico* ADMET prediction and (3) Advanced methods for determining protein-ligand binding are some major steps determined by computational methods (http://www.scfbio-iitd.res.in/tutorial/drugdiscovery.htm) which are briefly described in later sections. In this chapter there is brief information regarding the pharmaceutical properties of CSM which enable them as candidates for drug development. Then we have discussed various methods of computer-aided drug designing (CADD) and lastly, some advantages and challenges are discussed in relation to CADD.

## Pharmaceutical Potential of Cyanobacterial Secondary Metabolites

Secondary metabolites from cyanobacteria show exciting biological activities such as antibiotics, anticancer, antiviral, anti-inflammatory, anti-HIV (human immunodeficiency virus), antibacterial, antifungal, antimalarial and antiprotozoal activities which are depicted in (Figure 1) (Gademann and Portmann 2008; Wase and Wright 2008). Some properties of CSM are described below.

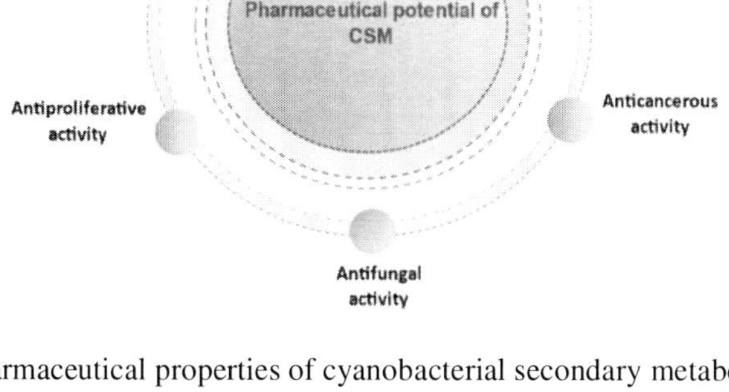


**Figure 1**. Pharmaceutical properties of cyanobacterial secondary metabolites (CSM).

## Anticancerous Activity

Chemotherapy is one amongst the most widely used treatments for cancer despite having life-threatening side effects. The effectiveness and side effects vary from person to person. Some drugs with maximum efficacy cause extreme side effects. Also, the drugs may have a limited supply which makes them more expensive. Due to the above stated reasons drugs derived from natural sources with fewer side effects are gaining more interest in this modern era.

Metabolites secreted by cyanobacteria have excellent durability and it is therefore, high time to explore CSM and their potential. The biosynthetic information on the chemical structures unique to these organisms will be very valuable for finding out new anticancerous agents. There are some bioactive compounds that are derived from cyanobacteria known to possess anticancer activities. Those compounds can be used as leading compounds to develop novel drugs against cancer with fewer side effects and treatment of cancer can be improved. In the 1990s, Moore lab of Oregon State University and Gerwick lab from the University of Hawaii started screening cyanobacterial isolates to identify new natural anticancerous compounds (Singh et al. 2011) Here are some examples of cyanobacterial metabolites which show promising anticancerous activities.

Cryptophycins are produced by cyanobacteria as secondary metabolites which is a potential anticancer agent. Cryptophycin 1 with anticancerous properties was isolated in Moore's lab from *Nostoc* sp. GSV224 (Patterson et al. 1991). The $IC_{50}$ value of Cryptophycin 1 against human nasopharyngeal cancer cells and LoVo human colorectal cancer cells are 5 $pgml^{-1}$ and 3 $pgml^{-1}$ respectively. It shows inhibitory activity against DMS 273 lung cancer cell lines and adriamycin-resistant M17 breast cancer. Cryptophycin 1 shows 100-1000 times more anticancerous potential in comparison to currently available anticancerous drugs such as taxol or vinblastine. It has the ability to obstruct the G2 to M phase transition. There are several natural and chemically isolated analogues of cryptophycin. The chemical analogue Cryptophycin-52 (LY 355073) which has entered phase 1 clinical trials shows only marginal activity (Shin and Teicher 2001). Cryptophycin 249 and 309 are highly stable and have a greater solubility with water. This is the reason these two analogues are being considered as second-generation clinical contenders (Liang et al. 2005). Also, these compounds show improved efficacy when compared with Cryptophycin-52 (LY 355073). These are encouraging signs to start clinical trials with compounds.

Curacin A is a potent tubulin interactive compound that is isolated from *Lyngbya majuscula* by Gerwick et al. (1994a). However, due to its insolubility, the bioactivity of this compound cannot be demonstrated under *in vivo* conditions. A number of semi-synthetic derivatives of curacin A have been generated using combinatorial techniques with improved solubility. Efforts are ongoing on these compounds for preclinical evaluations to verify their potential as anti-cancer drugs.

Dolastatin 10 is an anticancer cyanobacterial metabolite isolated from *Symploca* sp. (Leusch et al. 2001). Structurally it is composed of five peptide chains and four distinctive amino acids dolavaline, dolaisoleucine, dolaproline and dolaphenine. It blocks the G2/M phase transition in the cell by binding to the rhizoxin-binding site of the microtubule which affects microtubule assembly. It showed symptoms of peripheral neuropathy in 40% of the patients in the phase-II trial. The body of patients suffering from hormone-refractory metastatic adenocarcinoma and recurrent platinum-sensitive adenocarcinoma does not have any significant sensitivity towards this compound (Vaishampayan et al. 2000). All these negative results led to the discontinuation of further trials (Hoffman et al. 2003). TZT-1027 (auristatin PE or soblidotin) is another analogue of dolastatin 10 and antagonistic to two human xenograft models i.e., LX-1 lung carcinoma and MX-1 breast carcinoma in mice. It differs from dolastatin 10 in lacking thiazoline ring in dolaphenine residue (Kobayashi et al. 1997). It showed equal effectiveness against both p53 normal and mutant cell lines (Singh et al. 2011).

Dolastatin 15, a member of dolastatin family blocks the transition of G2 to M phase by binding to vinca alkaloid sites on tubulin. Due to its structural complexities, low synthetic yield and less solubility in water, no clinical trials have been done with these compounds so far. Water-soluble derivatives of dolastatin-15 cematodin (LU-103793) have high cytotoxicity in laboratory conditions. Phase-I trial carried out by BASF Pharma (Varanasi, India) showed its effectiveness against breast and other cancers but due to unexpected results in Phase-II, the trial has been discontinued (Singh et al. 2011). Several studies have suggested that synthetic and semisynthetic derivatives of dolastatins such as auristatin PE (TZT–1027) (Gradishar et al. 2012), synthadotin (ILX651) and cematodin (LU–103793) (Ercolak et al. 2015) show better pharmaceuticals and pharmacokinetic properties (Rafiei and Haddadi 2017). An inhibitor of cell growth i.e., Curacin-A was isolated from *Lyngbya majuscula* (Gerwick et al. 1994b). It interacts with tubulin and thus interferes with mitosis and cell growth (Yasin et al. 2019).

A cyclodepsipeptide metabolite known as apratoxin A was extracted from *Lyngbya* sp. of Guam. This compound has $IC_{50}$ ranging from 0.36 to 0.52 Nm against human tumour cell lines *in vitro*. However, this compound shows very low activity against early-stage adenocarcinoma *in vivo*. It induces cell cycle arrest at G-1 phase and apoptosis (Luesch et al. 2006). Apratoxin D, an analogue of apratoxin was isolated from *Lyngbya majuscula* and is cytotoxic to human lung cancer cells (Masuda et al. 2014). Cytotoxic activity towards lung cancer and neuroblastoma cells was also shown by symplocamide A extracted from *Symploca* sp. (Linington et al. 2008; Tan 2010).

Certain cyanobacterial bioactive compounds inhibit the excess growth of cancerous cells by inducing apoptosis and autophagy in them (Kim et al. 2012). When human leukaemia cell line HL-60 was incubated with an extract from *Synechocystis* and *Synechococcus* sp. cell shrinkage and apoptosis has been noted (Martins et al. 2008). Cell death is induced by Biselyngbyaside (a macrolide glycoside) extracted from *Lyngbya* sp. Cell death in mature osteoclasts and AML cell line was found to be induced by an extract from marine benthic *Anabaena* sp. (Yonezawa et al. 2012).

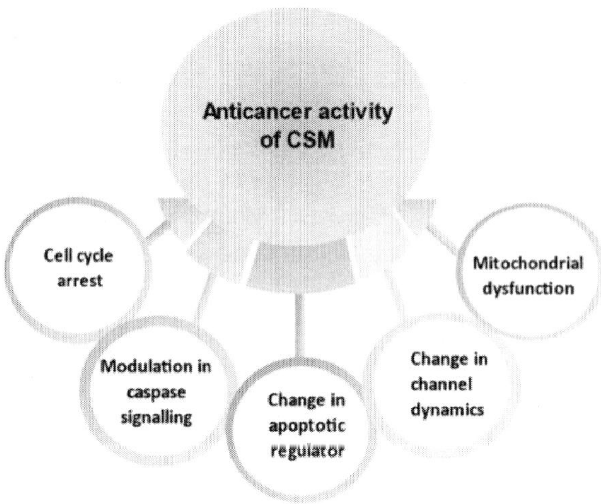


**Figure 2**. Anticancer activity associated with cyanobacterial secondary metabolites.

Another CSM scytonemin is also known to possess antiproliferative activity. It is a photoprotective compound that inhibits phosphorylation of cdc25C mediated by a polo-like kinase 1 (PLK1) in concentration dependent manner and inhibits transition between the G2/M transition in the cell cycle (Stevenson et al. 2002). Scytonemin obstructs cell proliferation and induces apoptosis in human T cell leukaemia Jurkat cells, with $IC_{50} = 7.8$ μM. At a concentration of 3 μM, 24% of the cell shows apoptosis. Among the small biomolecules, scytonemin was the first molecule to be known as an inhibitor of human PLK1 (Strebhardt and Ullrich 2006). Scytonemin can inhibit the growth of cells through the arrest of the cell cycle. Scytonemin can prevent cell growth in multiple myeloma and renal cancer through down-regulation of the PLK1 activity (Zhang et al. 2013). Various anticancerous activities of CSM are shown in Figure 2.

## Antiviral Activity

Several studies are there which support antiviral activity of cyanobacteria (Luescher-Mattli 2003). In 1997, Nowotny et al. reported antiviral activity of crude extract of *Microcystis aeruginosa* against Influenza A virus. Isolates of *Lyngbya lagerheimii* and *Phormidium tenue* are reported to have anti-HIV activity (Gustafson et al. 1989; Schaeffer and Krylov 2000; Vijayakumar and Menakha 2015). Certain research reports suggest antiviral activity of *Spirulina* extract against measles and HIV-1, which inhibits the replication process (Jha and Zi-Rong 2004). Anti-viral properties of cold-water extract of *A. platensis* are studied by Chen et al. in 2016. The extract inhibits a wide range of influenza viruses including oseltamivir-resistant strains. They also observed that extracts of *Spirulina* were able to increase the lifespan of mice infected with influenza virus. The extract works against the haemagglutination of influenza virus at an early stage of infection and decreases the yield of virus (Chen et al. 2016). According to Patterson et al. (1994) Chroococcales commonly produce antiviral agents. Studies have been conducted with the extract of 334 different cyanobacterial species to prove their inhibitory activity against HSV-2, RSV and HIV-1. Approximately 2% of the extracts show inhibitory activity against RSV and about 10% of the extracts show restrictive activities against HSV-2 or anti-HIV-1. Cyanobacterial extracts with lipophilic nature are more eminent as compared to hydrophilic extracts (Patterson et al. 1994). Reports of Hernández-Corona et al. (2002) showed that hot water extract from *Spirulina maxima* can antagonistically act on Herpes simplex virus type 2 (HSV-2). Acidic polysaccharides, like calcium spirulan (Ca-SP) is a potent inhibitor of herpes simplex virus-1 which is commonly released by cyanobacterium *Arthrospira platensis.* Calcium spirulan (Ca-SP) do so by choking the replication process (Kanekiyo et al. 2005). Apart from HSV-2 Ca-SP inhibit HSV-1, influenza virus, mumps, human immunodeficiency virus type 1 (HIV-1), human cytomegalovirus (HVMV) virus and measles by interfering with their replication (Hayashi 2008; Simpore et al. 2005). Reverse transcriptase activity of HIV-1 is also restricted by Ca-SP. The fusion between HIV-infected and uninfected $CD_4^+$ lymphocytes cells is also obstructed by the Ca-SP (Feldman et al. 1999). Two cyclic depsipeptides such as ichthyopeptins A and B also show antiviral activity against influenza A. Both the chemicals are released from *Microcystis ichthyoblabe*. These compounds retard the attachment and fusion of virus 1 with the host cell. As per the study of El-Baz et al. in 2013 50%, 53.3%, 56.7%, 66.7%, and 76.7% ethanol extract of *Arthrospira platensis* reduces infectious units of coxsackievirus B4, astrovirus type 1, adenovirus type 7, adenovirus type 40 and rotavirus Wa strain respectively in laboratory condition. Five inhabitants of freshwater such as *Anabaena sphaerica, Chroococcus turgidus, Oscillatoria limnetica,* and *Arthospira platensis* exhibit antiviral activity against adenovirus type 40 (Abdo et al. 2012). An acidic polysaccharide called nostoflan from *Nostoc flagelliforme* exhibits an excellent antagonistic effect with herpes simplex virus-1 (Kanekiyo et al. 2005). Cyanovirin-N and scytovirin which are categorised under carbohydrate-binding proteins interfere with the viral fusion which makes them potential virucidal drugs. Cyanovirin-N is an 11 KDa polypeptide actively synthesized by *Nostoc ellipsosporum* and characterized by having 101 amino acids. This protein has potent virucidal activity against HIV and other lentiviruses even in its nanomolar concentration (Boyd et al. 1997). It inhibits the fusion of HIV with the $CD_4$ cell membrane by interfering with the binding of gp120 proteins, $CD_4^+$ receptors and the coreceptor chemokine CCR5 or CXCR4. Cellegy Pharmaceuticals, San Francisco, CA, USA has developed a vaginal gel using cyanovirin-N to restrict the sexual transfer of the HIV virus (Klasse et al. 2008). It also shows antagonistic effects toward herpes simplex virus-6 and measles virus in laboratory conditions (Dey et al. 2000). Scytovirin, a 9.7 kDa polypeptide was isolated for the first time from the aqueous extract of *Scytonema varium*. This compound has 95 amino acids with five intrachain disulphide bonds (Bokesch et al. 2003). Even in low nanomolar concentration it binds to the glycoprotein gp41, gp120, and gp160 of HIV and paralyzes the virus.

## Antibacterial Activity

Rising resistance among bacteria against particular antibiotics and even multi-drug-resistant bacteria, like *Staphylococcus aureus* with methicillin-resistant, vancomycin-resistant *Enterococci* and AmpC b-lactamase producing Enterobacteriaceae, have posed huge therapeutic challenges. Noscomin (Mundt et al. 2003) isolated from *Nostoc commune*, showed more vigorous antibacterial activity as compared to standard drugs against *Staphylococcus epidermidis* (MIC 8 p.p.m.), *Bacillus cereus* (MIC 32 p.p.m.) and *Escherichia coli* (MIC 128 p.p.m.). As per the report of Bhateja et al. an extract of *Anabaena* showed antibacterial activity against *S. aureus* (vancomycin-resistant) with MIC of 32–64 $mgml^{-1}$ (Bhateja et al. 2006). Paracyclophanes like carbamidocyclophanes from *Nostoc* sp. CAVN 10 shows modest activity in opposition to *Staphylococcus aureus* (Bhateja et al. 2006). Nine ambiguines with antimicrobial activity were isolated from *Fischerella* sp., by Raveh and Carmeli 2007. If we compare the antibacterial activity of streptomycin and Ambiguine-I isonitrile against *Bacillus subtilis* and *Staphylococcus albus* the latter is more potent than the former. Two new norbietane compounds with antibacterial activity against *Salmonella Typhi, S. epidermidis, S. aureus, Vibrio cholarae, B. cereus, B. subtilis, E. coli* and *Klebsiela pneumoniae* are recently isolated from *Micrococcus lacustris* (Gutierrez et al. 2008).

## Antiprotozoal Activity

The World Health Organization estimated that around the world more than one billion people are suffering from diseases caused by *Trypanosoma, Plasmodium, Leishmania* and *Schistosoma* (Simmons et al. 2008). Failures in treating these diseases are majorly due to development of drug resistance in these organisms (Prioto 2007) and the lack of advancement in the field of new drug discoveries for these organisms (Sheifert et al. 2007).

An effort has been made by Anamanian International Cooperative Biodiversity to encourage the development of efficacious and inexpensive treatment of these diseases. They are screening extracts from terrestrial and marine sources. There are five classes of antiprotozoal compounds that have been reported from cyanobacteria. Protease inhibitor nostocarboline (Barbaraus et al. 2008), an alkaloid isolated from *Nostoc* sp. shows inhibitory activity against *Trypanosoma cruzi, Trypanosoma brucei, Plasmodium falciparum* and *Leishmania donovani* with $IC_{50}$ values ranging from 0.5 to

0.194 mM. Antiprotozoal activity against *T. brucei* is shown by Aerucyclamide C (Portmann et al. 2008) present in the extract of *Microcystis aeruginosa* PCC 7806. With submicromolar $IC_{50}$ values aerucyclamide B is active against *P. falciparum*. Clark et al. isolated six new acyl proline derivatives i.e., tumonoic acids D-I from *Blennothrix cantharidosmum* where tumonoic acid I showed moderate antimalarial activity ($IC_{50}$ 2 mM) (Clark et al. 2008).

From the above discussion, we can see that there are a number of metabolites from cyanobacteria that possess pharmaceutical properties ranging from anticancerous to antiprotozoal activity.

### Antifungal Activity

Fischerellin A, carazostatin, phytoalexin, hassallidin/balticidins, scytophycin, toyocamycin, tolytoxin, hapalindole, tjipanazole, and nostocyclamide are some antifungal compounds which are reported so far (Kajiyama et al. 1998; Abed et al. 2009). Several species of cyanobacteria like *Anabaena* sp. HAN21/1, *Anabaena cf. cylindrica* PH133, *Nostoc* sp. HAN11/1 and *Scytonema* sp. HAN3/2 produces an effective antifungal compound macrolide scytophycin. An unknown antifungal compound was produced by three different species of cyanobacteria namely *Fischerella* sp. CENA 298, *Scytonema hofmanni* PCC 7110 and *Nostoc* sp. N107.3 (Shishido et al. 2005). Most of these compounds are synthesized either by ribosomal pathways or by nonribosomal pathways. Suspension assay shows that norharmane (9H-pyrido (3,4-b) indole) and 4,4'-dihydroxybiphenyl synthesized by cyanobacterial antifungal activity (32-40 mcg ml-1) (Volk and Furkert 2006).

Some metabolites have already undergone phase I and Phase II trials however vast numbers of CSM are still unexplored due to experimental gaps.

## Certain Computational Analysis with CSM

From the above sections it was clear that CSM can be a suitable candidate for new drug development. With reference to the above-mentioned bioactivity, researchers have performed several computational studies on numerous CSM. Here we have enlisted some computational studies which have been done so far with CSM.

Through virtual screening, a new PKBb inhibitor was identified from marine algae. The compound has a unique scaffold compared to existing PKBb inhibitors. Computational studies revealed the antipathogenic nature of microcystin against pathogenic *X. vesicatoria* (Davis and Vasanthi 2015).

Polo-like kinase 1 (PLK-1) is an important target for the development of anticancer drugs as it plays an important role in the cell cycle process and is highly expressed in the human tumour cell. Previously molecular docking studies have been performed with cyanobacterial bioactive compound Scytonemin and PLK-1. Scytonemin acts as a non-competitive and competitive inhibitor of PLK-1. Other kinases such as Myt1, cyclin-dependent kinase 1/cyclin B, checkpoint kinase 1 and protein kinase C are also inhibited by scytonemin (Pathak et al. 2019). A molecular docking study was carried out between the cyanobacterial drug, cryptophycin-F and different types of cancer target proteins. This study compared the docking score between available commercial drugs cabazi-taxel and cryptophycin-F where both the cryptophycin-F and cabazitaxel showed more or less similar docking scores with the target cancer protein. This study confirmed cryptophycin-F can be a good alternative to cabazitaxel without any side effects. However, the study needs further validation through in vitro/in vivo studies (Sangeetha et al. 2014).

For discovering potential drugs against lung cancer *in silico* docking has been performed between 33 different cyanobacterial compounds and receptors responsible for lung cancer. Out of all the compounds tiglicamide A, an isolate of *Lyngbya confervoides* shows the highest docking score among all the compounds. Apart from that Hex software was used to compare the efficacy of tiglicamide A, apraclonidine and cabazitaxel where tiglicamide A was found to be most effective (Vijayakumar and Menakha 2014).

Recently numerous computational studies have been performed to know the potential of CSM against the viral disease SARS-CoV-2. Naidoo et al. performed computational docking studies to explore the inhibitory effects of 23 cyanobacterial metabolites against the SARS-CoV-2 Mpro and PLpro proteases. Among all the studied metabolites cylindrospermopsin, deoxycylindrospermopsin, carrageenan, cryptophycin 52, eucapsitrione, tjipanazole, tolyporphin and apratoxin A shows potential to inhibit SARS-CoV-2 Mpro. Compounds such as cryptophycin 1, cryptophycin 52 and deoxycylindrospermopsin displayed notable inhibitory activity against the PLpro of SARS-CoV-2. All the results are interpreted according to their binding energy score. Deoxycylindrospermopsin was found to be the most effective inhibitory candidate against SARS-CoV-2 proteases through further analysis of physicochemical features, potential toxicity, molecular dynamics simulations and MMPBSA energy scoring function (Naidoo et al. 2021).

An *in silico* study has been performed with a novel compound 9-Ethyliminomethyl-12-(morpholin - 4 - ylmethoxy) -5, 8, 13, 16–tetraaza–hexacene - 2, 3 dicarboxylic acid (EMTAHDCA) to identify its antibiotic property. Docking studies have been performed to assess the binding affinity of this novel compound with familiar drug targets such as OmpF porin protein and 30S ribosomal fragment. Results of the docking study reveal a strong binding affinity of EMTAHDCA with the targets in comparison to available drugs (Verma et al. 2016).

## Computer Aided Drug Designing (CADD)

CADD uses a wide targeted search as compared to the traditional HTS and combinational chemistry which results in an uprising in the hit rate of novel drug compounds. CADD not only explains the therapeutics activity at the molecular level but also predicts possible derivatives that would improve activity. Generally, CADD is used for three major motives. Firstly, it screens the large compound libraries into smaller groups of active compounds for

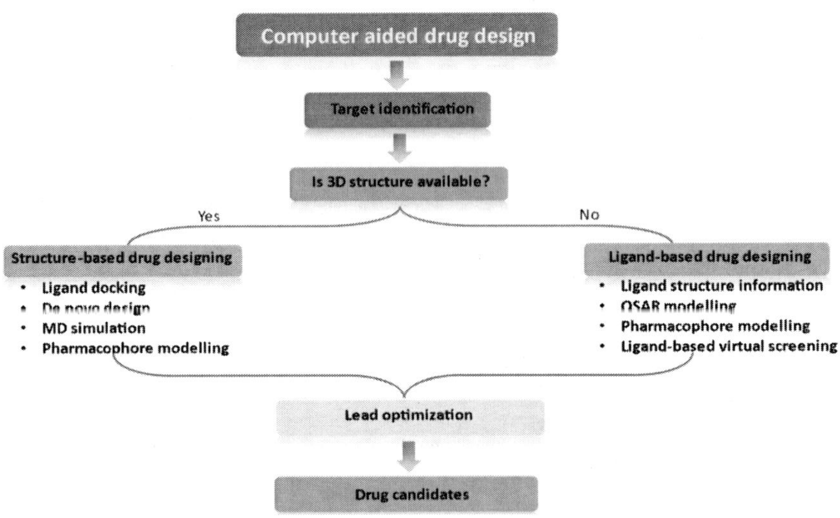


**Figure 3**. Basic steps involved in computer-aided drug designing.

experimental testing. Secondly, it's helpful for optimizing lead compounds to increase their affinity and other pharmacokinetics properties such as absorption, distribution, metabolism, excretion, and the potential for toxicity (ADMET); third is to design a new compound by growing one function of the starting molecule or by assembling fragment to form new chemotypes. Figure 3 illustrates different programmes and steps that are used in CADD.

## Structure-Based Drug Design (SBDD)

Structure-based drug design (SBDD) is centred upon the obtainability of the three-dimensional structure and inspection of binding site cavities of the target (Batool et al. 2019). This is a specific, fast and effective approach for the identification and optimization of lead molecules. It allows us to analyse a disease at its molecular level (Lionta et al. 2014).

Structure-based virtual screening (SBVS), molecular docking, and molecular dynamics (MD) simulations are some of the common methods used by SBDD. Through these methods, numerous studies can be possible such as analysis of binding energetics, the interaction between protein and ligand and assessment of conformational swap in the target receptor when a ligand binds to it (Kalyaanamoorthy and Chen 2011). SBDD is used by several pharmaceutical industries and medicinal chemists in the development of new drugs. For example, protein modelling and MD simulations are two core computational methods used in the development of human immunodeficiency virus (HIV) protease amprenavir (Clark 2006). SBDD approach has been used in the discovery of thymidylate synthase inhibitor raltitrexed against HIV has been done (Anderson 2003). Norfloxacin is an antibiotic used to treat urinary tract infections and is identified as topoisomerase II and IV inhibitor via SBVS (Batool et al. 2019). Fragment-based screening played a great role in identifying carbonic anhydrase inhibitor dorzolamide which is used against glaucoma, cystoid macular oedema (Grover et al. 2006). Apart from the above-mentioned examples, there are several other drugs which are now commercially available and SBDD played a major role in their discovery.

The basic steps involved in SBDD consist of the preparation of target receptor, identification of ligand binding site, preparation of compound library, molecular docking studies, molecular dynamic simulation, and binding free energy calculation.

## Preparation of the Target Structure

We can find the structure of the target protein from the protein data bank (PDB). As there is rapid development in structural elucidation techniques such as NMR and X-ray, deposition and the availability of target protein structures have increased over the last few decades. Certain protein structures still have not been completely solved due to experimental limitations (Wang et al. 2018). However, there are certain computational methods such as comparative homology modelling (Vyas et al. 2012), threading (Lemer et al. 1999), and ab initio modelling (Lee et al. 2017) which are used for the prediction of protein structure from their sequence. For accurate determination of three-dimensional protein structure, homology modelling is widely used where the structure of the protein is predicted from the amino acid sequence of the protein using a suitable template structure (Muhammed and Aki-Yalcin 2019).

Protein homology modelling is a multistep process such as (a) identification of template, (b) sequence alignments, (c) model building of the target, (d) model refinement, and (e) model validation (Vyas et al. 2012).

Protein threading is used to predict the protein structure when the sequence of the target protein shows low similarities (<25% sequence identity) with other proteins in the PDB. In case of protein threading along with sequence similarity, structural information like pairwise interactions, secondary structure and solvent accessibility are also considerable to increase prediction accuracy (Xu et al. 2008).

When no template structure is present in existing biological database, targeted ab initio modelling is used to predict the structure of protein (Lee et al. 2017). It finds the dihedral angle values for a given protein by considering global optimization which contribute to the structure's stability (possessing the global or near global minimum potential energy) (Yousef et al. 2019).

## Identification of the Ligand Binding Site

Valid information about the active site of protein is an important requirement to carry out definite docking. Site-directed mutagenic studies or X-ray crystallographic structures of proteins crystallized with substrates or inhibitors are the informative sources to find binding sites (Pan et al. 2017). For those who do not have any experimental information for their binding site, plenty of software and web servers such as CASTp (Binkowski et al. 2003), DoGSite Scorer (Volkamer et al. 2012), DEPTH (Tan et al. 2013), MSPocket (Zhu and

Pisabarro 2011), MetaPocket (Huang 2009) and Q-SiteFinder (Laurie and Jackson 2005) are available through which conjectural binding sites of the target protein can be predicted. Compounds that do not fit in the active site of the protein are rejected in the lead identification stage.

## Compound Library Preparation

Chemical databases such as ZINC (Irwin et al. 2005), PubChem (Kim et al. 2016), MCULE (https://mcule.com/), ChEMBL (Gaulton et al. 2012), DrugBank (Wishart et al. 2008), and ChemSpider (Pence and Williams 2010) are used to find and select chemical compounds. Lipinski's rule of five (Lipinski 2004), ADMET (absorption, distribution, metabolism, excretion, and toxicity) parameters and other risk parameters such as carcinogenicity, hepatotoxicity and acute rat toxicity are used for the screening drug-like compounds (Lipinski 2004). Lipinski's rule of five has set certain rules of physicochemical properties to identify the suitability of compounds as a drug. According to the rule, a compound should have a molecular weight of $\leq 500$, Hydrogen bond donors should be $\leq 5$, Hydrogen bond acceptor should be not more than 10, and an octanol–water partition coefficient log P should not be greater than 5 (Lipinski 2004). Human gastrointestinal absorption (HIA), blood-brain barrier (BBB) permeation, Pglycoprotein (P-gp) inhibition, cytochromes P450 (CYP) inhibition and plasma protein binding are some commonly used ADMET properties to identify the suitability of drug leads (Daina et al. 2017). Apart from the ADMET properties, the safety of drugs and their synthetic accessibility should be considered during the computational studies.

## Molecular Docking and Scoring Functions

Molecular docking allowed us to study the molecular interaction between the target receptor and ligand. Through docking, we can assess the binding affinity of the ligands with the target protein and rank them according to their binding affinity (Huang and Zou 2010). Mainly two factors such as (a) consideration of broad conformational space for different binding poses and (b) error-free prediction of binding affinity between the ligands and target in each pose are responsible to find the promising binding poses of the ligands with the active site of the target molecule (Kapetanovic 2008). A list of frequently used molecular docking programs is listed in Table 1.

Molecular docking can be categorised into two types: flexible-protein docking and flexible-ligand search docking. Three algorithms; simulation method, systematic method and stochastic method are commonly used in the flexible-ligand search docking (Sousa et al. 2006). In case of flexible-protein docking involves Monte Carlo (MC) and molecular dynamic (MD) methods (Hart and Read 1992).

**Table 1.** List of programmes/softwares used for molecular docking studies

| Programmes | Features | References |
|---|---|---|
| AutoDock | Lamarckian GA | Morris et al. 2009 |
| AutoDock Vina | Lamarckian GA | Trott and Olson 2009 |
| LigandFit | Shape matching (moments of inertia) | Venkatachalam et al. 2003 |
| GLIIDE | Descriptor matching/MC | Friesner et al. 2004 |
| GOLD | genetic algorithm | Jones et al. 1997 |
| DOCK6 | Shape matching (sphere images) | Allen et al. 2015 |
| Surflex | Surface-based molecular similarity | Jain 2003 |
| FlexX | Incremental construction | Kramer et al. 1999 |

## Molecular Dynamic (MD) Simulation

In the late 1970s, the MD simulation of a protein was performed for the first time (McCammon et al. 1977). Through this powerful technique, we can predict the position of each atom in a molecular system with respect to time which is based on Newton's laws of motions governing interatomic interactions (Karplus and McCammon 2002). A suitable force field is used to estimate the forces between interacting atoms through which the overall energy of the system can be determined (De Vivo et al. 2016). At every point in time, the motion and position of each atom of the system are collected in MD simulation which is quite difficult in any experimental technique. This technique unravels atomistic details such as conformational swap, binding and unbinding in the receptor at a fine resolution and is thus extensively used for structure-based drug design (Zhao and Caflisch 2015). MD simulation allows us to study the association and dissociation between receptor-and ligand and quantify the free energy landscape, thermodynamics and kinetics (Huang and Caflisch 2011). AMBER, CHARMM, GROMACS, NAMD and Desmond are some programmes to perform MD simulation. A complete list of programmes used in the molecular dynamic simulation is enumerated in Table 2.

**Table 2.** List of commonly used molecular dynamic (MD) simulation programmes/softwares

| Programmes | Simulation system | Reference/web address |
|---|---|---|
| GROMACS | Proteins, lipids, carbohydrate, nucleic acids | Van Der Spoel et al. 2005 |
| Desmond | Proteins, lipids | https://www.schrodinger.com/desmond |
| AMBER | Proteins, nucleic acids, carbohydrates | Case et al. 2005 |
| NAMD | Proteins, lipids, carbohydrates, nucleic acids | Phillips et al. 2002 |
| Tinker | Proteins, nucleic acids | Rackers et al. 2018 |
| CHARMM | Proteins, lipids, carbohydrates, nucleic acids | Brooks et al.2009 |
| LAMMPS | Proteins, lipids, carbohydrates, nucleic acids | https://lammps.sandia.gov/ |

## Ligand-Based Drug Design (LBDD)

When the three-dimensional structure of the target molecule is not available ligand-based drug design approach is used in CADD. This method is based on the fact that structural similarities are proportional to biological function. The information obtained from a batch of active compounds against a specific target receptor is helpful in the identification of physicochemical and structural properties responsible for the given biological activity (Gurung et al. 2021).

Pharmacophore modelling, quantitative structure-activity relationships (QSARs), and artificial intelligence (AI) are the techniques used in ligand-based virtual screening.

## Pharmacophore Modelling

To interact with the target receptor, spatial arrangements of chemical features in ligands are explained by the pharmacophore model (Schaller et al. 2020). Hydrogen bond acceptors, hydrogen bond donors, aromatic ring systems, hydrophobic areas, and positively and negatively charged ionizable groups are some chemical features used in pharmacophore modelling (Gurung et al. 2021). Pharmacophore based virtual screening can be used to identify ligands with similar spatial arrangements of key interacting moieties even if with different scaffolds. The molecule with its bioactive conformation binding site within the target binding site can be integrated into the pharmacophore model. In the QSAR studies pharmacophore modelling is used during the molecular alignment stage (Melo-Filho et al. 2014). Catalyst, PHASE, LigandScout, GALAHAD, and PharmMapper are some common software that allows the automatic construction of the pharmacophore model. A good pharmacophore model provides spatial constraints to the regions engaged with inactive molecules and makes the model less restrictive through proper optimizations. If a molecule does not show a pharmacophoric feature consistently then that feature is removed from the final model (Gurung et al. 2021). To minimize the false positive and false negative results there must be an optimum sensitivity and specificity in the generated pharmacophore model. Apart from this, the generated model must be validated with an independent external test set (Gurung et al. 2021).

In case of unavailability of proper information about receptor 3D structure and a set of known active compounds, a sequence-derived 3D pharmacophore model is used. For example, the software called Pharma3D uses information of the 3D crystal structures and homology models to procure common sequence motifs responsible for receptor and ligand interaction (Klabunde et al.2009).

## Quantitative Structure-Activity Relationships (QSARs)

QSAR methods are extensively used for lead identification and lead optimization in the drug discovery process. Change in the molecular structure can be proportional to the variation in biological activity and this principle forms the basis of QSAR studies. The biological activity of a new molecule is predicted by constructing a statistical model through QSAR correlation studies (Melo-Filho et al. 2014).

To generate a reliable QSAR model some prerequisites are the following: (a) appropriate selection of training and test set, (b) the selected data set must be inadequate number and with their bioactivities procured from common experimental protocols, (c) the physicochemical properties of the ligands must not be correlated with each other as it may overfit, (d) internal and external models must be used to validate the final model (Gurung et al. 2021).

QSAR can be categorised into six types which are enlisted here (a) 1D-QSAR related to the study of the interrelationship between the molecular properties such as logP, pKa and biological activities, (b) 2D-QSAR based on the study of correlation with the biological activities and structural patterns like 2D-pharmacophores and connectivity indices, (c) 3D-QSAR studies the relationships between the biological activities and noncovalent interaction fields surrounding the ligands, (d) 4D-QSAR modified version of 3D-QSAR with the addition of an ensemble of ligand configurations, (e) 5D-QSAR which integrates various induced-fit models in 4D-QSAR, and (f) extension of 5D-QSAR is named as 6D-QSAR which includes different solvation models (Patel et al. 2014). The comparative similarity indices analysis (CoMSIA) (Klebe et al. 1994), comparative molecular field analysis (CoMFA) (Cramer et al. 1988), HypoGen module of Catalyst (Li et al. 2000) and PHASE (Dixon et al. 2006) are some examples of 3D QSAR programs. Based on the chemometric method QSAR technique is categorised into two types such as linear and nonlinear methods. Examples of nonlinear QSAR methods are k -nearest neighbours (kNN), artificial neural networks (ANN), and Bayesian neural nets. The linear method includes linear regression (LR), multiple linear regression (MLR), partial least squares (PLS), principal component analysis (PCA), and principal component regression (PCR) (Gurung et al. 2021).

## Artificial Intelligence (AI) and Drug Discovery

AI involves machine intelligence which depends on the existing data of a given computer. Various *in silico* modelling methods use AI models to predict the toxicity and biological activity of lead compounds (Gurung et al. 2021). AI is also applicable for the studying of protein folding, virtual screening, ADMET etc. (Gurung et al. 2021) Machine learning (ML) and deep learning (DL) are two core factors that play a crucial role in new discoveries. Naive Bayesian (NB) (Sammut and Webb 2011), Random Forest (RF) (Breiman 2001) and support vector machine (SVM) (Cortes and Vapnik 1995), restricted Boltzmann machine (RBN), deep neural network (DNN), convolutional neural network (CNN) are few examples of ML learning and deep learning methods (Gurung et al. 2021). For the first time in 2012 machine learning was introduced in QSAR studies for drug development (Gurung et al. 2021).

Basically, for all the compounds which are considered for new drug development, it has to undergone three stages as discussed below.

## Advantages of CADD

Computational method is not only used to discover new drug lead/bioactive molecules against a particular target but it can be also applicable to find metabolic pathways of those bioactive compounds. *In silico* metabolic analysis provides experimental information which can be used to modify a given bioactive compound to increase its biological activity. This method guides to improvement of experimental procedures to get more comprehensive biological information (Yamazaki et al. 2004).

Computational methods are able to safely replace animal experimentation for testing drug safety. In all new computational methods, further animals would not have to suffer from chemicals (Gurung et al. 2021).

Certain models of Structural Activity Relationship (SAR) and QSAR have been developed which are dedicated to the prediction of carcinogenicity of a compound. However, studies reveal that *in silico* prediction of carcinogenicity is rarely possible (Cronin et al. 2003). There are some QSAR studies that have been done for genotoxicity, mutagenicity, skin irritation and eye irritation of naturally and synthetically derived cosmetics (Cronin et al. 2003). Toxicity of a compound can be directly predicted from the chemical structure of some computational packages. As this method is very simple and rapid in nature such packages are encouraged.

Through ADME prediction software one can able to identify the pharmacokinetic properties such as absorption, distribution, metabolism and excretion (ADME) of a compound. Such predictions enable us to localize the fate of the molecule in the body along with its toxicity. Information from the ADME can help us to find out the toxicological profile of lead compounds before the testing (Raunio 2011). Broadly use of computational tools in drug design makes the process more efficient, quick and cost-effective.

## Limitations of CADD

It has been observed that generally *in silico* studies are performed on chemical compounds which have already been synthesized. So, mostly computational methods are not applicable for hypothesized bioactive compounds. *In silico* studies only give a rough illustration of predetermined and calculated data which need proper validation. A number of papers have been published for pharmacological activity without any *in vitro* and *in vivo* experimental verification. The data generated by computational methods are purely hypothetical and not purely authenticated in nature. It is readily apparent that in a minority of papers we have found that computational approaches have resulted in predicted lead compounds for testing without the authors providing further experimental verification of biological activity (Amin and Welsh 2006).

Numbers of molecules which are screened through virtual screening are not experimentally tested to ensure that they are not false positives which causes molecular aggregation (Seidler et al. 2003).

In conventional QSAR techniques we can easily predict physicochemical properties like logP and solubility, but for predicting sophisticated biological dimension such as effectiveness and adverse effect of a drug this method is not exquisite (Roy et al. 2008). It is very challenging for traditional QSAR and machine learning techniques to process the huge data generated through high-throughput screening (HTS) technique. To overcome this problem AI has been developed which flexibly predict drug efficiency and adverse side effects in organisms. AI techniques can more accurately predict the ADMET properties as compared to the traditional method (Gurung et al. 2021).

For cost effective, time efficient and correct prediction of binding affinity between receptor and drug molecule AI is one of the explicit methods. However, there are certain reasons which make AI challenging. Performance of AI is dependable upon quantity and quality of available data as it is a data mining method (LeCun et al. 2015). Variations in available data mainly derived from biological assay and deficiency of qualitative data from public database is another hurdle in AI learning (LeCun et al. 2015; Rotemberg et al. 2019).

A computer system, even if it is the most systematic and complex, will never be able to replace the complexity of a biological system. Biological systems are unpredictable in nature and show 'chaotic behaviour.' Biological systems cannot be fully modelled by an algorithm shorter than themselves.

## Conclusion

In this chapter, we have tried to tie an integration between the computational methodologies and their application in the field of drug discovery by using CSM. Cyanobacterial compounds are well known for their pharmaceutical activity ranging from antimicrobial to anticancerous properties. They can be safely exploited in the pharmaceutical industries to develop drugs against microbes and deadly diseases such as cancer. Some computational studies have been performed by some researchers for the production of novel drugs from CSM and have produced good results. However, they need proper validations through *in vivo* and *in vitro* experimentation to succeed in the pharmaceutical industries. Nowadays *in silico* drug discoveries are gaining so much interest as it is less time consuming and it gives us a rough idea about the suitability of a studied compound as a drug against biological target before *in-vitro* and *in-vivo* experimentation. However, the future of computational methodologies in the drug discovery field is difficult to predict. It's a major question that the computational method will ever be able to be used for *in vitro* experimental studies. However, the answer is clearly no due to some major reasons stated in the above section. Here inference is that drug discovery from natural CSM through computational methods, *in vivo* and *in vitro* experimentation can bring a great revolution in pharmaceutical industries.

## Acknowledgments

Niharika Sahu (09/013(0927)/2020-EMR-I) is thankful to CSIR, New Delhi, India, for the award of Junior Research Fellowship. Amit Gupta (09/013(0912)/2029-EMR-I) extend his sincere appreciation to CSIR, New Delhi, India, for financial support in the form of Senior Research Fellowship. Varsha K. Singh (09/0013(12862)/2021-EMR-I) is thankful to CSIR, New Delhi, India, for the financial assistance in the form of Junior Research Fellowship.

## References

Abdo SM, Hetta MH, El-Senousy WM, Din RA, Ali GH (2012) Antiviral activity of freshwater algae. *J Appl Pharm* 2(2): 21-25.

Abed RM, Dobretsov S, Sudesh K (2009) Applications of cyanobacteria in biotechnology. *J Appl Microbiol* 106(1): 1-12.

Allen WJ, Balius TE, Mukherjee S, Brozell SR, Moustakas DT, Lang PT, Case DA, Kuntz ID, Rizzo RC (2015) DOCK 6: Impact of new features and current docking performance. *J Comput Chem* 36(15): 1132-1156.

Amin EA, Welsh WJ (2006) A preliminary in silico lead series of 2-phthalimidinoglutaric acid analogues designed as MMP-3 inhibitors. *J Chem Inf Model* 46(5): 2104-2109.

Anderson AC, (2003) The process of structure-based drug design. *Chem Biol* 10(9): 787-797.

Barbaraus D, Kaiser M, Brun R, Gademann K (2008) Potent and selective antiplasmodial activity of the cyanobacterial alkaloid nostocarboline and its dimers. *Bioorg Med Chem Lett* 18: 4413-4415.

Batool M, Ahmad B, Choi S, (2019) A structure-based drug discovery paradigm. *Int J Mol Sci* 20(11): 2783.

Bhateja P, Mathur T, Pandya M, Fatma T, Rattan A (2006) Activity of blue green microalgae extracts against in vitro generated *Staphylococcus aureus* with reduced susceptibility to vancomycin. *Fitoterapia* 77(3): 233-5.

Binkowski TA, Naghibzadeh S, Liang J (2003) CASTp: computed atlas of surface topography of proteins. *Nucleic Acids Res* 31(13): 3352-5.

Bokesch HR, O'Keefe BR, McKee TC, Pannell LK, Patterson GM, Gardella RS, Sowder RC, Turpin J, Watson K, Buckheit RW, Boyd MR (2003) A potent novel anti-HIV protein from the cultured cyanobacterium *Scytonema varium*. *Biochemistry* 42(9): 2578-2584.

Boyd MR, Gustafson KR, McMahon JB, Shoemaker RH, O'Keefe BR, Mori T, Gulakowski RJ, Wu L, Rivera MI, Laurencot CM, Currens MJ (1997) Discovery of cyanovirin-N, a novel human immunodeficiency virus-inactivating protein that binds viral surface envelope glycoprotein gp120: potential applications to microbicide development. *Antimicrob. Agents Chemother* 41(7): 1521-30.

Breiman L (2001) Random forests. *Machine learning* 45(1): 5-32.

Brooks BR, Brooks III CL, Mackerell Jr AD, Nilsson L, Petrella RJ, Roux B, Won Y, Archontis G, Bartels C, Boresch S, Caflisch A (2009) CHARMM: the biomolecular simulation program. *J Comput Chem* 30(10): 1545-1614.

Case DA, Cheatham III TE, Darden T, Gohlke H, Luo R, Merz Jr KM, Onufriev A, Simmerling C, Wang B, Woods RJ (2005) The Amber biomolecular simulation programs. *J Comput Chem* 26(16): 1668-1688.

Chen YH, Chang GK, Kuo SM, Huang SY, Hu I, Lo YL, Shih SR (2016) Well-tolerated Spirulina extract inhibits influenza virus replication and reduces virus-induced mortality. *Sci Rep* 6(1): 1-11.

Clark DE (2006) What has computer-aided molecular design ever done for drug discovery? *Expert Opin Drug Discov* 1(2): 103-110.

Clark BR, Engene N, Teasdale ME, Rowley DC, Matainaho T, Valeriote FA, Gerwick WH (2008) Natural products chemistry and taxonomy of the marine cyanobacterium *Blennothrix cantharidosmum*. *J Nat Prod* 71(9): 1530-7.

Cortes C, Vapnik V (1995) Support-vector networks. *Machine Learning* 20(3): 273-297.

Cramer RD, Patterson DE, Bunce JD (1988) Comparative molecular field analysis (CoMFA). 1. Effect of shape on binding of steroids to carrier proteins. *J Am Chem Soc* 110(18): 5959-5967.

Cronin MT, Jaworska JS, Walker JD, Comber MH, Watts CD Worth AP (2003) Use of QSARs in international decision-making frameworks to predict health effects of chemical substances. *Environ Health Perspect* 111(10): 1391-1401.

Daina A, Michielin O, Zoete V (2017) SwissADME: a free web tool to evaluate pharmacokinetics, drug-likeness and medicinal chemistry friendliness of small molecules. *Sci Rep* 7(1): 1-13.

Davis GD, Vasanthi AH (2015) QSAR based docking studies of marine algal anticancer compounds as inhibitors of protein kinase B (PKBβ). *Eur J Pharm Sci* 30(76): 110-8.

De Vivo M, Masetti M, Bottegoni G Cavalli A (2016) Role of molecular dynamics and related methods in drug discovery. *J Med Chem* 59(9): 4035-4061.

Dey B, Lerner DL, Lusso P, Boyd MR, Elder JH, Berger EA (2000) Multiple antiviral activities of cyanovirin-N: blocking of human immunodeficiency virus type 1 gp120 interaction with CD4 and coreceptor and inhibition of diverse enveloped viruses. *J Virol* 74(10): 4562-4569.

Dixon SL, Smondyrev AM, Rao SN (2006) PHASE: a novel approach to pharmacophore modeling and 3D database searching. *Chem Biol Drug Des* 67(5): 370-372.

Ercolak V, Sahin B, Gunaldi M, Duman BB, Afsar CU (2015) Efficacy of paclitaxel in the treatment of Kaposi sarcoma. *Eur Rev Med Pharmacol Sci* 19: 4095-4100.

Feldman SC, Reynaldi S, Stortz CA, Cerezo AS, Damonte EB (1999) Antiviral properties of fucoidan fractions from *Leathesia difformis. Phytomedicine* 6(5): 335-340.

Friesner RA, Banks JL, Murphy RB, Halgren TA, Klicic JJ, Mainz DT, Repasky MP, Knoll EH, Shelley M, Perry JK, Shaw DE (2004) Glide: a new approach for rapid, accurate docking and scoring. 1. Method and assessment of docking accuracy. *J Med Chem* 47(7): 1739-1749.

Gademann K, Portmann C (2008) Secondary metabolites from cyanobacteria: complex structures and powerful bioactivities. *Curr Org Chem* 12(4): 326-341.

Gaulton A, Bellis LJ, Bento AP, Chambers J, Davies M, Hersey A, Light Y, McGlinchey S, Michalovich D, Al-Lazikani B, Overington JP (2012) ChEMBL: a large-scale bioactivity database for drug discovery. *Nucleic Acids Res* 40(D1): D1100-7.

Gerwick WH, Proteau PJ, Nagle DG, Hamel E, Blokhin A, Slate DL (1994a) Structure of curacin A (1994) a novel antimitotic, antiproliferative and brine shrimp toxic natural product from the marine cyanobacterium *Lyngbya majuscula. J Org Chem* 59(6): 1243-5.

Gerwick WH, Roberts MA, Proteau PJ, Chen JL (1994b) Screening cultured marine microalgae for anticancer-type activity. *J Appl Phycol* 6: 143-149.

Gradishar WJ, Krasnojon D, Cheporov S, Makhson AN, Manikhas GM, Clawson A, Bhar P, McGuire JR, Iglesias J (2012) Phase II trial of nab-paclitaxel compared with docetaxel as first-line chemotherapy in patients with metastatic breast cancer: final analysis of overall survival. *Clin Breast Cancer* 12(5): 313-21.

Grosdidier A, Zoete V Michielin O (2011) SwissDock, a protein-small molecule docking web service based on EADock DSS. *Nucleic Acids Res* 39(suppl_2): W270-W277.

Grover S, Apushkin MA, Fishman GA (2006) Topical dorzolamide for the treatment of cystoid macular edema in patients with retinitis pigmentosa. *Am J Ophthalmol* 141(5): 850-8.

Gurung AB, Ali MA, Lee J, Farah MA, Al-Anazi KM (2021) An Updated Review of Computer-Aided Drug Design and Its Application to COVID-19. *Biomed Res Int.*

Gustafson KR, Cardellina JH, Fuller RW, Weislow OS, Kiser RF, Snader KM, Patterson GM, Boyd MR (1989) AIDS-antiviral sulfolipids from cyanobacteria (blue-green algae). *J Natl Cancer Inst* 81(16): 1254-8.

Gutierrez J, Barry-Ryan C, Bourke P (2008) The antimicrobial efficacy of plant essential oil combinations and interactions with food ingredients. *Int J Food Microbiol* 124(1): 91-97.

Han SI, Kim Y S, Kim TH (2008) Role of apoptotic and necrotic cell death under physiologic conditions. *BMB Rep* 41: 1-10.

Hart TN, Read RJ (1992) A multiple-start Monte Carlo docking method. *Proteins: Struct Funct Bioinfo* 13(3): 206-222.

Hayashi T (2008) Studies on evaluation of natural products for antiviral effects and their applications. *Yakugaku zasshi: J Pharm Soc Jpn* 128(1): 61-79.

Hernández-Corona, A., Nieves, I., Meckes, M., Chamorro, G., & Barron, B. L. (2002). Antiviral activity of Spirulina maxima against herpes simplex virus type 2. *Antiviral Research*, *56*(3), 279-285.

Hoffman MA, Blessing JA, Lentz SS (2003) A phase II trial of dolastatin-10 in recurrent platinum-sensitive ovarian carcinoma: a Gynecologic Oncology Group study. *Gynecol Oncol* 89(1): 95-98.

Huang B (2009) MetaPocket: a meta-approach to improve protein ligand binding site prediction. *OMICS J Integr Biol* 13(4): 325-330.

Huang D, Caflisch A (2011) The free energy landscape of small molecule unbinding. *PLoS Comput Biol* 7(2).

Huang SY, Zou X (2010) Advances and challenges in protein-ligand docking. *Int J Mol Sci* 11(8): 3016-3034.

Irwin JJ, Shoichet BK, (2005) ZINC− a free database of commercially available compounds for virtual screening. *J Chem Inf Model* 45(1): 177-182.

Jain AN (2003) Surflex: fully automatic flexible molecular docking using a molecular similarity-based search engine. *J Med Chem* 46(4): 499-511.

Jaki B, Orjala J, Sticher O (1999) A novel extracellular diterpenoid with antibacterial activity from the cyanobacterium *Nostoc commune*. *J Nat Prod* 62(3): 502-3.

Jha RK, Zi-Rong X (2004) Biomedical compounds from marine organisms. *Mar Drugs* 2: 123-146.

Jones G, Willett P, Glen RC, Leach AR Taylor R (1997) Development and validation of a genetic algorithm for flexible docking. *J Mol Biol* 267(3): 727-748.

Kajiyama SI, Kanzaki H, Kawazu K, Kobayashi A (1998) Nostofungicidine, an antifungal lipopeptide from the field-grown terrestrial blue-green alga *Nostoc commune*. *Tetrahedron Lett* 39(22): 3737-3740.

Kalyaanamoorthy S, Chen YPP (2011) Structure-based drug design to augment hit discovery. *Drug Discov* 16(17-18): 831-839.

Kanekiyo K, Lee JB, Hayashi K, Takenaka H, Hayakawa Y, Endo S, Hayashi T (2005) Isolation of an antiviral polysaccharide, nostoflan, from a terrestrial cyanobacterium, *Nostoc flagelliforme*. *J Nat Prod* 68(7): 1037-1041.

Kapetanovic I (2008) Computer-aided drug discovery and development (CADDD): *in silico*-chemico-biological approach. *Chem Biol Interact* 171(2): 165-176.

KarplusM, McCammon JA (2002) Molecular dynamics simulations of biomolecules. *Nat Struct Mol Biol* 9(9): 646-652.

Kim MK, Suh DH, Seoung J, Kim HS, Chung HH, Song YS (2012) Autophagy as a target for anticancer therapy and its modulation by phytochemicals. *J Food Drug Anal* 20: 241-245.

Kim S, Thiessen PA, Bolton EE, Chen J, Fu G, Gindulyte A, Han L, He J, He S, Shoemaker BA, Wang J (2016) PubChem substance and compound databases. *Nucleic Acids Res* 44(D1): D1202-13.

Klabunde T, Giegerich C, Evers A (2009) Sequence-derived three-dimensional pharmacophore models for G-protein-coupled receptors and their application in virtual screening. *J Med Chem* 52(9): 2923-2932.

Klasse PJ, Shattock R Moore JP, (2008) Antiretroviral drug–based microbicides to prevent HIV-1 sexual transmission. *Annu Rev Med* 59: 455-471.

Klebe G, Abraham U, Mietzner T (1994) Molecular similarity indices in a comparative analysis (CoMSIA) of drug molecules to correlate and predict their biological activity. *J Med Chem* 37(24): 4130-4146.

Kobayashi M, Natsume T, Tamaoki S, Watanabe JI, Asano H, Mikami T, Miyasaka K, Miyazaki K, Gondo M, Sakakibara K, Tsukagoshi S (1997) Antitumor activity of TZT-1027, a novel doiastatin 10-derivative. *Jpn J Clin Oncol* 88(3): 316-27.

Kramer B, Rarey M, Lengauer T (1999) Evaluation of the FLEXX incremental construction algorithm for protein–ligand docking. *Proteins*: *Struct Funct Bioinfo* 37(2): 228-241.

Lanzer MR, Rohrbach P (2007) Subcellular pH and Ca2+ in *Plasmodium falciparum*: implications for understanding drug resistance mechanisms. *Curr Sci* 92: 1561-1570.

Laurie AT Jackson RM (2005) Q-SiteFinder: an energy-based method for the prediction of protein–ligand binding sites. *Bioinformatics* 21(9): 1908-1916.

LeCun Y, Bengio Y, Hinton G (2015) Deep learning. *N1Z ature* 521(7553): 436-44.

Lee J, Freddolino PL, Zhang Y (2017) Ab initio protein structure prediction. In: Rigden DJ, editor. *From protein structure to function with bioinformatics*: Springer, Dordrecht, pp. 3-35.

Lemer CMR, Rooman MJ, Wodak SJ (1995) Protein structure prediction by threading methods: evaluation of current techniques. *Proteins: Struct Funct Bioinfo* 23(3): 337-355.

Li H, Sutter J, Hoffmann R (2000) HypoGen: An Automated System for Generating 3D Predictive Pharmacophore. *Pharmacophore perception, development, and use in drug design* 2: 171.

Liang J, Moore RE, Moher ED, Munroe JE, Al-awar RS, Hay DA, Varie DL, Zhang TY, Aikins JA, Martinelli MJ, Shih C (2005) Cryptophycins-309, 249 and other cryptophycin analogs: preclinical efficacy studies with mouse and human tumors. *Investigational New Drugs* 23(3): 213-24.

Linington RG, Edwards DJ, Shuman CF, McPhail KL, Matainaho T, Gerwick WH (2008) Symplocamide A, a potent cytotoxin and chymotrypsin inhibitor from the marine Cyanobacterium *Symploca* sp. *J Nat Prod* 71: 22-27.

Lionta E, Spyrou G, K Vassilatis D, Cournia Z (2014) Structure-based virtual screening for drug discovery: principles, applications and recent advances. *Curr Top Med Chem* 14(16): 1923-38.

Lipinski CA (2004) Lead-and drug-like compounds: the rule-of-five revolution. *Drug Discov Today Technol* 1(4): 337-341.

Luesch H, Chanda SK, Raya RM, DeJesus PD, Orth AP, Walker JR, Izpisúa Belmonte JC, Schultz PG (2006) A functional genomics approach to the mode of action of apratoxin A. *Nat Chem Biol* 2(3): 158-167.

Luesch H, Moore RE, Paul VJ, Mooberry SL, Corbett TH (2001) Isolation of dolastatin 10 from the marine cyanobacterium *Symploca* species VP642 and total stereochemistry and biological evaluation of its analogue symplostatin 1. *J Nat Prod* 64(7): 907-10.

Luescher-Mattli M (2003) Algae a possible source of new antiviral agents. *Curr Med Chem Anti-Infect Agents* 2: 219-225.

Martins RF, Ramos MF, Herfindal L, Sousa JA, Skærven K, Vasconcelos VM (2008) Antimicrobial and cytotoxic assessment of marine cyanobacteria-*Synechocystis* and *Synechococcus. Mar Drugs* 6(1): 1-11.

Masuda Y, Suzuki J, Onda Y, Fujino Y, Yoshida M, Doi T (2014) Total synthesis and conformational analysis of apratoxin C. *J Org Chem* 79: 8000-8009.

McCammon JA, Gelin BR, Karplus M (1977) Dynamics of folded proteins. *Nature* 267(5612): 585-590.

Melo-Filho CC, Braga RC, Andrade CH (2014) 3D-QSAR approaches in drug design: perspectives to generate reliable CoMFA models. *Curr Comput Aided Drug Des* 10(2): 148-159.

Morris GM, Huey R, Lindstrom W, Sanner MF, Belew RK, Goodsell DS, Olson AJ (2009) AutoDock4 and AutoDockTools4: Automated docking with selective receptor flexibility. *J Comput Chem* 30(16): 2785-2791.

Muhammed MT, Aki-Yalcin (2019) Homology modeling in drug discovery: Overview, current applications, and future perspectives. *Chem Biol Drug Des* 93(1): 12-20.

Mundt S, Kreitlow S, Jansen R (2003) Fatty acids with antibacterial activity from the cyanobacterium *Oscillatoria redekei* HUB 051. *J Appl Phycol* 15(2): 263-7.

Naidoo D, Roy A, Kar P, Mutanda T, Anandraj A (2021) Cyanobacterial metabolites as promising drug leads against the Mpro and PLpro of SARS-CoV-2: An *in-silico* analysis. *J Biomol Struct Dyn* 39(16): 6218-6230.

Oshiro CM, Kuntz ID, Dixon JS (1995) Flexible ligand docking using a genetic algorithm. *J. Comput. Aided Mol Des* 9(2): 113-130.

Pan L, Gardner CL, Pagliai FA, Gonzalez CF, Lorca GL (2017) Identification of the tolfenamic acid binding pocket in PrbP from Liberibacter asiaticus. *Front Microbiol* 8: 1591.

Patel HM, Noolvi MN, Sharma P, Jaiswal V, Bansal S, Lohan S, Kumar SS, Abbot V, Dhiman S. Bhardwaj V (2014) Quantitative structure–activity relationship (QSAR) studies as strategic approach in drug discovery. *Med Chem Res* 23(12): 4991-5007.

Pathak J, Mondal S, Ahmed H, Rajneesh, Singh SP, Sinha RP (2019) *In silico* study on interaction between human polo-like kinase 1 and cyanobacterial sheath pigment scytonemin by molecular docking approach. *Biointerface Res Appl Chem* 9: 4374-4378.

Pathak J, Singh PR, Sinha RP and Rastogi RP (2021) Evolution and Distribution of Cyanobacteria. In: *Ecophysiology and Biochemistry of Cyanobacteria*; Springer, Singapore. pp. 1-30.

Patterson GM L, Larsen LK, Moore RE (1994) Bioactive natural products from blue-green algae. *J Appl Phycol* 6: 151-157.

Patterson GM, Baldwin CL, Bolis CM, Caplan FR, Karuso H, Larsen LK, Levine IA, Moore RE, Nelson CS, Tschappat KD, Tuang GD (1991) Antineoplastic activity of cultured blue-green algae (cyanophyta). *J Phycol* 27(4): 530-6.

Pence HE, Williams A (2010) *ChemSpider: an online chemical information resource*.

Pérez Gutiérrez RM, Martínez Flores A, Vargas Solís R, Carmona Jimenez J (2008) Two new antibacterial norabietane diterpenoids from cyanobacteria, *Microcoleous lacustris*. *J Nat Med* 62(3): 328-31.

Phillips JC, Zheng G, Kumar S, Kalé LV (2002) November. NAMD: Biomolecular simulation on thousands of processors. In *SC'02: Proceedings of the 2002 ACM/IEEE Conference on Supercomputing*: 36-36.

Portmann C, Blom JF, Kaiser M, Brun R, Jüttner F, Gademann K (2008) Isolation of aerucyclamides C and D and structure revision of microcyclamide 7806A: heterocyclic ribosomal peptides from *Microcystis aeruginosa* PCC 7806 and their antiparasite evaluation. *J Nat Prod* 71(11): 1891-6.

Prioto G (2007). Nifurtimox-eflornithine combination therapy for second-stage *Trypanosoma brucei gambiense* sleeping sickness: a randomized clinical trial in Congo. *Clin Infec Dis* 45: 435-1444.

Rackers JA, Wang Z, Lu C, Laury ML, Lagardère L, Schnieders MJ, Piquemal JP, Ren P, Ponder JW (2018) Tinker 8: software tools for molecular design. *J Chem Theory Comput* 14(10): 5273-5289.

Rafiei P, Haddadi A (2017) Docetaxel-loaded PLGA and PLGA-PEG nanoparticles for intravenous application: Pharmacokinetics and biodistribution profile. *Int J Nanomedicine* 12: 935.

Rastogi RP, Sinha RP (2009) Biotechnological and industrial significance of cyanobacterial secondary metabolites. *Biotechnol Adv* 27(4): 521-539.

Raunio H (2011) *In silico* toxicology non testing methods. *Front Pharmacol* 2: 33.

Raveh A, Carmeli S (2007) Antimicrobial ambiguines from the cyanobacterium *Fischerella* sp. collected in Israel. *J Nat Prod* 70: 196-201.

Rotemberg V, Halpern A, Dusza S, Codella NC (2019) The role of public challenges and data sets towards algorithm development, trust, and use in clinical practice. In *Semin Cutan Med Surg* 38: E38-E42.

Roy PP, Leonard JT, Roy K (2008) Exploring the impact of size of training sets for the development of predictive QSAR models. *Chemometr Intell Lab Syst* 90(1): 31-42.

Sammut C, Webb GI (Eds.) (2011) Encyclopedia of machine learning. In: Sammut C, Webb GI, editors. *Springer Science & Business Media Boston*.

Sangeetha M, Menakha M, Vijayakumar S (2014) *In silico* prediction of anticancer cyanobacterial drug from *Nostoc*. *Biomed Prev Nutr* 4(1): 71-73.

Schaeffer DJ, Krylov VS (2000) Anti-HIV activity of extracts and compounds from algae and cyanobacteria. *Ecotoxicol Environ Saf* 45: 208-227.

Schaller D, Šribar D, Noonan T, Deng L, Nguyen TN, Pach S, Machalz D, Bermudez M, Wolber G (2020) Next generation 3D pharmacophore modeling. *Wiley Interdiscip. Rev Comput Mol Sci* 10(4): 1468.

Seidler J, McGovern SL, Doman TN Shoichet BK (2003) Identification and prediction of promiscuous aggregating inhibitors among known drugs. *J Med Chem* 46(21): 4477-4486.

Sheifert K, Lemke A, Croft SL, Kayser O (2007) Antileishmanial structure-activity relationships of synthetic phospholipids: *in vitro* and *in vivo* activities of selected derivatives. *Antimicrob Agents Chemother* 51(12): 4525-8.

Shin C (2001) Cryptophycins: a novel class of potent antimititotic antitumor depsipeptide. *Curr Pharmaceutical Design* 13: 1259-1276.

Shishido TK, Humisto A, Jokela J, Liu L, Wahlsten M, Tamrakar A, Fewer DP, Permi P, Andreote AP, Fiore MF, Sivonen K (2015) Antifungal compounds from cyanobacteria. *Mar Drugs* 13(4): 2124-2140.

Simmons TL, Engene N, Ureña LD, Romero LI, Ortega-Barría E, Gerwick L, Gerwick WH. Viridamides A and B (2008) lipodepsipeptides with antiprotozoal activity from the marine cyanobacterium *Oscillatoria nigro-viridis*. *J Nat Prod* 71(9): 1544-50.

Simpore J, Zongo F, Kabore F, Dansou D, Bere A, Nikiema JB, Pignatelli S, Biondi DM, Ruberto G, Musumeci S (2005) Nutrition rehabilitation of HIV-infected and HIV-negative undernourished children utilizing *spirulina*. *Ann Nutr Metab* 49(6):373-80.

Singh RK, Tiwari SP, Rai AK, Mohapatra TM (2011) Cyanobacteria: an emerging source for drug discovery. *J Antibiot Res* 64(6): 401-12.

Sousa SF, Fernandes PA, Ramos MJ (2006) Protein–ligand docking: current status and future challenges. *Proteins: Struct Funct Bioinfo* 65(1): 15-26.

Stevenson CS, Capper EA, Roshak AK, Marquez B, Eichman C, Jackson JR, Mattern M, Gerwick WH, Jacobs RS, Marshall LA (2002) The identification and characterization of the marine natural product scytonemin as a novel antiproliferative pharmacophore. *J Pharmacol Exp Ther* 303(2): 858-866.

Stouch TR, Kenyon JR, Johnson SR, Chen XQ, Doweyko A, Li Y (2003) *In silico* ADME/Tox: why models fail. *J Comput Aided Mol Des* 17(2): 83-92.

Strebhardt K Ullrich A (2006) Targeting polo-like kinase 1 for cancer therapy. *Nat Rev Cancer* 6(4): 321-330.

Tan KP, Nguyen TB, Patel S, Varadarajan R, Madhusudhan MS (2013) Depth: a web server to compute depth, cavity sizes, detect potential small-molecule ligand-binding cavities and predict the pKa of ionizable residues in proteins. *Nucleic acids Res* 41(W1): W314-21.

Tan LT (2010) Filamentous tropical marine cyanobacteria: A rich source of natural products for anticancer drug discovery. *J Appl Phycol* 22: 659-676.

Trott O, Olson AJ (2010) AutoDock Vina: improving the speed and accuracy of docking with a new scoring function, efficient optimization, and multithreading. *J Comput Chem 31*(2): 455-461.

Vaishampayan U, Glode M, Du W, Kraft A, Hudes G, Wright J, Hussain M (2000) Phase II study of dolastatin-10 in patients with hormone-refractory metastatic prostate adenocarcinoma. *Clin Cancer Res* 6(11): 4205-4208.

Van Der Spoel D, Lindahl E, Hess B, Groenhof G, Mark AE, Berendsen HJ (2005) GROMACS: fast, flexible, and free. *J Comput Chem* 26(16): 1701-1718.

Venkatachalam CM, Jiang X, Oldfield T, Waldman M (2003) LigandFit: a novel method for the shape-directed rapid docking of ligands to protein active sites. *J Mol Graph Model* 21(4): 289-307.

Verma E, Mishra AK, Singh AK, & Singh VK (2016) Structural elucidation and molecular docking of a novel antibiotic compound from cyanobacterium *Nostoc* sp. MGL001. *Front Microbiol* 7:1899.

Vijayakumar S & Menakha M (2014) Prediction of new cyanobacterial drug for treating lung cancer. *Biomed Aging Pathol 4*(1): 49-52.

Vijayakumar S, Menakha M, (2015) Pharmaceutical applications of cyanobacteria—A review. *J Acute Med* 5(1): 15-23.

Volk RB, Furkert FH (2006) Antialgal, antibacterial and antifungal activity of two metabolites produced and excreted by cyanobacteria during growth. *Microbiol Res* 161(2): 180-186.

Volkamer A, Kuhn D, Rippmann F, Rarey M (2012) DoGSiteScorer: a web server for automatic binding site prediction, analysis and druggability assessment. *Bioinformatics* 28(15):2074-5.

Vyas VK, Ukawala RD, Ghate M, Chintha C (2012) Homology modeling a fast tool for drug discovery: current perspectives. *Indian J Pharm Sci* 74(1): 1.

Wang X, Song K, Li L, Chen L (2018) Structure-based drug design strategies and challenges. *Curr Top Med Chem* 18(12): 998-1006.

Wase NV, Wright PC (2008) Systems biology of cyanobacterial secondary metabolite production and its role in drug discovery. *Expert Opin Drug Discov* 3(8): 903-929.

Wishart DS, Knox C, Guo AC, Cheng D, Shrivastava S, Tzur D, Gautam B, Hassanali M. DrugBank (2008) a knowledgebase for drugs, drug actions and drug targets. *Nucleic Acids Res* 1;36(s): D901-6.

Xu J, Jiao F, Yu L (2008) Protein structure prediction using threading. In: Zaki MJ, Bystroff C, editors. *Protein structure prediction* Humana Press, New Jersey, pp. 91-121.

Yamazaki Y, Kitajima M, Arita M, Takayama H, Sudo H, Yamazaki M, Aimi N, Saito K (2004) Biosynthesis of camptothecin. *In silico* and *in vivo* tracer study from [1-13C] glucose. *Plant Physiol* 134(1): 161-170.

Yasin D, Zafaryab M, Ansari S, Ahmad N, Khan NF, Zaki A, Rizvi MM, Fatma T (2019) Evaluation of antioxidant and anti-proliferative efficacy of *Nostoc muscorum* NCCU-442. *Biocatal Agric Biotechnol* 1; 17: 284-93.

Yonezawa T, Mase N, Sasaki H, Teruya T, Hasegawa SI, Cha BY, Yagasaki K, Suenaga K, Nagai K, Woo JT (2012) Biselyngbyaside, isolated from marine cyanobacteria, inhibits osteoclastgenesis and induces apoptosis in mature osteoclasts. *J Cell Biochem* 113(2): 440-8.

Yousef M, Abdelkader T, El-Bahnasy K (2019) Performance comparison of ab initio protein structure prediction methods. *Ain Shams Engr J* 10(4): 713-719.

Zhang G, Zhang Z, & Liu Z (2013) Polo-like kinase 1 is overexpressed in renal cancer and participates in the proliferation and invasion of renal cancer cells. *Tumor Biol 34*(3):1887-1894.

Zhao H, Caflisch A (2015) Molecular dynamics in drug design. *Eur J Med Chem* 91: 4-14.

Zhu H, Pisabarro MT (2011) MSPocket: an orientation-independent algorithm for the detection of ligand binding pockets. *Bioinformatics* 27(3):351-358.

# About the Editor

**Dr. Rajeshwar P. Sinha, PhD, FSAB**
Professor
Centre of Advanced Study in Botany
Banaras Hindu University, Varanasi, India
Email: rpsinhabhu@gmail.com


*Prof. Dr. Rajeshwar P. Sinha*, DAAD Fellowship Awardee and Fellow, Society for Applied Biotechnology, India, is a Professor of Molecular Biology, Centre of Advanced Study in Botany, Banaras Hindu University (BHU), Varanasi, India. He received his PhD (Biotechnology) form BHU, Varanasi and visited countries such as Argentina, Austria, Belgium, Canada, China, Germany, Greece, France, Italy, Japan, Luxembourg, Norway, Poland,

Republic of Korea, Spain, Switzerland, The Netherlands, UK and USA, in the field of academics/research. He is working on Physiological, Biochemical, Molecular, Nanobiotechnological and Computational biology aspects of cyanobacteria. His primary research focus is on UV-B radiation impacts on DNA damage and repair, Phycobiliproteins, Mycosporine-like amino acids, and Scytonemin etc. He has published over 450 research papers/reviews/book chapters/conference proceedings and edited/authored 10 books. He is a life time member of several national and international scientific societies and an editorial board member of various national and international journals. He has over 13115 citations with an h-index of 54 and i10-index of 134. He is among the top 02% most cited scientists in the world.

# Index

## A



## B



## C

## P



## R



## S



## T